Fifth Edition

The Canadian North
Issues and Challenges

Robert M. Bone

OXFORD
UNIVERSITY PRESS

OXFORD

UNIVERSITY PRESS

Oxford University Press is a department of the University of Oxford.
It furthers the University's objective of excellence in research, scholarship,
and education by publishing worldwide. Oxford is a registered trade mark of
Oxford University Press in the UK and in certain other countries.

Published in Canada by
Oxford University Press
8 Sampson Mews, Suite 204,
Don Mills, Ontario M3C 0H5 Canada

www.oupcanada.com

Copyright © Oxford University Press Canada 2016

The moral rights of the author have been asserted

Database right Oxford University Press (maker)

First Edition published in 1992
Second Edition published in 2003
Third Edition published in 2009
Fourth Edition published in 2012

Library and Archives Canada Cataloguing in Publication
Bone, Robert M., author
The Canadian north : issues and challenges / Robert
M. Bone. — Fifth edition.

Includes bibliographical references and index.
ISBN 978-0-19-901941-0 (paperback)

1. Canada, Northern—Geography—Textbooks. 2. Canada, Northern—Economic
conditions—Textbooks. 3. Natural resources—Canada, Northern—Textbooks.
4. Native peoples—Canada, Northern—Textbooks. I. Title.

HC117.N5B65 2016   917.19   C2015-906386-8

Cover image: Peter Mather/Getty Images

Oxford University Press is committed to our environment.
This book is printed on Forest Stewardship Council® certified paper
and comes from responsible sources.

MIX
Paper from
responsible sources
FSC
www.fsc.org   FSC® C004071

Printed and bound in Canada

1 2 3 4 — 19 18 17 16

Contents

List of Figures

IIIIIIIIIIIIIIIIIIIIIIIIIIIIIIIIIIIIIIIIIIIIIIIIIIIIIIIIIIIIIIIIIIIIIIIIIIIIIIIIIIIIIIIIIIIIIIIIIIIIIIIIIIIIIIIIIIIIIIIIIIIIIIIIIIIIIIIIII

List of Tables

List of Vignettes

Preface

The North is a fascinating region of Canada. I hope that the students reading this book will agree. In the twenty-first century new challenges have emerged, including climate change and Arctic sovereignty. Yet, the old challenges remain—the resource economy, the fragile environment, and the place of Aboriginal peoples in the Canadian North.

What is the northern image? Northerners and southerners perceive resource development differently. The first group sees the North as a homeland. They look at resource development from two perspectives: What benefits do northerners receive? And what effect does it have on the environment, especially wildlife? Southerners, particularly CEOs of resource companies, regard the North as a frontier open for development that will serve world markets and provide economic profits.

The first edition of this book was published in 1992—well before climate change was fully upon us and just at the start of comprehensive land claim agreements. What is remarkable to me is the magnitude of change that has taken place over this time period; even more astounding is the prospect for greater change in the future. Social change is one element, especially the advances made by the Aboriginal community, but the gap between Aboriginal and non-Aboriginal Canadians remains wide. Political change is another element with the emergence of Nunavut and the prospects for Nunavik. Still, Nunavut is haunted by a fiscal deficit and its dependency on Ottawa. Economic change is triggered by the growing world demand for resources, especially energy. However, the global economic cycle continues to rear its ugly head in the form of resource booms followed by resource busts. Glancing back to 1992, who would have imagined that changes in the climate, especially in the frigid Arctic, would reignite an interest in Arctic sovereignty and Arctic shipping?

The fifth edition, like the previous ones, is built on close relations with geographers and other northern scholars. My recent involvement with the students in the Master of Northern Governance and Development program at the University of Saskatchewan provided the spark to complete this edition, and their presentation to me of a Métis sash was very, very special.

The staff at Oxford University Press again provided the support so necessary for authors. More specifically, Lauren Wing served as the developmental editor while my old friend, Richard Tallman, again guided the manuscript into its final form. Over the years, Richard has become familiar with my writing and we work well together. Thank you, Richard. Anonymous readers selected by Oxford University Press made a number of constructive suggestions and brought a number of issues into focus.

Finally, a special note of appreciation goes to my wife, Karen, who creates that magical space so necessary for an author.

Robert M. Bone
October 2015

Northern Perceptions

As Canada's last frontier, the North holds a prominent place in the hearts and minds of Canadians. Yet, most Canadians have never visited this remote region. Its significance flows from its unique geography and Aboriginal peoples, its vast but untapped resources, and, most importantly, the sense that Canada's future lies in its northern lands and seas.

But what exactly is the North? For most, this region represents the colder lands of North America where winters are extremely long and dark, where permafrost abounds, and where few people live. For Aboriginal peoples, however, the North is their homeland where comprehensive land claim agreements have provided them with more control over their destiny and where the **duty-to-consult doctrine** ensures their future involvement in resource megaprojects. In places where no treaties exist, such as much of British Columbia, the June 2014 Supreme Court of Canada *Tsilhqot'in* **decision** goes beyond consultation to the need for Aboriginal approval, thus raising the bar for developers but strengthening the hand of First Nations.

The North is also the place where global warming is having a greater impact than elsewhere in Canada. The retreat of permafrost is one sign; another is the growing number of ships traversing the formerly ice-choked Northwest Passage. Most importantly, the issues of global warming, the Northwest Passage, and Arctic sovereignty have enhanced Canada's geopolitical role in the **Circumpolar World** and, more specifically, in the **Arctic Council**.

Most Canadians acknowledge the North as a resource frontier as well as the homeland of Aboriginal northerners. Nunavut represents the most politically sophisticated example of an Aboriginal homeland, while the oil sands development in northern Alberta illustrates its resource potential. Yet, those holding these seemingly diametrically opposed visions are compelled to search for common ground. For resource companies, the duty to consult forces them to inform affected Aboriginal communities about their projects and then to negotiate with First Nations who hold **Aboriginal title** to lands sought by the developers. For First Nations, the duty-to-consult doctrine provides an arena to argue for their culture and environment as well as to seek economic accommodation. Within this context, the search for accommodation is one more reason why the North symbolizes the nation's sense of purpose, the course of its

Vignette 1.1   What Is Development?

Development is based on a set of underlying assumptions about the world. In the case of Aboriginal tribes, their original assumptions were based on a spiritual relationship with the land and its animals that provided their food and clothing. With the coming of Europeans, the European approach to development held sway. But what is this type of development? Unlike the spiritual link to the land, Westerners' assumptions are rooted in a belief that humans can (and should) control nature rather than be in relationship with it. This Western view has formed the political and economic context of global capitalism where market forces shape the historical process of long-term change in society. In this sense, development is linked to **modernization theory**. Driven by economic growth flowing out of industrialization, development includes not only the notion of economic growth but also social and political changes that involve people's health, education, housing, security, civil rights, and other social characteristics. Seen in this light, development is a normative concept involving Western values, goals, and beliefs.

It follows that development in the Canadian North occurs within the market economy and is supported by the Canadian political system. Unfortunately, this concept of development views traditional values and ways as "hindrances" to economic and social progress. For that reason, until the late twentieth century, the place of Aboriginal peoples within the Canadian version of development was often ignored or dismissed. Since then, the legal validation of Aboriginal rights, the establishment of a modern land claim negotiation process, and the participation of Aboriginal organizations in the market economy signal the "necessary" inclusion of Aboriginal peoples within the Western development process. The 27 June 2014 Supreme Court of Canada ruling in *Tsilhqot'in* determined that this First Nation has Aboriginal title to its ancestral lands and control over those lands (Supreme Court of Canada, 2014). Notwithstanding these gains, what might seem to be the "inevitable" Aboriginal journey towards inclusion within Canadian society is far from complete (Schaefli and Godlewska, 2014: 111), not least because some groups do not want to be included.

future, and the evolving place of Aboriginal Canadians in that future (Vignette 1.1). An early example of economic accommodation was the Aboriginal Pipeline Group that had an option to acquire one-third ownership in the proposed Mackenzie Gas Project.[1] Looking to the future, could the highly controversial Northern Gateway project be converted into an Aboriginal pipeline along the lines of the Eagle Spirit proposal or will it suffer the fate of the failed Mackenzie Gas Project?[2]

Defining the North

The North is both easy and difficult to define. In a narrow sense, the North refers to the three territories. But such a definition ignores the natural conditions, such as permafrost, found on both sides of the sixtieth parallel. In this text, the spatial extent of the North extends far south of the Territorial North to include Arctic and Subarctic lands found in the northern reaches of all provinces except the three Maritime provinces (Figure 1.1).

The Arctic and Subarctic extend over a vast area—nearly 80 per cent of Canada. A generalized map of these two natural regions is presented in Figure 1.1. The North's

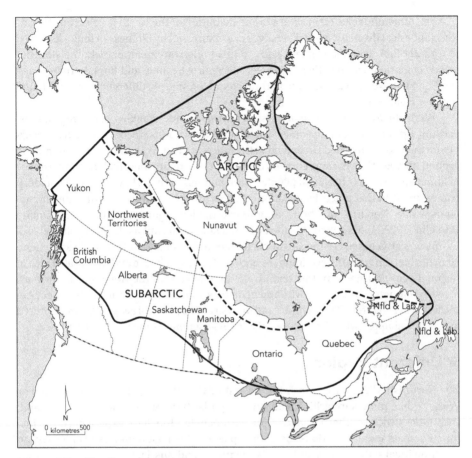

Figure 1.1   The Canadian North

The Canadian North consists of the Arctic and Subarctic natural regions. From an ecological point of view, these two regions are best described as the Tundra and Boreal biomes. As well, biomes emphasize the ecological aspect of flora and fauna and their adaption to that particular environment. The southern boundary of the Canadian North corresponds with the southern limit of the boreal forest and permafrost. Until recently, these physical boundaries were considered "fixed" or stable. But the impact of global warming is causing these natural boundaries to slowly but surely shift further north. Permafrost boundary is much more susceptible to change than natural vegetation boundaries because the frozen ground is relatively quickly thawed whereas plant life requires many years to become established in a new environment (see Chapter 2 for a fuller discussion).

southern edge closely corresponds to the southern limit of permafrost, indicating the close relationship of climate to permafrost. Permafrost, for example, only exists when the mean annual temperature is below zero. Furthermore, the geographic extent and depth of permafrost increases as the mean annual temperature decreases (Figures 1.3 and 2.10).

The Arctic exists in the three territories and in four of the seven provinces (Quebec, Newfoundland and Labrador, Ontario, and Manitoba), while the Subarctic occurs in all seven provinces with a northern landscape (British Columbia, Alberta, Saskatchewan, Manitoba, Ontario, Quebec, and Newfoundland and Labrador) and two territories (Northwest Territories and Yukon).

Yet, often understated in such a broad definition of the North is its extremely varied biophysical diversity. Such diversity is best expressed by biomes—a topic discussed in Chapter 2. A dynamic element adds to this understatement, namely, the slow but steady climatic warming affecting sea ice, permafrost, fauna, and flora. In these ways, the Canadian North has a unique natural character, sharply different from that found in other regions of Canada.

While the North is defined in physical terms, an understanding of this region must account for the Aboriginal peoples who arrived in Canada's North thousands of years before the Vikings reached Greenland in the late tenth century and then Newfoundland around the year 1000. An appreciation of the North therefore must include the existence of homelands where Aboriginal peoples, born and raised there, have a special and deep commitment to that place. First Nation reserves provide examples of truncated homelands while Nunavut functions at a territorial-level government. Louis-Edmond Hamelin (1979: 9) described this idea of homeland as "a trait as deeply anchored as a European's attachment to the site of his hamlet or his valley." In this single statement, Hamelin has captured the geographer's notion of a sense of place and the parallel idea of regional consciousness.[3] In this text, these three images of the North—as a northern resource Eldorado, as a cold environment, and as an Aboriginal homeland—reoccur in the following chapters. The richness and diversity of its physical geography are captured in the next chapter, while its cultural/historical geography is the focus of Chapter 3.

A Different Region

The North is a different region from the rest of Canada. As part of the Circumpolar World, it has more similarities with other high-latitude regions, such as Greenland, than it does with southern Canada. Northern peoples, too, have experienced the historic encroachment of nation-states, thus placing this region and its peoples within larger political settings, such as Canada, Denmark, and Russia.

But what are the differences between the North and the rest of Canada? To begin with, the North has the coldest environment in Canada, with permafrost—permanently frozen ground—abounding, the Arctic ice pack covering much of the Arctic Ocean, and the flashing of the aurora borealis across its winter skies. Second, the North is by far the largest region in Canada, accounting for 76 per cent of Canada's geographic area. A third revelation is that the North is almost equally divided between the Territorial North and the Provincial North (Table 1.1). While most Canadians realize that the three territories occupy a vast area of Canada, the sheer size of the northern area found in the seven provinces is surprising to many people. Of all the provinces, Quebec has the largest northern area, which makes up 81 per cent of that province. Two provinces—Newfoundland and Labrador and Manitoba—are next, with 74 per cent of their territory classified as "northern." Northern lands make up 65 per cent of Ontario and 50 per cent of Saskatchewan. Alberta and British Columbia trail behind, with 47 and 40 per cent, respectively.

A fourth fact marking the North as a different region from the rest of the country is its small population and its strong Aboriginal composition. Lying beyond Canada's ecumene, this thinly populated landscape contains less than 5 per cent of Canada's population—fewer than 1.5 million people. Of the two northern biomes, the Arctic contains under 100,000 people, making its population density of 0.01 persons per km² one

Table 1.1 Geographic Size of the Canadian North, by Province/Territory

Province or Territory	Total Area (000 km²)	Northern Area (000 km²)	North (%)	Canada (%)
Newfoundland/Labrador	405	300	74	3.0
Alberta	662	310	47	3.1
Saskatchewan	651	325	50	3.3
British Columbia	945	375	40	3.8
Manitoba	648	480	74	4.8
Ontario	1,076	700	65	7.0
Quebec	1,542	1,250	81	12.5
Provincial North	*5,929*	*3,740*	*63*	*37.5*
Yukon	482	482	100	4.8
Northwest Territories	1,346	1,346	100	13.5
Nunavut	2,093	2,093	100	21
Territorial North	*3,921*	*3,921*	*100*	*39.3*
Canada	*9,985*	*7,661*	*77*	*76.8*

Note: Column 3 shows the percentage of northern lands in each province or territory. Column 4 indicates the percentage of Canada's total area in each territory and in the northern reaches of the provinces.

Source: Statistics Canada (2012).

of the lowest in the world. Furthermore, Inuit comprise over 80 per cent of those living in the Arctic; and for much of the Subarctic the First Nations and Métis represent a majority. This demographic fact and the associated cultural change taking place are at the heart of critical economic and political issues and challenges (see Vignette 1.2).

Lastly, global warming is affecting the North's physical geography and those changes have cultural, economic, and political implications. Global warming—most noticeably in the form of higher summer temperatures—is taking place more rapidly in the North than in other regions of Canada. This implies that physical geography is no longer a stable or fixed feature of the northern landscape. Assuming that the process of warming continues, by the end of this century the Tundra and Boreal biomes will be greatly diminished in size, the Northwest Passage will be free of ice in the summer, and permafrost will have retreated into the territories. These topics, along with climate change, are discussed more fully in forthcoming chapters. Other features unique to the North, such as the sense of isolation felt by newcomers to the North, are discussed below.

Sense of Place

One of the defining features of the North is the different reaction to the northern landscape between those born and raised in the North and those who have come to the North for economic opportunities. Those who originate in the North, mainly Aboriginal peoples, have a strong **sense of place**. In simple terms, the North is their home. On the other hand, newcomers frequently have a hard time adjusting to the northern lifestyle,

---

Vignette 1.2   Culture Clash

The culture of a society provides its members a frame of reference for interpreting and responding to events and ideas. What happens when that frame of reference is lost? An argument could be made that the pace of change for Aboriginal cultures in northern Canada has been overpowering and that their frames of reference are being replaced by the dominant Canadian frame of reference. One indicator is the demise of Aboriginal languages and the adoption of English. Another indicator is that the western framework for decision-making overpowers the Aboriginal one. As an illustration, political scientist Graham White (2006: 412) writes:

> When we [Nunavut Task Force] try to incorporate **Inuit Qaujimajatuqangit** (IQ) into the existing Nunavut Government we create a "cultural clash." And, as is usual in all cultural clashes, the dominant culture dominates. The Inuit culture is forced to take on the shape of the dominant, rather than the other way round.

An important question is: Are "economic accommodations" a matter of adaptation or domination? For example, the 2001 Paix des Braves Agreement between the Quebec Cree and the Quebec government allowed Hydro-Québec to complete the final phase of the original James Bay Hydroelectric Project. On the one hand, the Cree will receive $3.6 billion over 50 years, as well as full responsibility for their own welfare and economic development. On the other hand, what is given up in such agreements is a large part of a land base that has been central to cultural and spiritual identity for hundreds or thousands of years.

---

which is typified by small communities with limited urban amenities and with long travel connections to their places of origin and previous residence, often major cities in southern Canada. While some adjust and become true northerners, many leave within five years. These transplanted Canadians, who are often nurses and teachers, find it unappealing to live in remote northern centres, such as Aklavik (Figure 1.2), and in extreme cases the absence of urban amenities is so overwhelming that they relocate south at the first opportunity. This psychological effect reaches its peak in isolated communities where air transportation represents the only means of reaching southern Canada. Such communities are found in the three territories and, with the exception of British Columbia, in all provinces with northern areas.

Companies and governments often employ incentives to attract southern workers to live in remote centres. One such incentive is a federal income tax deduction—Northern Residents Deduction—for living in the North, but this deduction varies by tax region. In the more northerly (less accessible) tax region the deduction is pegged at 100 per cent, while in the less northerly (somewhat more accessible) tax region the deduction falls to 50 per cent. By reducing the amount of taxable income, northerners are able to reduce their personal income taxes (McNiven and Puderer, 2000).

The Political North

The political North has two aspects. From a geographic perspective, this region is divided into territorial and provincial norths. These two norths have two different types of

Figure 1.2   Aklavik, Near the Arctic Ocean

The small Native community of Aklavik lies at the northern edge of the boreal forest in the Mackenzie Delta at the latitude of 68° 15'. In this winter scene, the Peel Channel is frozen, the ice road is operating, and, in the distance, lakes are widespread, forming a high percentage of the boreal forest landscape. A short distance beyond is the coastline where tundra forms the natural vegetation. Access to the outside world is by air, by small boat to Inuvik in the summer, and by ice road in the winter. Foods, building materials, and other goods from southern Canada reach Aklavik each summer by river barges that depart from Hay River. Hay River, on the southern shore of Great Slave Lake, has road and rail links to Edmonton, which serves as the "gateway" city to the Northwest Territories. In the winter, emergency supplies come by air at a much higher cost. In 2011, the Aklavik population was recorded as 633.

Source: David Boyer/National Geographic/Getty Images.

government, one territorial and the other provincial. One shortcoming faced by northern areas of provinces is that, in regard to population, they form a minority within their respective provinces and have limited political power. Also, provincial governments, understandably perhaps, tend to treat their northern areas as hinterlands rather than as provincial cores. The opposite is true for those residing in the three territories.

The federal government plays a more central role in the three territories than in the provinces. For example, Ottawa provides most funding for territorial governments to provide public services to their residents. Provinces, on the other hand, receive much of their funding from royalties from resource development and from provincial taxes, thus making them less dependent on Ottawa. This form of fiscal **dependency** is changing as both Yukon (2001) and the Northwest Territories (2014) receive the royalties from their natural resources, leaving Nunavut under the fiscal wing of Ottawa. For Aboriginal governments, a significant policy decision was associated with the Northwest Territories Devolution Act, passed by the federal government in 2014, because this ensures that Aboriginal governments will receive up to 25 per cent of royalties.

The northern areas of provinces do not benefit from a direct share of royalties. Instead, they resemble resource hinterlands somewhat along the lines of the core/periphery model discussed later in this chapter. Provincial governments have unlocked their northern resource hinterlands by promoting railway and highway construction to the northern reaches of the provinces. The strategy was simple: by lowering the cost of transporting natural resources such as minerals and timber to world markets, hinterland development became possible with the royalties benefiting all residents of the province. British Columbia provides one example of this strategy. The provincial government built and operated BC Rail (formerly the Pacific Great Eastern Railway) until it was sold to CN Rail in 2003. This provincial railway linked the vast northern interior of British Columbia with the port of North Vancouver. As a result, the forest and other resources of the northern interior of British Columbia were exploited and their products exported to southern markets by means of BC Rail.

Most Canadians view the three territories as Canada's North. This mental divide has caught the attention of scholars. Margaret Johnston (1994: 1) correctly observed that "there has been a tendency to view much of this [provincial] part of the north in Canada as less northern than Yukon and the Northwest Territories, and consequently it has received considerably less attention as a northern region." Indeed, Coates and Morrison (1992) describe the northern areas of seven provinces as "the forgotten north." In this text, as noted earlier, the North includes the three territories and the cold lands found in the northern sectors of seven provinces. The political division between territories and provinces warrants identifying the three territories as the "Territorial North" and the northern parts of the seven provinces as the "Provincial North."

Common Characteristics

Even though the geographic extent of the Canadian North is vast, this region has many common characteristics (Table 1.2). A cold environment, sparse population, and extensive wilderness areas with limited but unique biophysical diversity are the rule. Another characteristic is the high percentage of Aboriginal people in the northern population, unlike in southern Canada, where the Aboriginal population forms a very small proportion of the total population. In some areas, they form a majority of the local population. In Nunavut and Nunavik (the Inuit homeland in northern Quebec), the Inuit make up around 85 per cent of the population. Like other hinterlands within the core/periphery structure of the world economy, the North has a resource economy. While forestry and mining activities account for most of the value of economic production in the North, these industries employ relatively few workers. In fact, the vast majority are employed by public agencies. In the next pages, three common characteristics—the North's winter, the concept of nordicity, and the core/periphery model—are explored more fully.

The Arctic Circle and Seasonal Variation in Sunlight

Of all the regions of Canada, the variation in sunlight from summer to winter is most pronounced in the North. This phenomenon is due to the change in the angle of the sun rays striking the Earth. In more simple terms, the daily duration of sunlight is related to latitude. In the summer, higher latitudes experience longer periods of daylight. The opposite is true for the winter. In fact, winter in the North is a time of darkness while summer is a time of light (for more on this subject, see Vignette 2.6, "The Polar Night").

Table 1.2 Common Characteristics of the North

Physical Characteristics	Human Characteristics
Cold environment	Sparse population
Limited biophysical diversity	Population stabilization
Wilderness	High cost of living
	Few highways
Permafrost	Aboriginal population
Vast geographic area	Settling of land claims
Fragile environment	Financial dependency
Slow biological growth	Resource economy
	Reliance on imported foods
Global warming	Country food
Continental climate	Economic hinterland
	Remote and isolated
	Importance of wildlife to humans

The **Arctic Circle** (see Figure 1.3) marks the location on the Earth's surface where, at the time of the winter solstice (21 December), night lasts for 24 hours because the sun's rays do not rise above the horizon. The Arctic Circle represents the northward limit of the sun's rays at the time of the winter solstice. At this latitude (66° 33'N), the sun does not rise above the horizon for one day of the year. However, the **polar night** does not begin immediately because twilight occurs for a short period of time when the sun is just below the horizon (Burn, 1995: 70).

This phenomenon of darkness increases towards the North Pole, so those living in higher latitudes are subject to an even longer period without seeing the sun. Within this "zone of winter darkness," the number of days in which the sun does not rise above the horizon varies. At Alert (82° 29'N) there are nearly five months of continuous darkness, while at Inuvik (68° 21'N) there is just over one month of continuous darkness. The reappearance of the sun on the horizon usually provokes a favourable response from "transplanted" southern Canadians. In Inuvik, residents celebrate the return of the sun each January with a "Sunrise Festival" featuring a huge bonfire and a variety of community events.

The reverse conditions occur in the summer: Alert has nearly five months of continuous sunlight and Inuvik has just over one month of continuous sunlight—providing there is no cloudy weather. At the North Pole, the sun appears on the horizon at the spring equinox (21 March) and does not set until the fall equinox (21 September). With each additional day, the sun rises higher in the sky until the summer solstice (21 June), when it reaches its maximum height above the horizon.

Nordicity

While the North has a cold environment, this environment varies from place to place (Slocombe, 1995: 161). The human landscape also varies. **Nordicity**, a concept introduced by the Quebec geographer Louis-Edmond Hamelin (1979), combines human and

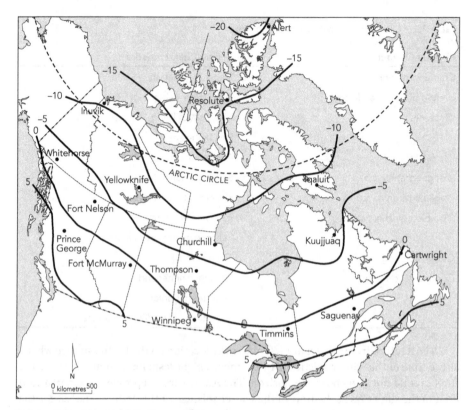

Figure 1.3 The Cold Environment

A measure of the North's cold environment is provided by mean annual air temperatures in degrees Celsius. These data, recorded from 1980 to 2010, indicate a warming trend (Table 2.2). Therefore, the isotherms are positioned further north than those based on data for the previous 20 years (1960–80). For more on this subject, see the section "A Warming North" and Table 2.2 in Chapter 2.

Source: Environment Canada (2014).

physical factors to measure the degree of "northernness" at specific places. It provides a quantitative measure of "northernness" for any place based on 10 selected variables that aim to represent all facets of the North. These variables, called polar units, are a combination of physical and human elements such as summer heat (or the lack of) and accessibility (or the lack of). The North Pole has a nordicity value of 1,000 polar units, which is the maximum value possible. The southern limit of areas in Hamelin's classification system occurs at 200 polar units. At that point, the North ends (as defined by nordicity) and the South begins.

The physical elements measure "coldness" while the human elements measure accessibility/development (see Appendix I for the complete list of variables, the assigned values, and the method of calculating nordicity). This approach permits the classification of the North into three regions (Middle North, Far North, and Extreme North) as shown in Figure 1.4.

Hamelin is describing the North from a southern Canadian perspective that reflects attitudes, beliefs, and values held by people residing in the more populated

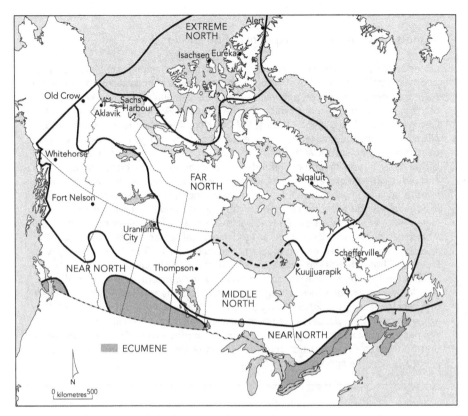

Figure 1.4   Canadian Nordicity

For geographers, "nordicity" permits the creation of three northern regions: the Middle North, the Far North, and the Extreme North. The southern limit of nordicity is marked by the 200 polar unit line. Yet, do the concept of nordicity, the spatial place names in the North, and the notion of a population ecumene provide a southern, northern, or balanced interpretation of the North?

Source: Adapted from Hamelin (1979: 150).

regions of Canada. An underlying assumption of such an ethnocentric viewpoint is that "development" of the North reduces its nordicity and therefore makes it more like southern Canada. Another assumption is that the North is viewed as a hinterland of southern Canada. Southern Canadians see the North quite differently from those living in the North. For instance, each descriptive label—Middle, Far, and Extreme Norths—may seem strange and out of place to a Canadian born north of the sixtieth parallel. A northerner, for example, might have a different mental map of Canada, with the North described as the "centre" and southern Canada as the "distant land."

Core/Periphery Model

Northern development takes place within a global economy. The **core/periphery model**—a product of Western thinking—best describes the economic relationship between industrial cores and periphery hinterlands within a capitalist economy (Wallerstein, 1979). Friedmann (1966: 76–98) turned this abstract economic theory

---

Vignette 1.3 Another Perspective: Fourth World

|||||||||||||||||||||||||||||||||||||||||||||||||||||||||||||||||||||||||||||||||||||||||||||||||||||||||||||||||||||||||||||||||||||||||||||||

Canada's North has another core/periphery perspective, namely that its Aboriginal peoples must function within an imposed political system that places little value on their aspirations, languages, and values. This political relationship sparked the term the "Fourth World." It was introduced by the Canadian Aboriginal leader and author George Manuel (Manuel and Posluns, 1974) to describe the poverty, internal **colonialism**, and subsequent statelessness experienced by Indigenous peoples within developed nation-states. Within Canada, the relationship between the core political centre (Ottawa) and its northern hinterland was associated with the imposition of law and order, with the arrival of Royal Canadian Mounted Police (RCMP) detachments in remote northern communities, and with social engineering, such as the notorious residential schools and the relocation of Aboriginal peoples into larger settlements to facilitate service provision. Only towards the close of the twentieth century did Aboriginal peoples, largely through land claim settlements, gain a modicum of independence, the best example being the creation in 1999 of the new territory of Nunavut. These changes in governance have helped to break some aspects of the political and economic confinement intrinsic to the concept of a "Fourth World." Still, Aboriginal peoples have a long way to go before the settlers' grip on the levers of power is shared (Lutz, 2008). Along these lines, Schaefli and Godlewska (2013: 111) argue that: "In Canada, pervasive uncritical acceptance of the exclusionary laws and policies that continue to marginalize Aboriginal peoples is rooted in willful ignorance that obfuscates the ways such policies uphold settler interests and renders such strategies possible and acceptable."

---

into a geographic model with one core and three peripheral regions. In turn, Bone (2014: 17, 18) has applied these four regions—core, rapidly growing or upward transition, slow growing or downward transition, and resource frontier—to Canada. In this text, the Canadian North is classified as a resource frontier.

The economic heart of the core/periphery model is resource extraction. Resource hinterlands are found in many parts of the world. Each is far from the manufacturing centres of the world but is connected to these larger centres and the market economy through trade. In Wallerstein's theory, one key assumption was that prices for manufactured goods increased over time more rapidly than agriculture and resource prices. In this way, the terms of trade became more and more favourable to the industrial core, thus allowing wealth to accumulate in this favoured region. But with the rise of China as an economic superpower, prices for resource commodities have risen sharply, suggesting that a "Super Cycle" for resource commodities has emerged, with higher commodity prices reversing the terms of trade, i.e., resource-rich countries and regions are benefiting from unusually high and seemingly sustainable commodity prices (Cross, 2007). The so-called Super Cycle first affected oil. Since 1972, world prices for oil have increased from US$2 a barrel to more than $100 a barrel by June 2014. Other commodity prices also have experienced this increase due to growing demand from China, India, and other rapidly industrializing countries. While the business cycle dictates that prices will fluctuate, as they did with oil in late 2014 and into 2015, only time will tell if the current range of energy and commodity prices will hold. In the meantime, resource hinterlands have enjoyed an unprecedented "boom." In fact, Cross (2007) goes so far as to suggest that the fear of a return to lower commodity prices, as was experienced in the late twentieth century, is overstated because the expanding Chinese and Indian economies will require more and more energy and resources.

Two other important theoretical issues are embedded in the core/periphery theory. First, Wallerstein assumed a homogeneous labour force. Such an assumption means that the core/periphery model is silent on the critical issue of the place of Indigenous peoples in global developments and, more specifically, of Aboriginal peoples in northern development in Canada. Second, Wallerstein saw no hope for diversification of the world's periphery because the core extracted all the wealth from trade with hinterlands. On the other hand, Friedmann depicted the periphery differently, and one hinterland area in his typology—the upward transition region—showed promise. While his work predated some of the specifics in this theory, Harold Innis (1930) presented a Canadian perspective on regional development through the lens of history, i.e., by examining the economic history of resource (staple) development in each region of Canada. In this way, Innis saw regional development occurring as a consequence of resource exploitation because it triggered a series of related economic activities that eventually led to regional economic diversification. In the case of the Canadian North, however, economic diversification is virtually impossible because its economy is subject to such a high level of economic leakage (the outflow of dollars to purchase outside goods and services not available in the North) and because its economy is based on non-renewable resources. Because of these two factors, Watkins (1977) feared that the North would fall into a "staple trap," that is, when the non-renewable resources are exhausted, the North's economy would collapse.

For our purposes, the North is a resource hinterland, more specifically a resource frontier periphery, while the rest of Canada and the world are the industrial core. Companies in the industrial core dominate the economy in the resource hinterland and control the hinterland's pace of resource development. Northern development has been subject to "good times and bad times" due to the variation in demand for resources. Put differently, much of the economic destiny of resource hinterlands is controlled by external forces and is extremely sensitive to fluctuations in world commodity prices. These fluctuations magnify the **economic cycle**, and may lead to boom/bust conditions in resource hinterlands. So far Cross has been correct about the Super Cycle. Even so, government intervention in the marketplace is warranted to ensure a portion of **northern benefits** from resource extraction remains in the hands of northerners.

The Old and New Styles of Resource Development

Over the past 50 years, the balance of power in the northern development equation has changed. No longer can resource companies and Crown corporations ignore the presence of Aboriginal peoples and simply proceed with their projects. For example, in 1949, the Iron Ore Company of Canada began to develop its iron mines, rail lines, and the townsite for Schefferville in northeastern Quebec, on the border with Labrador, without any concern for the local Innu and Naskapi or for their traditional hunting grounds (Table 1.3). By the early 1970s, however, after the Quebec government began the first phase of the James Bay Project on La Grande Rivière, the old-style approach to resource

Table 1.3 Evolution of Megaproject Proposals

Year	Megaproject	Aboriginal Issues	Environment Issues
1949–54	**Iron Ore Project:** Iron Ore Company of Canada developed the ore deposits in northern Quebec and Labrador by constructing a resource town, railway, power plant, and extensive port facilities at Sept-Îles.	Ignored	Ignored
1971–85	**James Bay Hydroelectric Project:** A 1973 court ruling forced Hydro-Québec to negotiate with the Cree and Inuit of northern Quebec over Aboriginal title before completing the construction of the first phase of the James Bay Project. The result was the James Bay and Northern Quebec Agreement (1975).	James Bay and Northern Quebec Agreement (JBNQA)	Recognized in the JBNQA
1974–7	**Mackenzie Valley Pipeline Proposal:** The Berger Inquiry and the National Energy Board examined the social and environmental aspects. Construction project did not proceed.	Discussed	Discussed
1981–8	**Oil Exploration in the Beaufort Sea:** As part of the National Energy Program, Ottawa announced the Petroleum Incentive Program, which provided $7 billion in federal funds for Canadian firms drilling for oil in frontier areas. Over 200 wells were drilled in the Beaufort Sea.	Ignored	Ignored
1982–6	**Norman Wells Oil Expansion and Pipeline Project:** The construction of this megaproject followed the 1981 federal Environment Assessment Panel report, *Norman Wells Oilfield Development and Pipeline Project.*	Federal social funding	Specific federal requirements
1991–2008	**NWT Diamond Mines:** Each mining proposal underwent a detailed environmental review and impact and benefits agreements were reached with local Aboriginal groups. More proposals for diamond mines are expected.	Impact and benefits agreements	Specific territorial and federal requirements
1993–2005	**Voisey's Bay Nickel Mine:** The Voisey's Bay mine proposal underwent a detailed environmental review, and impact and benefits agreements were reached with local Aboriginal groups.	Impact and benefits agreements	Specific provincial requirements
2000–11	**Mackenzie Delta Gas and Pipeline Proposal:** Extensive environmental and social reviews by the National Energy Board and the Joint Review Panel were completed in 2009. Aboriginal participation was strong, with the Aboriginal Pipeline Group set to own a third of the proposed pipeline. Construction did not begin because market conditions deteriorated.	Discussed	Discussed
2013–?	**Mary River Iron Mine:** Extensive environmental and social reports prepared by the company were accepted by the Nunavut Impact Review Board in 2013. Mining began in September 2014.	Impact and benefits agreement	Specific territorial requirements

development was challenged when the James Bay Cree and Inuit took the provincial government to court over this megaproject. The two sides were required by the Quebec court to settle their differences and the result was the 1975 James Bay and Northern Quebec Agreement (JBNQA), what is generally recognized as the first modern land claim agreement. The JBNQA marked the first formal acceptance that Aboriginal peoples occupying their traditional lands had a claim on resource development affecting those lands and that they deserved compensation. Around the same time, the extensive media coverage of the Berger Inquiry during the mid-1970s revealed the concerns of the Dene regarding the proposed Mackenzie Valley gas pipeline and its potential impacts on the environment and on their traditional livelihoods. With this coverage, the Canadian public began to realize that the state needed to intervene to protect the environment and to ensure northern peoples benefited from resource projects.

In spite of progress, however, the North still faces many issues and challenges related to the interaction of resource development, the environment, and the place of Aboriginal groups in resource development and Canadian society.[4] In broad terms, all of these challenges relate to the effect of the resource economy on the northern environment and peoples. Resolving disputes over a resource project is not easy. The Northern Gateway pipeline project is a case in point. This controversial pipeline, designed to transport bitumen from the Alberta oil sands to port facilities at Kitimat, BC, for overseas customers, was approved by the National Energy Board (NEB) in June 2014, subject to Enbridge satisfactorily dealing with 209 concerns identified by the NEB. Ottawa agreed with the NEB recommendation. Still, even though the federal Conservative government has viewed this project as in the "national interest," strong opposition from many First Nations and public concerns about oil spills damaging the coastline could prevent the project from proceeding.[5]

The next two chapters outline the physical and historical geography of the North. This background information equips the reader to better understand the physical limitations imposed on the resource economy, the historical process of northern development, and contemporary Aboriginal issues. Resource development and its impact on Aboriginal peoples and the fragile environment form the main thrust of the remainder of this text. To be sure, past human activities have damaged the northern environment, and the road to economic diversification is far from assured. We will see that the Aboriginal peoples of northern Canada face an uphill (but not impossible) struggle to find a place within Canada where they have a foot in both the market economy and their land-based economy.

Challenge Questions

1. While political scientists often define the North as consisting of the three territories of Yukon, Northwest Territories, and Nunavut, geographers do not. Why?
2. The duty-to-consult doctrine provides Aboriginal leaders with another tool to seek fair and just agreements with resource companies. But just what does this doctrine mean?
3. If you were to move from southern Canada to a small Inuit community in the Arctic, would you consider learning Inuktitut?

4. When Louis-Edmond Hamelin conceived of nordicity in the 1970s, he assumed that the North's physical geography was "stable" while its human geography was not. Is this still true? If not, why not?
5. Why did resource development begin in the Subarctic first?
6. Figure 1.3 illustrates the North's cold environment as measured by isotherms (mean annual air temperatures for many locations taken over a 30-year period). If you created a similar map based on mean annual air temperatures for the same set of locations but for only the last 10 years, would you expect a different set of isotherms?
7. How is the Super Cycle for resource commodities challenging the basic premise of the core/periphery theory and what are the implications for the North's economy?
8. From Table 1.3, by examining two megaprojects, the Iron Ore Project and the Voisey's Bay nickel mine, what have been the principal changes over the past 50 years in regard to Aboriginal and environmental issues?
9. Do you agree or disagree that what is given up in land claim agreements is not easily equated to a sum of money, however large? Explain.
10. Explain what Watkins meant by a "staple trap." Why did he believe the North would fall into this "trap"?

Notes

1. The "Producer Group" consisted of Imperial Oil, ConocoPhillips, Shell, and ExxonMobil. Each company had a subsidiary as a member of the Producer Group: Imperial Oil Resources Ventures Limited; ConocoPhillips Canada (North) Limited; Shell Canada Limited; and ExxonMobil Canada Properties. In July 2011, Shell decided to sell its gas reserves, indicating that, in the company's judgment, the Mackenzie Gas Project is no longer viable.
2. Eagle Spirit Energy Holdings and the Vancouver-based Aquilini Group have proposed a First Nations-led alternative pipeline to Enbridge's Northern Gateway project. The Aboriginal pipeline calls for a refinery in the interior of BC or Alberta to process the bitumen and then to transport the refined product to the BC coast. Eagle Spirit plans to file its proposal with the National Energy Board when it has the approval of the First Nations along the pipeline route.
3. "Sense of place" is a term used by geographers to denote the special and often emotional feelings that people have for the region in which they live. These feelings evolve over time and are derived from a variety of personal and group experiences that often occur in a calamity; some are due to natural factors, such as violent storms, while others result from cultural factors, such as an attack on the economy, language, or religion of the people. Whatever its origin, a sense of place is a powerful psychological bond between people and their region. A narrow view of this concept could see a rich resource frontier, such as the Klondike, associated with a mythical region abounding in great wealth known as "Eldorado."
4. In March 2014, the NWT government signed intergovernmental agreements on revenue-sharing and land/resource management with the Inuvialuit Regional Corp., Northwest Territory Métis Nation, Sahtu Secretariat Inc., Gwich'in Tribal Council, and Tlicho Government. Both the Akaitcho and Dehcho First Nations are expected to sign similar agreements in the near future.
5. Gitga'at First Nation claims ownership of the waters through which the oil tankers must pass en route to the proposed oil terminal at Kitimat. These ocean waters provide much food for the Gitga'at, and fear that an oil spill would destroy the marine life fuels their opposition to the Northern Gateway project.

References and Selected Reading

Assembly of First Nations (AFN). 2011. *Pursuing First Nation Self-Determination: Realizing Our Rights and Responsibilities.* www.afn.ca/uploads/files/aga/pursuing_self-determination_aga_2011_eng%5B1%5D.pdf.

Bone, Robert M. 2014. *The Regional Geography of Canada,* 6th edn. Toronto: Oxford University Press.

Burn, Chris. 1995. "Where Does the Polar Night Begin?" *Canadian Geographer* 39, 1: 68–74.

Cardinal, Harold. 1969. *The Unjust Society: The Tragedy of Canada's Indians.* Edmonton: Hurtig.

Coates, Ken, and William Morrison. 1992. *The Forgotten North: A History of Canada's Provincial Norths.* Toronto: James Lorimer.

Cross, Phillip. 2007. "The New Underground Economy of Subsoil Resources: No Longer Hewers of Wood and Drawers of Water." *Canadian Economic Observer* (Oct.). Statistics Canada Catalogue no. 11-010. www.statcan.ca/english/freepub/11-010-XIB/01007/feature.htm.

Environment Canada. 2014. "Canadian Climate Normals." http://climate.weather.gc.ca/climate_normals/index_e.html.

Fitzpatrick, Patricia. 2008. "A New Staples Industry? Complexity, Governance and Canada's Diamond Mines." *Policy and Society* 26, 1: 87–103.

Friedmann, John. 1966. *Regional Development Policy: A Case Study of Venezuela.* Cambridge, Mass.: MIT Press.

Hamelin, Louis-Edmond. 1979. *Canadian Nordicity: It's Your North, Too,* trans. William Barr. Montreal: Harvest House.

Helin, Calvin. 2006. *Dances with Dependency: Indigenous Success through Self-Reliance.* Vancouver: Orca Spirit.

Innis, Harold A. 1930. *The Fur Trade in Canada.* Toronto: University of Toronto Press.

Johnston, Margaret E., ed. 1994. *Geographic Perspectives on the Provincial Norths,* vol. 3. Centre for Northern Studies, Lakehead University. Mississauga, Ont.: Copp Clark.

Lutz, J.S. 2008. *Makuk: A New History of Aboriginal–White Relations.* Vancouver: University of British Columbia Press

McNiven, Chuck, and Henry Puderer. 2000. *Delineation of Canada's North: An Examination of the North–South Relationship in Canada.* Geography Working Paper Series No. 2000-3. Ottawa: Statistics Canada.

Manuel, George, and Michael Posluns. 1974. *The Fourth World: An Indian Reality.* Don Mills, Ont.: Collier Macmillan Canada.

Müller-Wille, Ludger. 2001. "Shaping Modern Inuit Territorial Perception and Identity in the Quebec–Labrador Peninsula." In Colin H. Scott, ed., *Aboriginal Autonomy and Development in Northern Quebec and Labrador,* 33–40. Vancouver: University of British Columbia Press.

Northwest Territories, Bureau of Statistics. 2007. "Aklavik." www.stats.gov.nt.ca/Infrastructure/Comm%20Sheets/Aklavik.html.

Rostow, W.W. 1960. *The Stages of Economic Growth: A Non-Communist Manifesto.* Cambridge: Cambridge University Press

Schaefli, Laura M., and Anne M.C. Godlewska. 2014. "Ignorance and Historical Geographies of Aboriginal Exclusion: Evidence from the 2007 Bouchard-Taylor Commission on Reasonable Accommodation." *Canadian Geographer* 58, 1:110–22.

Slocombe, D. Scott. 1995. "Understanding Regions: A Framework for Description and Analysis." *Canadian Journal of Regional Science* 18, 2: 161–78.

Statistics Canada. 2005. "Land and Freshwater Area, by Province and Territory." www40.statcan.ca/101/cst01/phys01.htm.

Supreme Court of Canada. 2014. *Tsilhqot'in Nation v. British Columbia,* [2014] S.C.C. 44. http://scc-csc.lexum.com/scc-csc/scc-csc/en/item/14246/index.do.

Wallerstein, Immanuel. 1979. *The Capitalist World-Economy.* Cambridge: Cambridge University Press.

Watkins, Melville H. 1977. "The Staple Theory Revisited." *Journal of Canadian Studies* 29: 160–9.

White, Graham. 2006. "Cultures in Collision: Traditional Knowledge and Euro-Canadian Governance Processes in Northern Land-Claim Boards." *Arctic* 59, 4: 401–14.

- 2 -

The Physical Base

The North is a vast polar area. Its geographic extent consists of more than three-quarters of Canada's **land mass** plus a sizable portion of the Arctic Ocean (Figure 1.1). An understanding of its complex physical geography is essential to appreciate its delicate balance with human populations, resource development, and a vulnerable environment. Added to this mix, global warming and climate change represent dynamic elements that are rapidly altering the North's physical geography. A warmer Arctic has the potential to radically transform global energy balance and global circulation systems, thus creating a much hotter world. Stone (2015) describes such unprecedented events with the metaphor, **Arctic Messenger**, warning the world of the global consequences of a warmer Arctic.

Overview

While "cold environment" provides a simple summary of the region's physical geography, polar climate, permafrost, and periglacial features provide a more accurate description of its cold nature. Figure 1.3, "The Cold Environment," expresses both the intensity and spatial dimensions of cold through mean annual temperatures. On this map, the coldest areas include the northern tip of Ellesmere Island at latitude 83°N, where the annual mean temperatures are –20°C or colder; the warmest areas are found near the zero annual mean isotherm at much lower latitudes. Fort McMurray, latitude 56°N, lies just south of the zero isotherm, which also marks the southern edge of permafrost. On the west and east coasts, two currents, the warm **Alaska Current** and the cold **Labrador Current**, affect the location of this annual zero isotherm by driving it northward to 60°N in southern Yukon and then veering southwards to 53°N along the Labrador coast.

Oddly enough, the real story of the North is not its cold environment but its rapid retreat due to global warming. Thus, countering this reality of a cold place is its relentless march to a warmer North caused largely by the **greenhouse effect**. The warming of the North began in earnest some 30 years ago and the details of this warming—as recorded in mean annual temperatures—is demonstrated for 15 northern communities in Table 2.2. Another sign of recent warming is revealed in the late summer increase in the extent of open water in the Arctic Ocean.

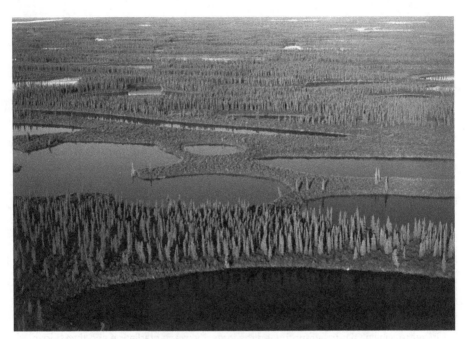

Figure 2.1   Northern Edge of the Boreal Forest in the Mackenzie Delta

The northern edge of the boreal forest forms the treeline, a narrow transition zone from forest to tundra. Here, wetlands (lakes, muskeg, and peat bogs) are interspersed with small, stunted black and white spruce trees. Given sufficient time, global warming could see the Mackenzie Delta become part of the closed boreal forest. Already, signs of ground-hugging tundra species growing into a shrub-like deciduous woodland have emerged (Stone, 2015: 256).

Source: Staffan Widstrand/Nature Picture Library.

The cold environment of the North, then, provides one focus for this chapter while the other is the warming effect. To simplify our task of understanding its interrelated physical nature, our attention is focused on the broad spatial patterns making up the North's cold environment. Biomes, by focusing attention on the adaptation of animals and plants to their particular landscape, provide a vital geographic framework for our discussion of the North's physical geography. The **Tundra biome** and **Boreal biome** offer a natural vegetation expression of the North's geographic extent (Figure 2.1). Climate and permafrost have played a key role in the formation of these two biomes while geomorphic processes unique to the North, such as **gelifluction**, add to our understanding of micro landforms found in this cold environment.

Encompassed within these two biomes are five geomorphic regions—the Canadian Shield, the Interior Plains, the Cordillera, the Hudson Bay Lowlands, and Arctic Lands—that capture the North's varied landscapes and their particular geological structure. The **physiography** of these regions varies due to their distinctive geological history. While each has its own particular landforms, glaciation and permafrost have marked and shaped the surface of each geomorphic region.

A key point stressed in this chapter is that the physical geography of the North is no longer a static phenomenon. Its physical base is slowly changing due to **global warming**. Over a relatively short span of time—three or four decades—warmer annual temperatures have melted sea ice, ice in permafrost, and glaciers. As reported in August 2010 (Smith, 2010), Professor John England observed a large piece of ice, perhaps the size of Bermuda, breaking off from the **Ward Hunt Ice Shelf** on Ellesmere Island in Nunavut. England concluded that the **ice shelf** itself is badly fractured, rapidly failing, and could eventually disappear.

On the other hand, natural vegetation takes much more time to adjust to the new temperature regime. For that reason, the Tundra and Boreal biomes show only minute signs of shifting their boundaries in our warming world. Within the **Holocene Epoch**, for instance, climatic conditions were sufficiently stable for long periods of time (thousands of years) to allow for the formation of Tundra and Boreal biomes (Vignette 2.1).

Vignette 2.1 The Holocene Epoch, Climate Change, and Global Warming

Following the Late Wisconsin Ice Age, the Holocene Epoch marks the beginning of a warmer and relatively stable climate (Table 2.1). The most likely causes of this remarkable **climate change** involve a combination of factors in a powerful feedback loop. The first step in this process of change began some 15,000 years ago. At that time, the Earth's orbit shifted, causing the angle of incoming solar radiation to increase, which, in turn, strengthened the intensity of solar radiation and thus began the warming of the Earth (**Milankovitch theory**). The second step saw this slight increase in the global temperature accelerated by the disappearance of the ice sheets and the duration of snow cover, thus reducing the **albedo** effect, that is, less solar radiation, perhaps as much as 90 per cent, was reflected back into outer space and more solar radiation was absorbed into the surface of the Earth no longer covered by ice and snow. The last step resulted in an increase in the percentage of greenhouse gases in the atmosphere, thus allowing the atmosphere to retain more solar energy and thereby heat the Earth's surface.

Two Powerful Forces

Global energy balance and the **global circulation system** form critical elements in our natural world. The global energy balance, for instance, refers to the balance between incoming energy from the sun and outgoing heat from Earth. The amount of solar energy emitted by the sun is constant, but the capacity to retain solar energy in the atmosphere varies depending on the amount of greenhouse gases in the atmosphere. Simply put, more greenhouse gases in the atmosphere lead to warmer surface temperatures. The reverse is also true. Human activities, such as the burning of fossil fuels, add greenhouse gases to the atmosphere. In this way, the North's temperature is rising and human activities, for the first time, are playing a part in global warming and climate change.

The amount of solar energy received at the Earth's surface varies. The global circulation system redistributes much of that energy. In short, the amount of solar energy received by the Earth varies by latitude.[1] For that reason, high-latitude areas, like the North, receive a relatively small amount of solar energy compared to low-latitude countries. The explanation for this spatial variation is found in its angle of incidence, that is, the angle at which the sun's rays strike the Earth's surface. In high latitudes, this angle is oblique, meaning that the rays are spread over a larger surface area, thus reducing the intensity of its warming capacity. The reverse is true for low latitudes, where the rays of the sun strike the Earth at right angles. The net result is energy surpluses in the tropics and energy deficits in the Arctic. Fortunately, the global circulation system ameliorates this imbalance somewhat through ocean currents and atmospheric winds that transfer heat from low to high latitudes, thus maintaining the present patterns of world temperatures and global climate zones. In recent decades, however, global warming has increased world temperatures, affected the global circulation system, and, as a result, initiated the process of altering the Arctic and other global climatic zones.

Vignette 2.2   The North's Energy Deficit

The North is an area of energy deficit. The principal factor contributing to this deficit is the low amount of solar energy received in high latitudes and the high loss of this energy through reflection from the ice/snow surface (the albedo effect). Williams (1986: 6–7) defines a cold environment as having a negative annual heat balance due to a greater amount of long-wave radiation emitted to outer space than the amount of incoming short-wave radiation and energy transfer by global winds and ocean currents. In general, the amount of the North's energy deficit increases with latitude but latitudinal fluctuations occur, and these fluctuations correspond closely with the mean annual isotherms shown in Figure 1.3. Interestingly, the decrease in ice/snow cover due to global warming is a major factor in the more rapid warming of the North than other areas of the world. Within the Canadian North, warming is more apparent in the western half due in part to the influence of the warm Pacific Ocean.

Northern Biomes

Biomes represent systems of natural communities within five natural vegetation zones (aquatic, desert, grassland, forest, and tundra) found in the world. Within these zones, individual plants and animals have adjusted to the specific environmental conditions. Each biome, therefore, is characterized by a distinctive kind of biological community adapted to those physical conditions.

Natural vegetation serves to distinguish the Tundra and Boreal biomes. The complexity of these two biomes, however, is reflected in their ecology and **biodiversity**. Because of its extremely cold environment, the Tundra biome has the smallest number and variety of flora and fauna—though some species, habitats, and ecological processes are unique, such as the ability of some plants to survive extreme cold and dryness, the physiological features that allow mammals to maintain body heat through an Arctic winter, and the presence of life within sea ice.

Human activities often have a negative impact on the North's fragile environment and therefore on its biodiversity (Vignette 2.3). Clear-cut logging in the boreal forest

---

Vignette 2.3   A Fragile Environment

The northern environment is often described as fragile, meaning that the risk of anthropogenic damage is much higher in the North than in other regions of Canada. The primary reason for this regional variation is the much greater length of time required for nature to repair human damage to a northern environment compared to the time required in more temperate regions of the world. One illustration of the delicate nature of this cold environment is revealed by the relationship of temperature and precipitation to plant growth. Plant growth varies with temperature and precipitation. A combination of low temperatures and meagre precipitation results in very low levels of biological activity and, hence, a longer period of time is required for nature to heal itself. The same rule applies within the two biomes, where the Boreal biome can recover more quickly from physical damage because it has a longer, warmer summer and receives more precipitation than the Tundra biome.

---

frequently results in new species taking hold, thus displacing the original forest. Mining activities discharge toxic effluents into water systems and emit air pollutants into the atmosphere, while hydroelectric projects replace a natural environment with an industrial landscape by disrupting the flow of rivers, building dams, reservoirs, and power stations, and constructing high-tension power lines.

How were these two biomes formed? When the great ice sheets of the last ice age still covered much of North America some 15,000 years ago, plants and animals had to wait for these ice sheets to melt before they could become established. This biological process took thousands of years before the Tundra and Boreal biomes were formed. While the climate was relatively stable during the Holocene Epoch, minor variations did occur. Once cooling took place during the **Little Ice Age** (*c.* 1450–1850), the treeline retreated southward because its major species, black spruce, requires an average summer monthly temperature of 10°C or higher to permit regeneration. However, over the last 30 years, the treeline has slowly edged northward due to warmer and longer summers.

Vignette 2.4   Ecology, Ecosystems, and Ecozones

Ecology is the study of the interactions of living organisms with one another and their physical environment. One of the characteristics of biological life is its high degree of complexity and interrelationship. For example, a group of individuals of the same species living and interacting in the same geographic area is defined as a population. Many populations may exist in the same geographic area or habitat and, collectively, these populations form a biological community. Ecosystems or biomes represent large, continental units while ecozones represent smaller units within ecosystems. Both units express the spatial extent of this complexity and interrelationship. In Canada, there are 15 terrestrial and five marine ecozones. A map of the terrestrial ecozones for Canada is found at http://sis.agr.gc.ca/cansis/publications/maps/eco/all/zones/index.html.

The Tundra Biome

Lying north of the treeline, the Tundra biome was formed under an Arctic climate where normal tree growth is prohibited by the cold temperature. This biome is found primarily in Nunavut and the Northwest Territories, but it also exists in Quebec, Newfoundland and Labrador, Ontario, Yukon, and Manitoba (Figure 1.1). It is associated with the Arctic climate where the warmest month has a mean temperature of less than 10°C and the coldest month is often less than –30°C (Figures 2.6 and 2.7). Precipitation is low, often under 300 millimetres annually (Figure 2.8). In spite of such dry conditions, wetlands, such as peat bogs, exist because permafrost prevents the water from draining away. As well, permafrost inhibits root development. Under such natural conditions, normal tree growth and soil formation are not possible. In fact, its short, cool growing season limits natural vegetation growth to low-lying vegetation known as tundra. This type of natural vegetation consists of shrubs that hug the ground to avoid the impact of Arctic winds while sedges, reindeer mosses, grasses, and a wide variety of flowers remain near ground level. Another reason for plants to hug the ground is because a strong air temperature gradient exists, so that a warm summer day could see a temperature of 20°C at ground level but 10 cm above ground level the air temperature could fall by half. These tundra plants have horizontal root systems because the thin **cryosolic soils** are underlain by permafrost. Tussock sedge and ground-hugging shrubs provide summer grazing for caribou, which are still a major source of food for Aboriginal peoples. Heath, herbs, and lichen are the typical plants. Reindeer moss, a grey-green, sponge-like lichen, is an important source of food for caribou, muskox, and other herbivores. While trees such as willows do grow, they reach a height of only a few centimetres.

Beyond the tundra vegetation, an even colder environment exists where even tundra plants cannot survive. Known as polar desert, its barren surface consists of rock and unconsolidated material. The principal life form is the primitive lichen commonly found on the south-facing slopes of boulders. It is not surprising, then, that the biodiversity of the polar desert is extremely low and vegetation structures are very simple.

The Arctic Ocean falls under the Tundra biome as a marine subset. Much of the Arctic Ocean is covered by a permanent floating ice cap. This ice cap, also known as **Arctic ice pack**,[2] consists of old ice and new ice that slowly moves around the Arctic Ocean. Old or **multi-year ice** represents a harder ice that forms the core of the Arctic ice cap. **New ice** or first-year ice refers to ice that was formed over the course of one year. For most of the year, the Arctic ice cap covers the entire Arctic Ocean. During the dark Arctic winter, new ice is formed, causing the ice cap to reach its maximum extent around early March. During the long summer days, solar radiation warms the ice and the open water, causing the melting of the Arctic ice cap. By September, the ice cap reaches its minimum extent. In October, the cycle repeats itself. Over the last decade and perhaps longer, the Arctic ice cap has not only lost much of its old ice, but its geographic extent is diminishing and its thickness is thinning. Stone (2015: 219) has shown graphically that not only is the geographic extent of the Arctic ice cap decreasing, but that its composition is increasingly becoming young ice (year-old ice), with less and less old, hard ice (four or more years old).

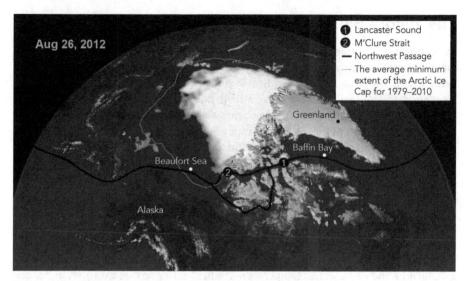

Figure 2.2   The Arctic Ice Cap, 26 August 2012

On 26 August 2012, the Arctic ice cap reached its smallest geographic extent over the past 30 years at 3.63 million km². In the following two years, however, cooler summers limited the ice melt and the geographic extent of the Arctic ice cap was larger than in 2012, at 5.35 and 5.28 million km² (NSIDC, 2015). These "cold" summers are thought by NASA to be anomalies based on its findings over a period of 30 years (1979–2010) that indicate a general increase in the area of open water and thus a retreat of the summer extent of the Arctic ice cap. As indicated in this NASA satellite image, the southern route of the Northwest Passage was ice-free while its northern route was blocked by pack ice at M'Clure Strait. In fact, M'Clure Strait has only been ice-free for a short period of time in 2007 and 2008.

Source: NASA/Goddard Space Flight Center Scientific Visualization Studio. The Blue Marble data is courtesy of Reto Stockli (NASA/GSFC).

Since young ice more quickly turns to water in the summer, the pace of ice loss and therefore more open water can only increase in the years ahead. As Barber and others (2008: 7) observed, this warming process affects physical, biological, and social systems associated with the Arctic Ocean.

Significantly, since 1979, the extent of the Arctic ice cap has decreased in most years (Figure 2.2). Its slow but uneven retreat each summer might lead to an ice-free Arctic Ocean by the end of the twenty-first century. Without a doubt, the effect of global warming resulting in more open water in the summer has increased interest in commercial shipping through the Northwest Passage, sparked a renewed interest in Arctic sovereignty by Ottawa, and caused Canada to claim the seabed lying in international waters. These geopolitical topics are discussed in Chapter 8.

The transition zone between the two biomes is represented by the treeline. Like many other natural features, the treeline in fact is a transitional zone between the closed boreal forest and exclusively tundra vegetation. While the treeline closely corresponds to the isotherm representing a 10°C **mean monthly temperature** for July, other natural factors, such as the depth of the **active layer of permafrost**, topography that protects trees from wind, south-slope radiation, and well-drained land may result in patches of trees growing north of this isotherm or, conversely, tundra occurring south of this isotherm. All of these factors have produced a narrow transition zone at the northern edge of the Boreal biome that can be described as "wooded tundra." In this transition zone, the proportion of tundra to forest varies. Towards the southern edge of the Tundra biome, patches of bush-size spruce and larch are found in sheltered, low-lying areas while high, more exposed lands are treeless. Stands of more mature trees mixed with open areas consisting of a thick ground cover of lichens are found (see Figure 2.5). This transition zone represents a biological and cultural boundary. Polar bears and Arctic foxes prevail in the Tundra biome, while beaver, black bears, and moose are restricted to

Vignette 2.5 Polar Desert

Polar desert exists in the higher latitudes of the Arctic where extremely cold, arid conditions occur throughout the year. Low summer temperatures combined with a very thin active layer of permafrost greatly limit biological activity. For that reason, little vegetation is found in polar deserts. Lichens are by far the most important group of primitive plants found in polar deserts. Since the principal geomorphic process is a freeze/thaw cycle, a sterile, barren-looking landscape consisting of shattered bedrock, patterned ground, and unconsolidated materials prevails. An example of polar desert is provided in the photograph below of a barren landscape with no visible vegetation. Located on Melville Peninsula (around 69°N), frost action has produced a rugged surface of shattered rock fragments.

Figure 2.3 Polar Desert

Frost-heaved rock fragments along jointing sites, near Hall Beach, Melville Peninsula, Nunavut, provide an example of physical weathering and frost heave. In this case, extreme temperature differences caused the rocks to split and an upward movement caused by ice in the active layer of permafrost causes the rock fragments to shift.

Source: Reproduced with permission of Natural Resources Canada 2015, courtesy of the Geological Survey of Canada.

Figure 2.4   Wooded Tundra Landscape

The transition zone between the Tundra and Boreal biomes consists of a blending of two natural vegetation types within a cold environment. Here, the Boreal biome is represented by a sparse stand of stunted black and white spruce while tundra ground cover consists of lichen, dwarf birch, dwarf willow, and labrador tea.

Source: Northwest Territories (2004). Photo: Dave Downey, Department of Environment and Natural Resources, Government of the Northwest Territories.

the forest lands of the Boreal biome. Barren ground caribou, on the other hand, migrate between the two biomes.

The treeline (and therefore the Arctic) does not follow a latitudinal direction, but has a distinct northwest to southeast direction due to two factors. First, the **continental effect** causes the interior of the North to warm in the summer, thus allowing the treeline to reach the mouth of the Mackenzie River, well north of the Arctic Circle, which is located at 66° 33′N and shown on Figure 1.3. Second, the cold waters of Hudson Bay and the Labrador Sea prevent tree growth along its coastline. The latitudes of three communities just south of the treeline are Aklavik (68° 13′N), Churchill (58° 48′N), and Cartwright (53° 36′N). Aklavik, located on a deltaic island at the mouth of the Mackenzie River in the Northwest Territories, has a warmest-month average mean temperature of 14°C; Churchill, situated near the shore of Hudson Bay in northern Manitoba, has an average July temperature of 11.8°C; and Cartwright, lying along the Labrador coast, has a mean July temperature of 12°. As these spring/summer temperatures are expected to rise in the years to come, the melting of snow will expose the tundra vegetation to sunlight needed for photosynthesis earlier, thus encouraging the treeline to advance further north.

The Boreal Biome

The Boreal biome is the largest natural region in North America. Its natural vegetation, the boreal forest or taiga, provides a clear indication of its geographic extent. The boreal forest extends in a continuous belt from Yukon to Labrador. While its long winters are cold, summers are short but warm, allowing for a much richer vegetation cover than that found in the Tundra biome. Over the year, average monthly temperatures are much higher than those found in the Tundra biome. January average temperatures are around –20°C, some 10°C warmer than those found in the Tundra biome. The result is a closed boreal forest as depicted in Figure 2.5. Similarly, July average temperatures often reach 15°C (Figure 2.7). Annual precipitation is much higher, particularly in Ontario, Quebec, and Labrador, where it is more than double that found in the Tundra biome (Figure 2.8). As well, the existence of discontinuous permafrost means that large areas are permafrost-free and that the depth of the active layer is much greater than in continuous permafrost, thus allowing deep root development and enhanced drainage. Outside of wetlands, these natural conditions promote the growth of the boreal forest where coniferous trees predominate.

The closed boreal forest forms the principal natural vegetation in the Boreal biome. The forest parkland is a transition zone between the Boreal biome and the Grassland biome. The closed boreal forest, a dense forest of mature fir, spruce, and pine, covers the northern areas of seven provinces stretching from British Columbia to Newfoundland and Labrador. This closed forest is also found in southern Yukon and a small part of the Northwest Territories (principally the upper Mackenzie Valley). Within this huge zone, the forest cover is broken by a variety of wetlands, including lakes, **muskeg**, and peat bogs. Black spruce and larch are the most common species in these wetlands. On well-drained land, however, species of spruce, fir, pine, and larch are common, along with stands of poplar and birch. Towards its southern limits, broad-leaf trees, particularly aspen and birch, are found.

Pacific air masses have a strong warming effect on the western section of the Boreal biome. During the winter, close proximity to the Pacific Ocean keeps temperatures of the western section of this biome relatively high while the frozen Hudson Bay reinforces the cold continental effect on the winter temperatures of the eastern part of the boreal forest. The average January temperatures for Prince George (–9.6°C) in BC, at a latitude of just under 54°, compared to Chibougamau (–18.4°C) in Quebec, at about the same latitude (just under 55°), demonstrate the powerful effect of marine influences on climate. Even a more telling effect of marine influences is that Prince George lies well south of the mean annual zero isotherm while Chibougamau falls to its north.

Podzolic and gleysolic soils are common in the Boreal biome. The thin, acidic **podzolic soils** are formed under cool, wet growing conditions where the principal vegetative litter is derived from a coniferous forest. **Gleysolic soils** are associated with extremely poorly drained and often waterlogged land such as marshes and bogs. A low evaporation rate, immature drainage, and permafrost ensure an excess of ground moisture, resulting in severely leached soils and the widespread occurrence of ponds, lakes, and bogs. Yet, the longer summer temperatures in the Boreal biome allow well-drained ground to thaw to a depth of several metres, promoting biological activity, plant growth, and chemical action. Unlike the cryosolic soils in areas of continuous permafrost, podzolic soils are associated with discontinuous and sporadic permafrost, both of which have relatively thick active layers (for more on the types of permafrost and their geographic extent, see the subsection on permafrost in this chapter).

Figure 2.5   Boreal Forest in the Northwest Territories

The boreal forest extends across Canada in a green swath, reaching into high latitudes in the Northwest Territories. Just south of the Arctic Circle in the vicinity of Great Bear Lake and the Mackenzie River Valley, the boreal forest consists of mature stands of black and white spruce. In this photograph, the boreal forest extends to the shore of Oscar Lake.

Source: The Boreal Songbirds Initiative, www.borealbirds.org/. © Ducks Unlimited, Canada.

The boreal forest is a rich zone of biodiversity with different types of wildlife: moose, deer, martens, rabbits, beaver, foxes, wolves, bears, eagles, and various birds. Local medicinal plants for treating colds, diabetes, and heart and skin problems are known and used by First Nations and Métis. The biological diversity of the boreal forest includes thick layers of moss, soil, and peat that store huge amounts of organic carbon and thus play a significant role in regulating the Earth's climate. Boreal wetlands also filter millions of gallons of water each day that fill Canada's northern rivers, lakes, and streams.

Polar Climate

In the classification of climates throughout the world, the polar climate is the coldest climatic type. The polar climate is divided into four climatic subtypes: the Arctic, **Subarctic**, mountain, and ice cap climates. In this text, however, the mountainous areas of northern British Columbia, Yukon, and the Northwest Territories are treated as part of the Subarctic. Similarly, the ice cap climatic type is merged with the Arctic climatic type. The ice cap climate is associated with glaciers found on Baffin, Devon, and Ellesmere islands. This climate has a mean temperature below freezing for all months. Those small areas covered by glaciers are considered part of the Arctic climate. Each climatic type is associated with an **air mass** that reflects the weather characteristics of that type. As these air masses move across the continent, they affect weather in other regions.

The polar climate is characterized by extreme seasonal variations in the amount of solar energy. Summer days, for example, are long while winter days receive little to no sunlight. The Arctic Circle marks the latitude where the sun remains above the horizon for one summer day (21 June, the summer solstice) each year and remains below the horizon for one winter day (21 December, the winter solstice) each year. At latitudes well beyond the Arctic Circle, summer days and winter nights can last for months. At the North Pole, the Sun is above the horizon for six months and below the horizon for the rest of the year. A day of continuous darkness is referred to as a "polar night."

Monthly receipts of solar energy vary widely throughout the year. On a yearly average, the Earth's poles receive 40 per cent less radiation than the equator (Lawford, 1988: 144). Within the Canadian North, the major dividing line, the Arctic Circle, marks the point where solar radiation is reduced to zero for one day (21 December), and at higher latitudes for longer periods of time. Such low levels of radiation result in continuous cooling of the land and the buildup of masses of frigid Arctic air, which are associated with daily high temperatures of –40°C or lower and strong surface winds. In fact, the Arctic coast is one of Canada's windiest places, with annual average wind speeds exceeding 20 km/hr. Cold polar air is often associated with extreme wind-chill conditions that will freeze exposed flesh in a matter of seconds. Arctic winds drive frigid air masses southward, causing stormy and sometimes blizzard conditions in the Canadian Prairies, sub-zero temperatures in eastern Canada, and freezing temperatures in the southern United States. In the spring, solar radiation increases but much of its effect is lost due to snow-covered surface. In fact, up to 80 per cent of the spring solar radiation is reflected into space by the snow-covered ground. Once the snow is gone, winter's grip is quickly broken. Temperatures recover rapidly and the ensuing warm weather quickly melts the ice from lakes, rivers, and the ocean. By early July, sea ice has disappeared from Hudson Bay, and a few weeks later the ice is gone from along the edge of the Arctic coast. The polar pack, no longer attached to the coastline, drifts around the Arctic Ocean. During the long summer days, massive amounts of solar radiation penetrate to the ground and open waters, causing their surface temperatures to rise sharply. In response, plants quickly appear and flower and the warm ocean water decays the Arctic ice pack more quickly. In turn, more open water means a decrease in surface albedo, which results in warmer ocean water and an acceleration in the melting of the ice pack.

Daily summer temperatures in the Mackenzie Valley and southern Yukon can reach into the low thirties Celsius. By late August, however, summer is over. Within another month, ice has formed on lakes and rivers, marking a return to a frozen landscape. By October, the surface of the open waters in the Arctic Archipelago has again frozen.

The wide latitudinal difference in mean monthly air temperatures is illustrated by the isotherms in Figures 2.7 and 2.8. The coldest **mean daily temperature** for January (–35°N) occurs in the northern extremes of Ellesmere Island (Figure 1.3). Such low January temperatures result from the absence of incoming solar radiation in January. As the most northern island in Canada, Ellesmere Island extends north of 80°N and therefore falls into the zone of polar nights. As an example, the climate station of Alert

---

Vignette 2.6   The Polar Night

||||||||||||||||||||||||||||||||||||||||||||||||||||||||||||||||||||||||||||||||||||||||||||||||||||||||||||||||||||||||||||||||||||||||||||||||||||||||||||||||||

The polar night is a period of continuous winter darkness. Twilight does not occur. Polar nights take place north of 72° 33'N, well beyond the Arctic Circle (66° 33'N). Why is this? We know that the sun does not rise above the horizon at the Arctic Circle at the winter solstice. Yet there are not 24 hours of continuous darkness because diffused light from the sky is caused by the sun's rays being reflected from a position below the horizon onto the atmosphere and then back down to Earth. Twilight may last for an hour or more at the time when the sun is below (but less than 6° below) the horizon, thus providing at that time sufficient light for outdoor activities. On 21 December, the latitude of 72° 33'N, not the Arctic Circle, marks the geographic point where 24 hours of continuous winter darkness occurs.

Sources: Burn (1995, 1996).

(82° 31'N), located on Ellesmere Island, has a mean daily January temperature of –29.4°C (Natural Resources Canada, 2013). In contrast, Fort McMurray in northern Alberta (56° 39'N) has a mean daily January temperature of –17.4°C (Natural Resources Canada, 2013a). This resource city based on the oil sands lies just south of the southern limit of the North where the mean annual temperature is slightly above zero. In comparison with Alert, Fort McMurray benefits from a daily average of just over seven hours of daylight each clear day in January. Even though the angle of these incoming solar rays has limited heating power because of their oblique angle, they do warm the surface of the ground. As well, Fort McMurray gains from its proximity to a warm ocean where the ameliorating effect of the occasional Pacific air mass moderates January temperatures. Surrounded by the frozen Arctic Ocean, Alert does not experience the ameliorating effect of marine influences.

A different configuration of the mean daily temperature for July demonstrates both the continental effect of the northern land mass where the land heats up quickly and the cooling effect of the Arctic Ocean, Hudson Bay, and the North Atlantic that keeps the July temperatures of adjacent coastal areas cool. As a result, the July isotherms show a distinct northwest-to-southeast direction. As well, a warm corridor extends down the Mackenzie Valley to Norman Wells, where July temperatures are similar to those experienced in the Canadian Prairies. For instance, Norman Wells (65°N) has a mean July temperature of 17.1° while Saskatoon, situated in the heart of the Canadian Prairies at 52°N, has a July temperature of 18.5°C (Natural Resources Canada, 2013b). No warm corridor exists in the eastern half of Canada's North. In fact, along the Atlantic coast, the Labrador Current, consisting of Arctic waters, exerts a chilling effect on July temperatures and extends the Tundra biome to latitudes in the low fifties (Figure 1.1).

Canada's North receives low amounts of precipitation except along its east coast. In fact, as latitudes increase, annual precipitation declines. The least amount of precipitation, in the Tundra, is due to the inability of Arctic air masses to absorb moisture from cold bodies of water. The isohyets shown in Figure 2.8 clearly indicate that the lowest

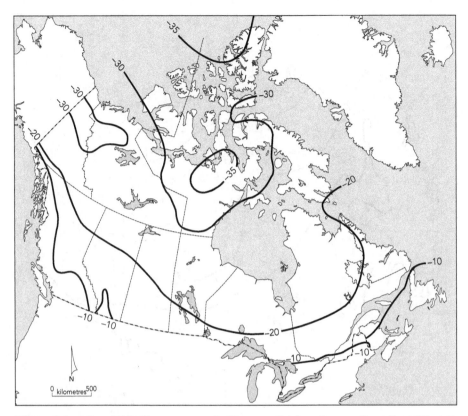

Figure 2.6 Mean Daily Temperatures in Degrees Celsius, January (based on Climatic Normals for 1941–71)

The Pacific and Atlantic oceans have a strong temperature-moderating influence on adjacent coastal areas but this **marine effect** lessens inland. Well beyond the marine influence, the continental effect takes hold resulting in much colder January temperatures than those experienced at the same latitudes along the coast. The reverse takes place in July, when inland temperatures are much higher than those along the coast. The –20°C January isotherm provides evidence of this phenomenon as this isotherm dips southward from Whitehorse, Yukon, at 61°N, to reach middle latitudes around 50°N in Ontario and Quebec and then bends northward to the southeast tip of Baffin Island at 63°N.

Source: Natural Resources Canada (2013a). Licensed under the Open Government Licence—Canada.

amounts of annual precipitation are found in the Arctic Archipelago, where some islands record less than 100 mm annually. The lowest annual precipitation occurs in these ice-locked islands (Vignette 2.7). Victoria Island, for example, has such scant rain and snowfall (often less than 140 mm annually) that the island is described as a "polar desert." In addition to precipitation, water is released in the summer by the melting of ice in the active layer of permafrost.

Closer to the northern edge of the boreal forest, annual precipitation in the Tundra biome increases. At Yellowknife, the annual precipitation is around 250 mm. The Boreal biome, in contrast to the Tundra, generally receives more precipitation, usually over 300 mm annually. Precipitation does vary, however. The greatest amount

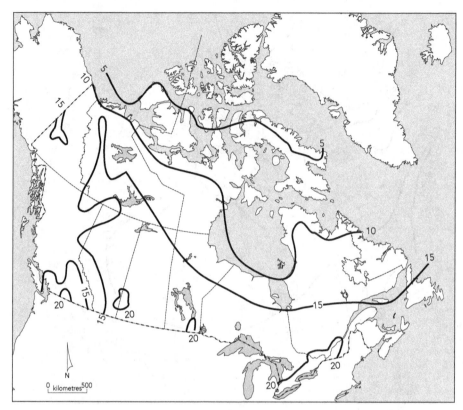

Figure 2.7 Mean Daily Temperatures in Degrees Celsius, July (based on Climatic Normals for 1941–71)

The most remarkable features of the July isotherms are (1) they ignore latitudes and follow a northwest–southeast trend; and (2) the 15°C July isotherm reaches beyond the Arctic Circle by following the Mackenzie River Valley. This warm temperature corridor allows the treeline to reach the Mackenzie Delta (see Figure 2.1). Both the Mackenzie River Valley and the Yukon River Valley provide unique natural conditions, including lower elevations, to allow such higher temperatures at these polar latitudes.

Source: Natural Resources Canada (2013b). Licensed under the Open Government Licence—Canada.

Vignette 2.7 Arctic Archipelago

The Canadian Arctic Archipelago is a group of islands in the Arctic Ocean. Covering over 1.3 million km², they form the largest group of islands in the world. The largest islands are Baffin Island (507,451 km²), Victoria Island (217,290 km²), and Ellesmere Island (196,236 km²). Most islands have elevations below 200 metres and few topographic features. Elevations do rise above 2,000 metres in the eastern islands of the Arctic Archipelago. Mount Barbeau on Ellesmere Island, for example, reaches an elevation of 2,616 metres. At these high elevations, ice shelves and glaciers exist, though both are retreating because the local climate is warming. The geological history of the Arctic Archipelago began some 3 billion years ago and Precambrian rock exists at the surface on Baffin, Devon, and Ellesmere islands. Most of the Arctic islands were formed much later and contain sedimentary rocks. Within these sedimentary rocks, vast oil and gas deposits exist.

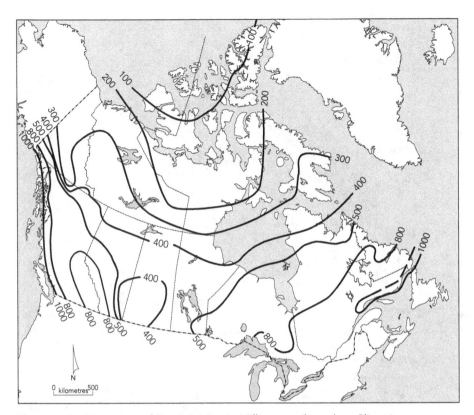

Figure 2.8 Mean Annual Precipitation in Millimetres (based on Climatic Normals for 1941–71)

Air masses originating over the Atlantic and Pacific oceans bring most moisture to the North. Air masses forming over the cold Arctic Ocean usually contain little moisture. As a result, the highest levels of annual precipitation are found along the western and eastern edges of the North while the Arctic Archipelago is extremely dry, causing scientists to refer to this area as a polar desert (Figure 2.3). But could the Arctic Archipelago become a warmer and wetter place? The reasoning goes as follows: first, as open water in the Arctic Ocean increases, air masses are expected to capture more moisture and therefore distribute it over the Arctic Archipelago as precipitation; second, with open water, cloud cover over the Arctic Archipelago could increase, thus trapping the energy below and effectively increasing surface temperatures.

Source: Natural Resources Canada (2013c). Licensed under the Open Government Licence—Canada.

of precipitation occurs in the Cordillera and along the Atlantic coast. In these two areas, the high terrain results in orographic precipitation (rain or snow caused when warm, moisture-laden air is forced to rise over hills or mountains and is cooled in the process). The south coast of Baffin Island receives around 400 mm annually, while the annual total for the southern coast of Labrador and northern Quebec exceeds 800 mm. Most precipitation falls as snow. In the spring, the runoff peaks when melting snow and ice flow into the streams and rivers. Some communities along the Mackenzie River are subject to spring flooding. Fort Simpson, situated on an island at the confluence of the Liard and Mackenzie rivers, has been inundated a number of times, and Aklavik, located in the delta of the Mackenzie River, is threatened by flood waters almost every spring. The occurrence of spring flooding at Aklavik is so regular that, in

Figure 2.9 The Arctic Archipelago

Source: Adapted from Infoplease, www.infoplease.com/atlas/region/nunavut.html.

the late 1950s, the federal government decided to create the new town of Inuvik rather than expand the community of Aklavik.

Climate Change

Since its existence, the Earth has experienced many climates. Each climate represents an average of relatively stable weather conditions over a significant geological time frame. In the **Pleistocene Epoch**, for example, climatic change was associated with the advance and retreat of continental ice sheets that span some 100,000 years. These advances and retreats were controlled by variations in the tilt and position of the Earth's orbit. In short, the total summer radiation received in northern latitudes varied from low to high in accordance with changes in the Earth's orbit, albedo effect, and the proportion of greenhouse gases in the atmosphere. Low summer radiation correlated with ice formation and glacial advance while high summer radiation corresponded with ice melt and glacial retreat. Milankovitch (1941) first discovered this correlation between the series of glacial advances and retreats in the Pleistocene Epoch and the position of the orbital tilt of the Earth. Huybert and Wunsch (2005) later confirmed Milankovitch's findings (for more on how the Earth moves around the sun and the Milankovitch cycles, see Stone, 2015: 318–19).

Climate change now is associated with global warming, whereby higher global temperatures are caused primarily by human activities that increase the amount of greenhouse gases in the atmosphere. Of all parts of the Earth, the North has endured the greatest increase in temperature in the last 30 years. One factor is its albedo or reflective quality. As its snow and ice cover melt, more and more of its surface, whether land or water, is able to absorb rather than reflect solar radiation.

What exactly is "climate"? Climate is an average of complex natural phenomena that make forecasting climate change or reconstructing past climatic changes very difficult. For instance, while the source of the Earth's heat is constant emission of solar radiation from the sun, many variables affect the warming of the Earth. One is the capacity of the ground to absorb (or reflect) solar energy; another is the reduction of solar radiation reaching the Earth's surface because of dust in the atmosphere caused by volcanic activity or large meteorites striking the Earth's surface. In addition, radiation from the sun may vary slightly and that variation can account for changes in the Earth's average temperature. For example, sunspots (flares of solar energy) might increase solar energy to the Earth, but this theory is unproven. Yet, the **Maunder Minimum**, a period of relatively few sunspots, occurred at the same time as the Little Ice Age, suggesting that a low level of sunspot activity on the sun's surface could result in slightly less solar radiation reaching the Earth.

The recording of surface temperatures and precipitation began around 1850 in Europe and North America and somewhat later in other parts of the world, and these measurements provide the basis for our understanding of climates. Reconstruction of ancient climates involving proxy methods such as interpreting the geological record and relating temperature change to variations in oxygen isotopes found in ice cores add to our understanding of climate change. As for predicting future climates, global climate models (GCMs) focus on estimating temperature increases based on the amount of greenhouse gases in the atmosphere. Since long-term variations in climate are measured in decades, centuries, and even millennia, ice-core analyses from glaciers and polar ice caps are our best indicators of climate change because they "average" out the short-term weather variations and reflect longer-term temperature trends. (See Chapter 8 for a more detailed account of the impact of global warming on the Arctic ice cap.)

According to proxy measurements, a general interpretation of climatic change for the current Holocene Epoch is revealed in Table 2.1. Houghton et al. (2001) estimated variations from the current global average annual temperature for key climatic periods, but their proxy time frame does not always agree with the time frame in Table 2.1. The Little Ice Age[3] is a case in point. Nevertheless, the following figures from Houghton et al., for the third assessment report of the **Intergovernmental Panel on Climate Change** (IPCC), reveal the range of temperature change in the Holocene Era and provide some perspective on our present and future situation. Their 2001 GCM prediction of a global temperature increase of $+4°$ by the end of this century was confirmed in 2013 in the Fifth IPCC *Assessment Report* (IPCC, 2013: 21). In viewing the global data

Table 2.1 Major and Minor Climatic Variations in the Holocene Epoch

Period	Name	Climate conditions
14,000 BP	Holocene warming	Slow warming from the last ice age; large ice melt.
12,000 BP–10,500 BP	Younger Dryas	Rapid cooling, prolonged cold period, then rapid warming.
7000 BP–5000 BP	Climatic Optimum	Warm conditions; temperatures were perhaps 1 to 2 degrees Celsius warmer than they are today. Great ancient civilizations began and flourished.
5000 BP–4000 BP		Cooling trend; drops in sea level and the emergence of many islands.
4000 BP–3500 BP		Short warming trend.
3500 BP–2750 BP		Colder temperatures and renewed ice growth; drop in sea level of between 2 to 3 metres below present-day levels.
2750 BP–2150 BP		Slight warming not as warm as the Climatic Optimum.
2150 BP–AD 900		Cooling trend; Nile River (AD 829) and Black Sea (AD 800–1) froze.
AD 1100–1300	Little Climatic Optimum or Medieval Optimum	Warm; warmest climate since the Climatic Optimum; Vikings established settlements on Greenland and Iceland.
1300–1550		Cool and more extreme weather; abandonment of settlements in the Southwest United States.
1450–1850	Little Ice Age	Coldest temperatures since the beginning of the Holocene;populations die from crop failure and famine in Europe.
1850–present	Contemporary climate	Warming trend.

Source: Ward (2012).

below, regional difference can be significant. For instance, the Arctic is warming much faster—at about twice the global rate (Stone, 2015: 196).

- −10°C: Late Wisconsin Ice Age, 30,000 BP to 15,000 BP
- +5°C: Climatic Optimum, 9000 BP to 5000 BP
- +3°C: Medieval Optimum AD 800 to 1300
- −1°C: Little Ice Age, 1450 to 1850
- +0.6°C: Anthropogenic warming, 1850 to 2000
- +4°C: Predicted rapid warming, 2000 to 2100

Around 15,000 BP, the Earth warmed again, causing the ice sheets to melt and thus allowing biological life to re-enter the ice-free terrain. Geologists refer to this current warm period as the Holocene, and within that very short geological period, minor climatic variations have taken place; the warmest period, known as the Climatic Optimum, occurred between 7,000 and 5,000 BP. During the Climatic Optimum global temperatures were considerably higher than those currently experienced. Kaufman and others (2004) reconstructed Climatic Optimum temperatures for Canada's western Arctic. Their findings support the notion that temperatures in higher latitudes rose during this period, largely due to the albedo effect. In more recent times, two minor variations in global temperatures took place—the Medieval Optimum (c. 800–1300) and the Little Ice Age (c. 1450–1850). The Little Ice Age did affect the Thule, who had previously hunted the bowhead whale in the open waters of the Arctic Ocean. Within this colder environment, more extensive ice cover took away their opportunity to harvest these huge sea mammals and the Thule (and their descendants, the Inuit) had to hunt seal and caribou (Fossett, 2001).

A Warming North

Temperature recordings over a 50-year period at climate stations across Canada provide data about temperature change. As noted earlier, this trend as revealed by proxy data such as ice melting is particularly noticeable in the Canadian North. Evidence of such temperature changes is available in the Canadian Climate Normals produced by Environment Canada as 30-year summaries of the average climatic conditions of a particular location. Our focus is on the **mean annual temperature**, which provides a snapshot of temperature change over time for a selected number of climate stations. For a 50-year period, from 1961 to 2010, Environment Canada has produced three overlapping 30-year periods: 1961 to 1990; 1971 to 2000; 1981 to 2010. By selecting the mean annual temperatures for 12 climatic stations across the Canadian North and three southern stations, a regional picture of temperature change is shown in Table 2.2, as well as by isotherms in Figure 1.3. From Table 2.2, the following observations are:

- From 1961 to 2010, the 12 northern climate stations averaged a temperature increase of 0.7°C while the three southern stations had a slightly lower increase of 0.5°C.
- Mean annual temperatures for the 12 northern climate stations increased over the three 30-year periods, but most of the increase took place in that last period, from 1981 to 2010.
- The greatest increases occurred in the western half of the North at Inuvik (1.3°C), Whitehorse (0.9°C), Yellowknife (0.9°C), Fort Nelson (0.7°C), and Fort McMurray (0.8°C), while lower increases took place in the Eastern Arctic with Iqaluit at 0.2°C and Kuujjuaq at 0.4°C. In the High Arctic, Resolute (0.9°C) experienced a significant warming trend.

Without a doubt, this climatic warming has consequences for the Aboriginal peoples. As the twenty-first century advances, the Arctic climate is anticipated to become much warmer and sea-ice cover is expected to be much less; in other words, the area of the Arctic will significantly decline. Sea ice, so important to Inuit communities and to Inuit hunters, is losing its grip on Arctic waters. Already, land-fast sea ice, so critical for Inuit hunters, lasts for a shorter period of the year and open water exists

Table 2.2 Temperature Increases across the Canadian North for 15 Climatic Stations (mean annual temperatures in Celsius)

Climate Station	Latitude	1961–90	1971–2000	1981–2010	Temperature Increase
Alert	83° 31'	−18.1°	−18.0°	−17.7°	0.4°
Resolute	74° 43'	−16.6°	−16.4°	−15.7°	0.9°
Inuvik	68° 18'	−9.5°	−8.8°	−8.2°	1.3°
Iqaluit	63° 45'	−9.5°	−9.8°	−9.3°	0.2°
Yellowknife	62° 27'	−5.2°	−4.6°	−4.3°	0.9°
Whitehorse	60° 42'	−1.0°	−0.7	−0.1°	0.9°
Fort Nelson	58° 50'	−1.1°	−0.7°	−0.4°	0.7°
Churchill	58° 44'	−7.1°	−6.9°	−6.5°	0.6°
Kuujjuaq	58° 06'	−5.8°	−5.7°	−5.4°	0.4°
Fort McMurray	56° 39'	+0.2	+0.7°	+1.0°	0.8°
Thompson	55° 48'	−3.4°	−3.2°	−2.9°	0.5°
Cartwright	53° 42'	−0.3°	−0.5°	0.0°	0.3°
Winnipeg	49° 55'	+2.4°	+2.6	+3.0°	0.6°
Timmins	48° 34'	+1.2°	+1.3°	+1.8°	0.6°
Petit Saguenay	48° 11'	+2.5°	+2.5°	+2.8°	0.3°
South Average		+2.0°	+2.1°	+2.5°	0.5°
North Average		−6.5°	−6.2°	−5.8	0.7°

Source: Environment Canada (2014).

for a longer period (Aporta, 2010; Laidler, 2010; also see Chapter 8). Inuit marine hunters are faced with a shorter hunting season on sea ice and, when travelling by snowmobile, the danger of falling through usually thin sea ice has become a serious problem (Ford, 2009; Laidler et al., 2009; Laidler, 2010). Food security for the Inuit depends heavily on accessing marine life from sea ice. Seals, for example, which are accessed over the ice, comprise a critical element in their diet and are essential for their health and for maintaining their traditional culture. On a more positive note, the return of the bowhead whales to the Arctic Ocean signals a dramatic potential increase in country food; indeed, perhaps harvesting, in the decades to come, will reach the levels of the Thule.

Permafrost

Permafrost, or permanently frozen ground, is found in almost all of the Canadian North (Mackay, 1972). In many ways, permafrost epitomizes the true nature of the North. As a product of the extremely cold period known as the Pleistocene, permafrost is defined as ground remaining at or below the freezing point for at least two years. It reaches its maximum depth in high latitudes, where it extends to several hundred metres or more into the ground; at more southerly sites its depth may be less than 10 metres. The greatest recorded thicknesses in Canada, on Baffin and Ellesmere islands, are over 1,000 metres. Pockets of unfrozen ground, called **taliks**, exist within the permafrost landscape. Taliks are usually found alongside or under lakes and rivers that do not freeze to the bottom.

At the height of the Pleistocene, permafrost occupied a much greater area of North America. In the post-glacial Holocene period, the geographic extent of permafrost has decreased due to warming. In recent decades, global warming has accelerated the melting and retreat of permafrost (Smith, 2011: 6). By the end of the twenty-first century, its extent will be greatly diminished.

Permafrost consists of two layers. The upper layer of permafrost (called the active layer) thaws each summer and then refreezes each winter. The thickness of the active layer varies from a few centimetres in the Tundra biome to five or more metres in the Boreal biome. The lower layer of permafrost never thaws, but towards its southern limits pockets of permafrost exist where the active layer may reach five metres. The southern extent of permafrost is associated with a mean annual air temperature of 0°C (Williams, 1986: 3). Along that annual isotherm, the July mean temperature is well above 15°C, which means that the active layer reaches its maximum thickness.

Since permafrost varies in depth and geographic extent, it is divided into four types—continuous, discontinuous, sporadic, and alpine (Figure 2.10). An area is classified as having continuous permafrost when over 80 per cent of the ground is permanently frozen; for discontinuous permafrost, 30–80 per cent must be frozen; and for sporadic permafrost, less than 30 per cent of the ground in an area is permanently frozen. The sporadic permafrost zone represents a transition area between permanently frozen and unfrozen ground. Alpine permafrost is not defined by the percentage of permanently frozen ground but by its presence in a mountainous setting. Such permafrost is found in British Columbia and Alberta. In terms of geographic distribution, continuous permafrost is found where the mean annual temperature is around –7°C, while discontinuous permafrost lies between –5°C and –7°C (see Figure 1.3). Sporadic permafrost is found between 0°C and –5°C. It occurs in small patches.

Permafrost affects the biophysical environment by slowing the growth of vegetation, impeding surface drainage, and creating periglacial landforms such as pingos (small, conical mounds or hills with an ice core). More specifically, active layer thickness influences root systems and soil moisture conditions. Both are critical for vegetation succession. In this way, permafrost shapes the landscape, vegetative communities, and ecosystems (Lewkowicz and Harris, 2005; Burn and Kokelj, 2009). Permafrost also affects northern hydrology through its influence on infiltration, runoff, groundwater storage, and rivers; and it indirectly affects the hydrologic cycle through its influence on evapotranspiration. Permafrost plays an active role in shaping the local landscape. For instance, **thermokarst landscapes** are characterized by very irregular surfaces caused by the melting of ice-rich frozen ground. Another example of permafrost altering the landscape is through a special form of solifluction, known as gelifluction. Sloping land is prone to slumping when the active layer becomes saturated with meltwater and the entire mass slowly slides down the permanently frozen slope. As the climate continues to warm, the degradation of permafrost is expected to accelerate, resulting in more thermokarst landscapes (Kokelj and Jorgenson, 2013).

Permafrost and periglacial landforms are often found in the same geographic area. Periglacial features are created by the action of freezing and thawing. The melting of permafrost has only become a problem since efforts have been made to develop the North by building roads, pipelines, and towns, and, more recently, by warmer summer temperatures (Vignette 2.8). The design of these human-made features must take into account the presence of permafrost. Exposure of ice-rich ground during construction can result in retrogressive thaw slumps (Burn and Lewkowicz, 1990), and such slumps can prove costly to the project and to the environment. Permafrost has made northern construction "a matter of geotechnical science and engineering" (Williams, 1986: 27). Construction in permafrost areas has taken advantage of the frozen ground by placing piles within the frozen ground. These piles form the foundation for the various structures—government buildings, bridges, schools, and dwellings. If the permafrost melts, of course, those structures would be at risk.

Global warming presents a new concern for melting of ice in frozen ground. According to Smith (2011: 6), permafrost thermal monitoring sites indicate that warming of permafrost is occurring across the permafrost region, although the magnitude of

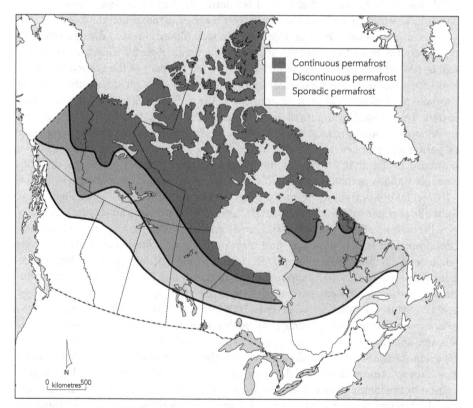

Figure 2.10 Permafrost Zones in the Canadian North

The three permafrost zones extend across the Canadian North. In addition, occurrences of permafrost are found at higher elevations in the Cordillera. In the early 1960s, the construction of the Great Slave Lake Railway from the rail head at Grimshaw, Alberta, to Hay River, Northwest Territories, took place in the zone of sporadic permafrost. Reports from the construction crew stated that little permafrost was encountered.

Source: After *The Atlas of Canada*. atlas.gc.ca.

Vignette 2.8 Permafrost, Vegetation Cover, and Subsidence

Disturbance or removal of vegetation cover by construction projects can result in the melting of permafrost and the resultant subsidence or sinking of the ground. How does this melting come about? The natural vegetation cover provides insulation from the sun's radiation. When removed, the darker ice-rich ground then becomes warmed by the sun and the higher soil temperatures melt the ice contained in the ground. The result is a shift from stable terrain to unstable terrain. One effect is subsidence, with the ground settling to a lower level. Another is gelifluction in hilly areas, where slumping takes place on the steep slopes. Finally, subsidence results in thermokarst topography, as the melting of ice embedded in the ground forms an irregular or hummocky landscape. For many decades, construction firms have taken special measures to ensure the thermal regime of the ground is not altered by the removal of vegetation. One measure is to cover steep slopes with wood chips and thus insulate the ground from the sun's rays. Another is to undertake construction projects in the winter, which minimizes the impact on the frozen ground.

---

this warming varies regionally (Kokelj et al., 2014; Way and Viau, 2014). Inevitably, the geographic extent of permafrost will diminish, with continuous permafrost changing into discontinuous permafrost and discontinuous permafrost into sporadic permafrost. Exposed ice, such as along the frozen shoreline at Tuktoyaktuk, is subject to ocean wave action and it disintegrates rapidly.

Northern Hydrology

Precipitation and temperature play key roles in northern hydrology. While precipitation is low, the long winter stores the snow, releasing it in the spring. The precipitation pattern for the North is lowest in its higher latitudes, where a polar desert exists because of extremely low precipitation. Figure 2.8 illustrates the spatial pattern of precipitation.

The pattern of hydrology in the North has distinct winter/summer components. For example, northern rivers and streams have three phases—very active (water released in the spring from snow/ice melt, as well as spring rainfall), active (rainfall, though limited, keeps the hydrological cycle active in the summer), and inactive (water is frozen in winter and precipitation falls as snow, very little of which evaporates into the atmosphere because of the cold temperatures and the lack of solar radiation). With such a sudden surplus of water in the spring, the banks of rivers and streams are challenged to contain these waters. Flooding often occurs in low-lying areas, such as the Mackenzie Delta. For summer months, the flow of water is well within the capacity of streams and rivers. Winter sees the hydrological cycle slow because precipitation falls as snow and ice covers the rivers and lakes.

Permafrost also plays an important role in the North's hydrology. Permafrost areas are often associated with immature drainage systems, largely because the permanently frozen terrain limits the infiltration of surface water into the ground, inhibits the development of stream channels, and prevents groundwater storage. In the spring, the saturation of the active layer with meltwaters is common. Along steeply sloping land, this saturated layer can slither downslope as a form of gelifluction. Otherwise, the

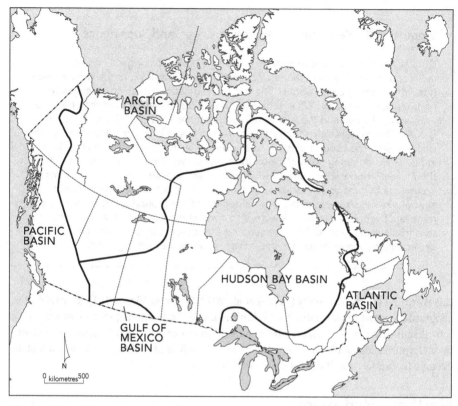

Figure 2.11   Drainage Basins of Canada

Canada has four major river basins and one minor basin, the waters of which flow to the Gulf of Mexico. A more detailed account of river basins is shown in *The Atlas of Canada*.

Source: After *The Atlas of Canada*. atlas.gc.ca.

water is ponded. In areas of immature drainage, when the ice in permafrost melts and subsidence occurs, thermokarst ponds dot the landscape.

Though annual precipitation in the North is low, water resources in the North make up the bulk of Canada's water reserves. These resources are found in four drainage basins—the Arctic, Hudson Bay, Atlantic, and Pacific basins (Figure 2.11). The river systems found in each drainage basin empty into the three oceans surrounding Canada. The Arctic and Hudson Bay drainage basins form nearly 75 per cent of the area of Canada and, despite relatively low precipitation over much of this area, these two basins account for almost 50 per cent of the streamflow (Table 2.3).

The longest river in Canada, the Mackenzie River, is just over 4,200 km in length, making it the tenth longest river in the world. The Mackenzie River illustrates the seasonal nature of streamflow common for all northern rivers that flow into the Arctic Ocean. In terms of runoff volume, the Mackenzie River is by far the most important river in the North. Since most of its headwaters are located in British Columbia, Alberta, and Saskatchewan, spring melting first occurs south of 60°N while the lower reaches of the Mackenzie River are still frozen. Ice jams frequently occur, causing widespread flooding. Yet, these ice jams affect the Mackenzie Delta in a positive way by

Table 2.3 Drainage Basins of Canada and Their Streamflows

Drainage Basin	Area (million km²)	Streamflow (m³)
Hudson Bay	3.8	30,594
Arctic	3.6	20,491
Atlantic	1.6	21,890
Pacific	1.0	24,951
Total	10.0	97,926*

*The drainage basin of the Gulf of Mexico extends into a small portion of southern Alberta and Saskatchewan, and accounts for an area of 21,600 km² or 0.003 per cent of Canada's territory. Accordingly, the Canadian section of the Gulf of Mexico drainage basin contributes a similarly small amount of streamflow: 209 m³.

replenishing tens of thousands of lakes, ponds, and wetlands in the region. The Mackenzie River drainage basin includes the Liard and Peace rivers. Since the upper reaches of these two rivers are below the sixtieth parallel, their spring melt takes place well before the ice on the lower Mackenzie River has thawed, thus adding to flooding in the Mackenzie Delta. Recent research into the release of ice jams in the Mackenzie Delta reveal that earlier river ice breakup is due to a combination of an earlier spring warming and lower winter snowfall (Lesack, Marsh, Hicks, and Forbes, 2014).

Northern rivers empty vast quantities of fresh water into northern seas. The amount of river water flowing into the Arctic Ocean is increasing due to the increased melting of permafrost. This additional runoff is increasing the flow of fresh water into major rivers and eventually to the Arctic Ocean and Hudson Bay. The Liard River, for instance, is one of the main tributaries of the Mackenzie River. Over the last decade, the Liard has seen an increase in its spring runoff due to additional runoff from melting permafrost (Connon, Quinton, Craig, and Hayashi, 2014). Of all Canadian rivers flowing into the Arctic Ocean, the Mackenzie River discharges the greatest quantities of fresh water into the Arctic Ocean. Other northern rivers, such as La Grande and Churchill, also empty fresh water into the sea. One impact of this fresh water is to stratify the ocean waters, that is, river water tends to overlie the colder, saltier, and denser ocean waters. This stratification of ocean waters results in surface layers having a lower salt content. The fresh water also provides a suitable estuarial habitat for marine life, such as bowhead whales.

The Arctic Ocean and Sea Ice

The Arctic Ocean is a critical element in the changing global climates (Barber et al., 2008; Stone, 2015). Despite its remoteness and frozen state, the Arctic Ocean is a critical component in the interconnected circulation system that regulates Earth's climate. As the Arctic ice pack diminishes, the impact of the Arctic Ocean on global climates increases. A NASA ice expert, Josefino Comiso (NASA, 2010), expresses this relationship:

The oceans are crucial to Earth's climate system, since they store huge amounts of heat. Changes in sea ice cover can lead to circulation changes not just in the Arctic Ocean, but also in the Atlantic and Pacific oceans. If you change ocean circulation, you change the world's climate.

By definition, the Arctic ice pack is "old" ice that has not melted for at least two years and is not attached to land and therefore moves in a clockwise motion within the Arctic Ocean. This circumpolar movement within the Arctic Ocean is called the Arctic Oscillation. Polar **pack ice** is much thicker and harder than new ice, which forms to join the Arctic ice pack but melts each summer. **Land-fast ice** is a common form of new ice. Land-fast ice is attached to coasts and extends offshore over shallow parts of the continental shelf. Unlike the polar pack ice, land-fast ice remains fixed in location while pack ice drifts, driven by currents and winds.

The seas of the Canadian North extend from the Beaufort Sea to the Labrador Sea and include the waters of Hudson Bay. These cold water bodies have less impact on the northern hydrological cycle than might be expected. The reasons are twofold: (1) cold seas and sea ice have a lower evaporation rate than warm seas, thus limiting moisture forming in Arctic air masses; and (2) the circulation of the Arctic Ocean is relatively isolated from the rest of the world's currents because the **Beaufort Gyre circulation system** tends to limit flows to the Atlantic and Pacific oceans and because of the cold water. However, with the warming of the waters of the Arctic Ocean and the loss of its ice cover, a major question facing climatologists is whether this circulation system is weakening, thus leading to greater interaction with the Atlantic and Pacific oceans.

Arctic ice and waters in relatively small amounts flow into the Atlantic Ocean around Baffin Bay, where they mix with the warmer Atlantic waters, forming Subarctic water. Baffin Bay is connected to the Arctic Ocean through Nares Strait and Jones and Lancaster sounds, and to the Labrador Sea through Davis Strait. Most icebergs are formed from the Greenland Ice Sheet. The Labrador Current carries them to the waters off Newfoundland, where they represent a hazard to both ocean shipping and offshore drilling rigs. During winter, when there is extensive ice cover, a **polynya** or open-water area (called "north water") exists in the northern part of Baffin Bay. The explanation for the natural factors that create open water in a frigid environment still eludes physical scientists (Vignette 2.9).

Vignette 2.9 Polynyas

The Arctic Ocean has a thick cover of ice, but a dozen or more areas of open water do occur each winter. Many recur each winter. The largest recurring open water lies between Baffin Island and Greenland. Such open water is known as a polynya, which is a derivation of the Russian word for open water surrounded by ice. The explanation remains unresolved but most hydrographers believe that the basic mechanism involves various combinations of tides, currents, ocean-bottom upwellings, and winds that keep the surface waters moving and thus prevent them from freezing. Polynyas may be as small as 50 metres across or as large as the famous North Water polynya, which often extends over 130,000 km². These biological "hot spots" serve as Arctic oases where marine animals, polar bears, and birds congregate. Not surprisingly, archaeologists have uncovered Thule settlement sites near the North Water. One "hot spot" is Lancaster Sound. The North Water polynya extends into Lancaster Sound, which is considered the "Serengeti of the Arctic" because most of the world's narwhals and a large number of whales, walruses, and seals frequent these waters. As well, polar bears roam the perimeter of the open water, hunting for ringed and bearded seals, though these bears also will feed on the carcasses of beluga whales, grey whales, walruses, narwhals, and bowhead whales.

Sea ice is one of the unique features of northern oceans. Another is icebergs. In late winter, sea ice extends from the Arctic Ocean to the North Atlantic waters offshore from Labrador and the island of Newfoundland. Icebergs are floating masses of freshwater ice that broke away from glaciers. Most icebergs in the North Atlantic have calved (broken away) from glaciers as the glaciers enter the sea. Each year, 10,000 to 15,000 icebergs enter the North Atlantic, with most originating from glaciers along the west coast of Greenland. A few come from glaciers in the eastern Canadian Arctic islands. Icebergs float in the Labrador Current beyond the edge of the land-fast sea ice. This zone, known as "Iceberg Alley," extends from Baffin Bay to the waters off the coast of Labrador and Newfoundland. Once the sea ice has melted, icebergs can be found close to land. Before disintegrating, icebergs may float as far south as 40°N, which is approximately the same latitude as New York City. Iceberg viewing is an important late spring tourist activity in Newfoundland.

Sea ice varies in thickness and duration. The most durable, oldest, and thickest ice is found in the Arctic ice pack. But as Stone (2015: 219) notes, the annual summer melting of the Arctic ice pack has greatly reduced the amount of old ice. In the summer, sea ice disappears first in the Great Lakes and offshore of Atlantic Canada but not in the Arctic Ocean. While a portion of the Arctic ice pack melts each summer, the length of time for open water in the Arctic Ocean is very short—perhaps only weeks before ice re-forms. On the other hand, in mid-latitude water bodies, such as Lake Athabasca in northern Saskatchewan and Alberta, open water exists for almost half of the year.

The increasing open water due to the summer reduction in ice cover in the Arctic Ocean has altered its currents, affected wave action, and thus the amount of ice/water/energy flowing into the North Atlantic through Baffin Bay. Satellite imagery and on-site measurements reveal that the ice cover has diminished in both extent and average thickness. The significance of the Arctic ice cover having a greater proportion of new ice is that this thinner ice melts more quickly in the summer, thus creating a longer period of open water. In September 2007, for instance, satellite imagery indicated that the extent of open water in the Arctic Ocean was more extensive than in the previous decades, with the extent of sea ice diminished to 4.3 million km^2 (NSIDC, 2007). This record was broken in September 2012 when sea ice cover declined to 3.6 million km^2 (NSIDC, 2015). The most recent data for September 2014 revealed that the extent of Arctic sea ice was 5.3 million km^2 (NSIDC, 2015). In spite of fluctuations in the annual minimum ice cover data, the trend indicates a gradual retreat of the sea ice cover on the Arctic Ocean. For more on this subject, see the section "The Arctic Ocean and Its Changing Ice Cover" in Chapter 8.

Northern Landscapes

The North's varied landscapes and their particular geological structures are captured in its five geomorphic regions—the Canadian Shield, the Cordillera, the Interior Plains, the Hudson Bay Lowlands, and the Arctic Lands. At a macro-scale, these five geomorphic regions provide insight into the North's heterogeneous nature.

With the exception of the Hudson Bay Lowlands these geomorphic regions were formed billions of years ago. Their surface has been etched by various geomorphic processes but especially by glaciation during the last ice age. Its impact on the landscape remains a key feature. In mountainous areas, alpine glaciation "sharpened" the features of mountains; the massive ice sheets pummelled and scarred the surface as they moved southward; the melting of these ice sheets created enormous runoffs and formed vast glacial lakes. On the other hand, a more subtle mark on the northern landscape found exclusively in areas of continuous permafrost takes the form of periglacial features. While its primary geomorphic processes are frost shattering (physical weathering of rocks) and mass wasting (mass movement of material down a slope), pingos provide the most dramatic example of a periglacial landform. With the degradation of permafrost due to a warming climate, the topographic features, thermokarst landscapes, and hydrological systems of the North are changing.

The five geomorphic regions in the North have strikingly difference landscapes (Figure 2.12). While each region has its unique features, they have three common characteristics—they cover a large, contiguous area with similar relief features; they have experienced similar geomorphic processes that shaped the terrain; and they have had a similar geological history and therefore possess a common geological structure.

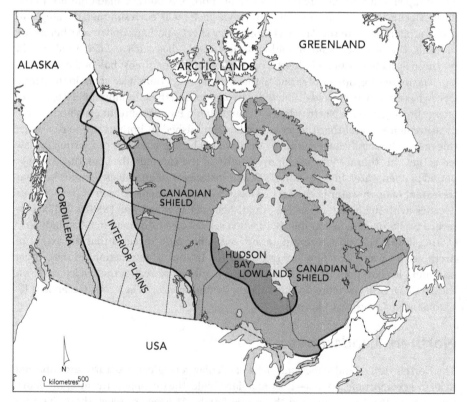

Figure 2.12 Geomorphic Regions of Canada

Five of Canada's seven geomorphic regions are found in the Canadian North. The other two are the Great Lakes–St Lawrence Lowland, located in southern Ontario and Quebec, and the Appalachian Upland in Atlantic Canada.

Source: After Slaymaker (1988).

These geomorphic regions extend over thousands of km^2, making them easily identified from a high-flying aircraft or from satellite photographs.

Glaciation during the Late Wisconsin period affected the five northern regions by modifying the surface either directly, by glacial erosion or deposition, or indirectly, by the formation of glacial lakes and by the invasion of depressed land by seawater. **Erratics**—huge boulders moved great distances and then deposited on the ground— are one example of the power of glacial erosion; **nunataks** are mountain peaks that rose above the alpine glaciers. The Canadian Shield, stretching from Labrador to the Northwest Territories, is the largest geomorphic region in the Canadian North. It is the geological core of the northern landscape. Its Precambrian rocks are more than 2.5 billion years old and lie beneath most of the more recently formed strata, such as the Interior Plains and the Hudson Bay Lowlands.

Over most of the northern areas of Manitoba, Ontario, Quebec, Labrador, the northern half of Saskatchewan, the northeast corner of Alberta, and about half of the Northwest Territories, the Canadian Shield is exposed at the Earth's surface. Shaped like a saucer, its highest elevations are found around its outer limits while the central area lies beneath the waters of Hudson Bay. Much of the exposed Canadian Shield consists of a rough, rolling upland. Along the east coast from Labrador to Baffin Island, the Shield has been strongly uplifted. Glaciers moving downslope to the sea deeply scoured valleys and created fjords, giving the coast spectacular scenery.

The Interior Plains, a flat to gently rolling landscape, lie between the Cordillera and the Canadian Shield. The sedimentary rocks of the Interior Plains were formed after the end of the Precambrian era (some half-billion years ago) and include Cretaceous-age rocks formed some 100 million years ago. Within these Cretaceous-age rocks, vast oil, gas, and potash deposits exist. At the end of the last glacial advance, a mantle of glacially deposited debris or glacial till covered these sedimentary rocks. In other places, glacial lakes drained, forming flat lowlands consisting of **glaciolacus-trine** material, while outwash plains resulted from meltwater streams that deposited **glaciofluvial** materials. At the same time, fast-flowing rivers fed by the melting ice sheet carved deep valleys known as glacial spillways. These and other glacial features on the landscape are indications of the last ice age. The Mackenzie River and its tributaries created a river valley system that extends to the Arctic Ocean. Two huge deltaic landforms exist. One is at the western end of Lake Athabasca, where the Peace and Athabasca rivers deposit their silt. The other is the Mackenzie Delta, located at the mouth of the Mackenzie River.

The Cordillera, a complex mountainous region, occupies much of British Columbia, Yukon, and a small portion of southern Alberta and the Northwest Territories west of the Mackenzie River. The Canadian Cordillera begins at the forty-ninth parallel, the Canada–United States border in the West, and this mountainous belt extends south into Mexico and north into Alaska. The Cordillera in Canada is about 800 km wide. Two geologic forces, known as faults and folds, created its basic landform features, and erosional agents, primarily ice and water, reshaped these features. The Cordillera contains a variety of mountainous terrain as well as plateaus, valleys, and plains. It also includes what are today remnant glaciers and ice fields. This geomorphic region has majestic mountains, including the world-famous Rocky Mountains. These mountains were formed by severe folding and faulting of sedimentary rocks. The highest mountain in Canada, Mount Logan, has an elevation of nearly 6,000 metres and is part of the St Elias Mountains in the southwest corner of Yukon.

During the Late Wisconsin Ice Age, the Cordillera was glaciated. Alpine glaciers created arêtes, cirques, and U-shaped valleys. Over most of the Cordillera, the land became free of ice some 10,000 years ago. In the post-glacial period, river terraces, alluvial fans, flood plains, and deltas have formed. While today many glaciers are retreating, some glaciers are still active, though their rate of movement is relatively slow.

The Hudson Bay Lowlands region consists of a low, flat coastal plain that, geologically speaking, emerged recently from the Tyrrell Sea, the name given to the much larger Hudson Bay that existed some 8,000 years ago. The surface of this lowland consists of recently deposited marine sediments combined with reworked glacial till. These deposits accumulated in the post-glacial period when the lowland was beneath the Tyrrell Sea. With the retreat of the ice sheet from this area, the process of isostatic rebound came into play. More and more of the seabed became land as the surface of the Earth gradually rose after having been depressed by the massive weight of the ice sheet. With the receding waters of the Tyrrell Sea, this process created a distinct geomorphic region known as the Hudson Bay Lowlands.

The Precambrian bedrock underlying the Hudson Bay Lowlands is masked entirely by glacial and marine sediments. The inland boundary of this region is marked by elevations of around 180 metres, which indicate that, with the removal of the weight of the ice sheet, the Earth's crust began to regain its former shape. This phenomenon is called **isostatic uplift** or isostatic rebound. Assuming that the rate of isostatic uplift remains around 70 to 130 cm/100 years, Professor Barr speculated that the Hudson Bay Lowlands will continue to increase in size and Hudson Bay will recede further. Within 12,000 years, Hudson Bay could become a fraction of its present size (Barr, 1972). However, this hypothesis, presented in 1972, does not account for the possibility of rising sea levels caused by global warming.

The Arctic Lands are a complex geomorphic area centred on the Arctic Archipelago. Here, geological events and geomorphic processes have created lowlands, hilly terrain, mountains, and even polar deserts. Polar deserts, a unique northern landscape, exist in the highest latitudes and driest conditions of the Canadian Archipelago. The map of precipitation indicates that much of the Canadian Archipelago receives less than 200 millimetres per year, making this area an ideal location for polar deserts (Figure 2.8). Added to the low precipitation, two other factors contribute to a polar desert: continuous permafrost and long periods when the sun does not appear inhibit plant growth.

In the cold, dry climate found in the Arctic Lands, the primary geomorphic process is frost action. This frost action results in many periglacial features, such as pingos and tundra polygons, which provide a distinctive appearance to the land. Scattered bedrock and patterned ground are widespread in the lowlands, while alpine glaciers occur in the more mountainous landscapes. Rolling to hilly terrain underlain by permafrost is affected by slumping, caused by **solifluction** or gelifluction: "soil flow" down a sloping frozen surface (Trenhaile, 1998: 84–5). A variety of landforms are found in the Queen Elizabeth Islands (those islands lying poleward of the Northwest Passage, including Prince Patrick Island in the west and Devon Island in the east). Most of Canada's glaciers are found in the mountainous terrain of the eastern section of the Queen Elizabeth Islands, especially Ellesmere Island. These glaciers are retreating, signalling the start of a major shift in their geographic extent. If the warming trend continues, glaciers could disappear in time. A similar retreat is taking place with sea ice and permafrost, though changes in the latter are more difficult to measure. As well, ice in the ground is less exposed to sunlight and is insulated from surface heat by vegetation cover.

Glaciation

Glaciation is one of the principal erosional agents that have fine-tuned a variety of landscapes in Canada's North. Glaciation involves the formation, advance, and retreat of glaciers. The most recent phase of continental glaciation began some 25,000 years ago. Glacial formation took place as the climate cooled, causing snow to accumulate and eventually forming glaciers. Glacial retreat commenced when the climate began to warm, so that the southern front of the continental glaciers began to melt. As the glaciers melt and recede, material contained in the ice is deposited on the ground. In the Canadian North, unglaciated terrain is limited to a small area of Yukon where there was not enough precipitation to nourish the expansion of glaciers.

During the last glacial advance of the Pleistocene Epoch, two huge ice sheets, the Laurentide and the Cordillera, covered most of Canada. These ice sheets reached a maximum thickness of 4,000 and 2,000 metres, respectively. In comparison, today the largest Canadian glaciers are found on Ellesmere Island and have a thickness of nearly 1,000 metres. Some time around 15,000 BP, the climate warmed and these huge ice sheets began to melt, i.e., retreat. By 14,000 BP, an ice-free corridor appeared along the eastern edge of the Cordillera ice sheet, connecting the unglaciated areas in Yukon with the rest of ice-free North America. By about 10,000 BP, most of these two ice sheets had melted.

During this process of advance and retreat, two geomorphic processes took place. First, the advancing ice sheet caused glacial erosion; later, the retreating ice sheet deposited debris on the land. Glacial erosion took various forms, such as scraping off the unconsolidated material and plucking out huge chunks of bedrock. Where the bedrock was highly resistant, the rock was scraped and scoured. In the mountains, alpine glaciers moved quickly downslope, which had the effect of "sharpening" mountain features, as shown in Figure 2.13. Alpine glaciers create mountain peaks formed by cirques (basins carved by the glaciers) and arêtes (razor-like ridges). Other evidence of the erosional power of alpine glaciers is found in U-shaped valleys in the Cordillera and the coastal fjords of Labrador and Baffin Island.

As the Earth warmed, both alpine and continental glaciers deposited vast amounts of debris. Glacial features of deposition consist of water-sorted deposits and unsorted deposits. As the ice sheets melted, some of the material contained in the ice was discharged into running meltwater and glacial lakes. The debris held in the ice sheets was deposited on the land; in some cases, glacial meltwaters sorted the debris or till into eskers and outwash plains. **Eskers**—long, narrow ridges of sorted sands and gravel— were deposited from melt streams within or beneath the decaying ice sheet (Figure 2.14). Some eskers are over 100 kilometres in length. The most common glacial deposit is **glacial till**, which consists of unsorted material deposited by a melting ice sheet or glacier. Glacial till extends over vast areas while **drumlins**, formed by massive subglacial flooding, appear as clusters of low, elongated, whale-backed hills, shaped by the flow of the ice. Most drumlins are believed to have been formed a short distance behind

Figure 2.13 Alpine Glaciers

The St Elias Mountains, located in southern Yukon, contain many alpine glaciers with an erosional force that sharpens the features of mountains. The three shown in this photograph are descending in three parallel valleys from a much larger glacier located in the bowl-shaped basin (called a cirque) at the top of the mountain. Between the valleys are sharp-edged ridges or arêtes.

Source: Reproduced with permission of Natural Resources Canada 2015, courtesy of the Geological Survey of Canada.

the ice margin just prior to deglaciation and therefore record the final direction of ice movements. As the massive ice sheets melted, enormous quantities of water were released. To the south, these meltwaters either overtaxed the existing southern-flowing drainage system or formed glacial lakes. Often, these glacial lakes were created when the northward-flowing rivers were still blocked from reaching the sea by the remaining ice sheet. The largest glacial lake, Lake Agassiz, covered much of Manitoba as well as parts of northwestern Ontario, Saskatchewan, Minnesota, and North Dakota.

Glaciation has created different landforms in each of the five geomorphic regions. In the Cordillera, mountain features have been sharpened into arêtes and peaks; in the Interior Plains, deposits of glacial debris are everywhere; the Canadian Shield has been subjected to ice plucking, scraping, and scouring; the Hudson Bay Lowlands region was depressed by the weight of the ice sheet that sea water from Hudson Bay entered when the ice sheet retreated; and, in the higher elevations and latitudes of the Arctic Lands, disintegrating glaciers, including the remnants of the Ellesmere Ice Shelf, stand out.

Periglacial Features

Periglacial landforms exist in a dry, cold environment where permafrost is wide-spread. Its landform features are often moulded by the process of freezing and thawing. The most distinctive periglacial landforms are ice-cored hills known as **pingos**.

Figure 2.14   Esker in Arctic Quebec

This esker is located near the Inuit settlement of Salluit, on Hudson Strait at the top of northern Quebec. In general, eskers range from small, sinuous ridges a few tens of metres long, to linear features up to 15 km long and even longer. They consist of post-glacial gravel and other sediments that were deposited by subglacial or englacial streams.

Source: Natural Resources Canada, "Esker," *Interpretation Guide of Natural Geographic Features*, 2008, Centre collégial de développement de matériel didactique, photo no. 6432, www.cits.mcan.gc.ca/site/eng/resoress/guide/esker/pg07.html.

These hills are associated with permafrost and a cold, dry Arctic climate where frost action (the freezing-and-thawing cycle) is the dominant geomorphic process. They occur in lowlands where continuous permafrost exists, and are formed when an ice lens in permanently frozen ground is nourished by extraneous water. Over time, the ice lens grows and pushes itself upward, forming a mound and eventually a hill. However, the most common periglacial feature is **patterned ground**—symmetrical forms, usually polygons, caused by intense frost action over a long period of time (Figure 2.15). Patterned ground includes frost-sorted circles of polygon patterns of stones and pebbles. The general process of frost heave causes coarse stones to move to the surface and outward.

During the **Late Wisconsin glacial period**, these conditions were associated with the southern edge of the Laurentide and Cordillera ice sheets. As these ice sheets retreated, more periglacial features were formed. Hugh French (1996: 5) has described this process:

During the cold periods of the Pleistocene, large areas of the now-temperate middle latitudes experienced intense frost action and reduced temperatures because of their proximity to the ice sheets. Permafrost may have formed, only to have been degraded during a later climatic amelioration.

Figure 2.15   Tundra Polygons

Tundra polygons rearrange the surface of the ground into polygon-like shapes. They are formed because of the freeze–thaw cycle, which is the principal erosional agent in areas of continuous permafrost. The freeze–thaw cycle causes the ground to contract and expand, thus rearranging the surface material into polygon-like forms. Relic tundra polygons still exist in areas in which continuous permafrost conditions once existed.

Source: Reproduced with permission of Natural Resources Canada 2011, courtesy of the Geological Survey of Canada (Photo KGS-791 by J.M. Harrison).

French estimates that periglacial features are found over one-fifth of the Earth's surface. Today, relic periglacial features exist great distances from Canada's cold, dry Arctic climate, but the formation of new patterned ground and other periglacial features continues to take place in the Arctic Lands. Tundra polygons are a common periglacial feature found in the Arctic Lands.

Grand Theme

From this discussion of the North's physical geography, we can see that the complex but changing nature of this cold environment provides many challenges and calls for continued research to monitor its changing nature, especially that of the Arctic Ocean and its potential impact on the global climate. Reflecting back to the impact of global warming on the Ward Hunt Ice Shelf mentioned at the beginning of this chapter, this warming effect symbolizes the enormous change taking place, which has the potential to affect the rest of the globe. At this juncture, the reader is alerted to a grand theme in geography, namely: *How will global warming affect our cold environment, especially the two northern biomes, permafrost, and the Arctic ice cap?* The answer to this question will determine to a large extent not only what happens to Canada's northern environment but, quite possibly, what happens to the global environment.

Challenge Questions

1. What is the relationship between latitude and the amount of incoming solar radiation?
2. Why is the North considered a fragile environment?
3. Figures 2.7 and 2.8 are maps showing mean monthly temperatures for January and July based on 1941 to 1970 Climate Normals. Based on the 1981 to 2010 Climate Normals provided for 15 climatic stations in Table 2.2, how would you redraw the isotherms on these two maps?
4. Temperature increases are predicted to be much greater in the Arctic than in southern Canada. What role does the albedo effect play in these temperature increases?
5. Based on the information in Table 2.1, which Holocene climate best describes our current climate?
6. Why are most periglacial features found in the High Arctic, where the primary geomorphic process is frost shattering?
7. What landforms caused by glaciation are found in the Cordillera but not in the Interior Plains?
8. How would the loss of the Arctic ice cap stand out as a pivotal natural change that would dramatically impact global climates?
9. The amount of greenhouse gases in the atmosphere has increased over the last century and is likely to continue to increase in the twenty-first century. Whether you agree with this statement or not, offer an explanation for your position.
10. Why is ice, in the form of glaciers, ice shelves, and Arctic pack ice, more likely to melt than ice contained in permafrost?

Notes

1. Latitude is an angle that ranges from 0° at the equator to 90° at the North and South poles. Keep in mind that latitude is an imaginary line created to determine the north–south position of a point on the Earth's surface. It also provides a measure of the amount of solar radiation striking the Earth at different latitudes. The reason is because the Earth is a sphere and therefore the incoming solar rays strike the Earth's surface at different angles. Both the angle of the sun's rays and the number of daylight hours in Canada's North change at each latitude throughout the year as the Earth revolves around the sun. For example, at the equator, these rays are at right angles throughout the year and thus provide the maximum heat per unit of the surface. In sharp contrast, at the North Pole, the sun's rays only reach the ground for six months of the year and during those six months, the angle of the sun's rays are low, meaning they add little heat to the Earth's surface.
2. The Arctic ice pack is sometimes referred to as the polar ice pack. However, since there are ice packs in both the Arctic and Antarctic oceans, the term "Arctic ice pack" is preferred.
3. The Little Ice Age took place over at least four centuries. The generally accepted time span is 1450 to 1850, with the period between 1600 and 1800, which had the coldest winters, marking the height of the Little Ice Age. Most evidence for the colder winters than those experienced in the twentieth century is based on historical records from Europe,

Iceland, and Greenland. Rivers and lakes froze and the growing season shortened; sea ice surrounded Iceland and Greenland, cutting off supply ships; the Greenland colony disappeared. While records indicate that colder winters began as early as the fourteenth century, consistently cold winters did not occur until the mid-point of the fifteenth century and ended in the middle of the nineteenth century, when the Industrial Revolution and the burning of large quantities of coal initiated the process of global warming. Some scholars believe that a slightly warmer period occurred in the 1500s, after which the climate deteriorated substantially. For more information on the Little Ice Age, see H.H. Lamb (1969: 236).

References and Selected Reading

Aporta, Claudio. 2010. "Inuit Sea Ice Use and Occupancy Project (*ISIUOP*)." International Polar Year Project. gcrc.carleton.ca/isiuop.

Barber, D.G., J.V. Lukovich, J. Keogak, S. Baryluk, L. Fortier, and G.H.R. Henry. 2008. "The Changing Climate of the Arctic."*Arctic* 61, 5:7–26.

Barr, William. 1972. "Hudson Bay: The Shape of Things to Come." *Musk-Ox Journal* 11: 64.

Birks, H.J.B. 2007. "Climate Change in the Holocene: An Overview." Indigenous Peoples and Climate Change Conference, Oxford, England, 12–13 Apr. http://www.eci.ox.ac.uk/news/events/indigenous/birks.pdf.

Boon, S., D.O. Burgess, R.M. Koerner, and M.J. Sharp. 2010. "Forty-Seven Years of Research on the Devon Island Ice Cap, Arctic Canada." *Arctic* 63, 1: 13–29.

Burn, Chris R. 1995. "Where Does the Polar Night Begin?" *Canadian Geographer* 39, 1: 68–74.

————. 1996. *The Polar Night*. Scientific Report No. 4. Inuvik: Aurora Research Institute, Aurora College.

———— and S.V. Kokelj. 2009. "The Environment and Permafrost of the Mackenzie Delta Area." *Permafrost and Periglacial Processes* 20: 83–105.

———— and A.G. Lewkowicz. 1990. "Retrogressive Thaw Slumps." *Canadian Geographer* 34, 3: 273–6.

Connon, R.F., W.L. Quinton, J.R. Craig, and M. Hayashi. 2014. "Changing Hydrologic Connectivity due to Permafrost Thaw in the Lower Liard River Valley, NWT, Canada." Special issue, *Canadian Geophysical Union*. doi: 10.1002/hyp.10206.

Environment Canada. 2002. "Canada Climate Normals 1971–2000." www.msc-smc.ec.gc.ca/climate/climate_normals/index_e.cfm.

————. 2007. "Ecosystems." www.mb.ec.gc.ca/nature/ecosystems/index.en.html.

————. 2014. "Canadian Climate Normals." 13 Feb. http://climate.weather.gc.ca/climate_normals/index_e.html.

Etkin, Dave. 1989. "Elevation as a Climate Control in Yukon." *Inversion: The Newsletter of Arctic Climate* 2: 12–14.

Ford, J., and L. Berrang-Ford. 2009. "Food Insecurity in Igloolik, Nunavut: A Baseline Study." *Polar Record* 45, 234: 225–36.

Fossett, Renée. 2001. *In Order to Live Untroubled: Inuit of the Central Arctic, 1550–1940*. Winnipeg: University of Manitoba Press.

French, Hugh M. 1996. *The Periglacial Environment*, 2nd edn. London: Longman.

————. 2007. *The Periglacial Environment*, 3rd edn. Toronto: Wiley.

Hare, F. Kenneth, and Morley K. Thomas. 1979. *Climate Canada*, 2nd edn. Toronto: Wiley.

Houghton, J.T., Y. Ding, D.J. Griggs, M. Noguer, P.J. van der Linden, X. Dai, K. Maskell, and C.A. Johnson. 2001. *Climate Change 2001: The Scientific Basis*. Contribution of Working Group I to the Third Assessment Report of the Intergovernmental Panel on Climate Change. Cambridge: Cambridge University Press. www.ipcc.ch/pub/reports.htm.

Huybert, Peter, and Carl Wunsch. 2005. "Obliquity Pacing of the Late Pleistocene Glacial Terminations." *Nature* 434: 491–4.

Intergovernmental Panel on Climate Change. 2013. *Fifth Assessment Report: The Physical Science Basis*. https://www.ipcc.ch/report/ar5/wg1/.

Kaufman, D.S., et al. 2004. "Holocene Thermal Maximum in the Western Arctic." *Quaternary Science Reviews* 23: 529–60.

Kokelj, S.V., and M.T. Jorgenson. 2013. "Advances in Thermokarst Research." *Permafrost and Periglacial Processes* 24, 2:108–19.

Kokelj, S.V., T.C. Lantz, S.A. Wolfe, J.C. Kanigan, P.D. Morse, R. Coutts, N. Molina-Giraldo, and C.R. Burn. 2014. "Distribution and Activity of Ice Wedges across the Forest-Tundra Transition, Western Arctic Canada." *Journal of Geophysical Research: Earth Surface.* doi: 10.1002/2014JF003085.

Laidler, G.J., J. Ford, W.A. Gough, T. Ikummaq, A. Gagnon, A. Kowal, K. Qrunnut, and C. Irngaut. 2009. "Travelling and Hunting in a Changing Arctic: Assessing Inuit Vulnerability to Sea Ice Change in Igloolik, Nunavut." *Climatic Change* 94, 3–4: 363–97.

Laidler, Gita. 2010. *Inuit Siku (Sea Ice) Atlas.* app.fluidsurveys.com/surveys/gjlaidler/siku-atlas-consultation-1/?code=X45wR&l=en.

Lamb, H.H. 1969. "Climatic Fluctuations." In H. Flohn, ed., *World Survey of Climatology,* vol. 2, *General Climatology.* New York: Elsevier.

Lawford, R.G. 1988. "Climatic Variability and the Hydrological Cycle in the Canadian North: Knowns and Unknowns." *Proceedings of the Third Meeting on Northern Climate,* Canadian Climate Program, 143–62. Ottawa: Minister of Supply and Services.

Lesack, L.P., P. Marsh, F. Hicks, and D. Forbes. 2014. "Local Spring Warming Drives Earlier River-Ice Breakup in a Large Arctic Delta." *Geophysical Research Letters* 41:1560–6. doi: 10.1002/2013GL058761.

Lewkowicz, A.G., and C. Harris. 2005. "Morphology and Geotechnique of Active-Layer Detachment Failures in Discontinuous and Continuous Permafrost, Northern Canada." *Geomorphology* 69: 275–97.

Mackay, J.R. 1972. "World of Underground Ice." *Annals of the Association of American Geographers* 62: 1–22.

———— and C.R. Burn. 2002. "The First 20 Years (1978/79 to 1998/99) of Ice-Wedge Growth at the Illisarvik Experimental Drained Lake Site, Western Arctic Coast, Canada." *Canadian Journal of Earth Sciences* 39, 1: 95–111.

Milankovitch, M. 1941. *Kanon der Erdbestrahlungen und seine Anwendung auf das Eiszeitenproblem.* Belgrade. (New English translation, *Canon of Insolation and the Ice Age Problem.* Agency for Textbooks, 1998).

NASA. 2010. "Arctic Sea Ice." Earth Observatory, 19 Oct. earthobservatory.nasa.gov/Features/WorldOfChange/sea_ice.php.

————. 2012. "Arctic Sea Ice Shrinks to New Low in Satellite Era." http://www.nasa.gov/topics/earth/features/arctic-seaice-2012.html.

National Snow and Ice Data Center (NSIDC). 2007. "Overview of Current Sea Ice: September 2007." *Arctic Sea Ice News* (Fall). nsidc.org/news/press/2007_seaiceminimum/20070810_index.html.

————. 2013. "Arctic Sea Ice Reaches Lowest Extent for 2013." *Arctic Sea Ice News & Analysis,* 20 Sept. http://nsidc.org/arcticseaicenews/2013/09/draft-arctic-sea-ice-reaches-lowest-extent-for-2013/.

————. 2015. "State of the Cryosphere." 8 Jan. https://nsidc.org/cryosphere/sotc/sea_ice.html.

Natural Resources Canada. 2013a. "Mean January Daily Temperature." *Hydrological Atlas of Canada,* Plate 14. http://geogratis.gc.ca/api/en/nrcan-rncan/ess-sst/fe584f1a-c178-5be4-a316-c5df2793c883.html.

————. 2013b. "Mean July Daily Temperature." *Hydrological Atlas of Canada,* Plate 8. http://data.gc.ca/data/en/dataset/52672467-c1d0-5e64-87ac-993e67c5910c.

————. 2013c. "Annual Precipitation." *Hydrological Atlas of Canada,* Plate 10. http://geogratis.gc.ca/api/en/nrcan-rncan/ess-sst/c65855e6-fe60-51b4-bc7b-5743bb03581f.html.

Northwest Territories. 2004. "Vegetation on the Taiga Plains." forestmanagement.enr.gov.nt.ca/forest_resources/eco_land_classification/taiga07_vegetation.html.

Sharp, M., D.O. Burgess, J.G. Cogley, M. Ecclestone, C. Labine, and G. Wolken. 2011. "Extreme Melt on Canada's Arctic Ice Caps in the 21st Century." *Geophysical Research Letters* 38: L11501. doi: 10.1029/2011GL047381.

Slaymaker, Olav. 1988. "Physiographic Regions." In James H. Marsh, ed., *The Canadian Encyclopedia,* vol. 3, 1671. Edmonton: Hurtig.

Smith, Dan, comp. 2010. "U Alberta's John England on Fracture of Ward Hunt Ice Shelf." *GeogNews* (News Digest of the Canadian Association of Geographers, no. 94), 27 Aug. www.geog.uvic.ca/dept/cag/geognews/geognews.html.

Smith, S. 2011. "Trends in Permafrost Conditions and Ecology in Northern Canada." In *Canadian Biodiversity: Ecosystem Status and Trends 2010*. Technical Thematic Report No. 9. Ottawa: Canadian Council of Resource Ministers.

Stone, David P. 2015. *The Changing Arctic Environment: The Arctic Messenger*. New York: Cambridge University Press.

Trenhaile, Alan S. 2013. *Geomorphology: A Canadian Perspective*, 5th edn. Toronto: Oxford University Press.

Ward, Dale. 2012. "ATMO: Weather, Climate and Society." University of Arizona. http://www.atmo.arizona.edu/students/courselinks/fall12/atmo336/lectures/sec5/holocene.html.

Way, Robert G., and Andre E. Viau. 2014. "Natural and Forced Air Temperature Variability in the Labrador Region of Canada during the Past Century." *Theoretical and Applied Climatology*. doi: 1007/s0074-014-12 48-2.

Williams, Peter J. 1986. *Pipelines and Permafrost: Science in a Cold Climate*. Ottawa: Carleton University Press.

- 3 -

The Historical Background

History is essential for an understanding of present events taking place in Canada's North. For most Canadians, history is about nation-building—the creation of a nation (or nations) by European colonizers and settlers. Harold Innis (1930) provided an early interpretation of nation-building in his classic book, *The Fur Trade* (Buxton, 2013), which emphasized that resource staples drove exploration and then settlement.[1] For the North, the valued commodity (staple) was the beaver and the harvesting of beaver pelts by Indian trappers provided the basis for the fur trade. Towards the end of the nineteenth century the valued commodity was gold, and by the middle of the twentieth century it was iron ore.

For Aboriginal Canadians, two histories capture their past. One describes events prior to contact with Europeans and the other after contact. The pre-contact history begins some 30,000 years ago; the second one takes place at different times and places as contact is made with European and later Canadian nation-builders. Our interest focuses on the contact history with Europeans, the subsequent engagement of the two peoples through alliances and trade, and their changing relationship in the fur economy as new circumstances emerged. In the fur trade, for example, Indian trappers played a vital role by supplying fur and food to the French and then British fur traders. On the other hand, mining left little room for direct participation by Aboriginal peoples. Unlike trapping, the extraction of minerals from the ground had little place in Aboriginal experience. In both cases Ottawa attempted to minimize its financial commitment, but during the Klondike gold rush of the late 1890s the flood of gold seekers into the Yukon, many of them Americans, forced the federal government to ensure sufficient presence of the North West Mounted Police. Still, the federal stance on the North is best described as "laissez-faire" until the mid-twentieth century, when Prime Minister John Diefenbaker's "Northern Vision" was proclaimed.

The North's history is set into four phases:

1. The Beginning
2. The Fur Trade
3. Resource Development Begins
4. The Northern Vision

The Beginning

Old World hunters who crossed the Beringia land bridge into Alaska were in all likelihood the first people to set foot in North America (Vignette 3.1). But just when these paleolithic people reached Alaska and then the heartland of North America remains in doubt.[2] Archaeological evidence confirms that Paleo-Indians occupied what is today the southwest United States some 11,000 years ago, suggesting that their ancestors migrated along the ice-free corridor between the Cordillera and Laurentide ice sheets some 12,000 years ago or earlier (Vignette 3.2).

Yet, a nagging feeling exists among archaeologists that Old World hunters reached the interior of North America before the ice-free corridor appeared (Bonnichsen and Turnmire, 1999), possibly by island-hopping along unglaciated islands exposed by the lower sea level until they were south of the Cordillera Ice Sheet. Archaeological evidence suggests humans arrived in Yukon before the last ice advance—some 25,000 to 30,000 years ago. Archaeological findings such as those at Bluefish Caves near Old Crow in Yukon support this hypothesis, although the dating of the Bluefish Caves artifacts remains controversial (Cinq-Mars and Morlan, 1999).

One theory espoused by Wright (1995) suggests that two major migrations from Siberia to the temperate areas of North America took place—one before the last ice advance (35,000 years ago or more), the other when the ice sheets began to retreat about 14,000 years ago. In recent years, the coast migration theory has gained the most support from archaeologists. The basic idea is that Old World hunters spread along the Pacific coast by island-hopping before the ice-free corridor opened. This theory assumes that, with the lower ocean level, more offshore islands beyond the reach of the Cordillera Ice Sheet were available to these ancient peoples.

Vignette 3.1   Migration Routes into North America

During part of the last ice age, the sea level dropped, exposing the sea bottom in the Bering Strait area. The result was a land bridge several hundred kilometres wide and some 1,500 km in length linking Asia to North America. This land is called Beringia. Now submerged some 100 metres below sea level, the land bridge allowed Old World hunters to occupy these unglaciated lands and move into Alaska. Around 18,000 years ago, the last ice advance had reached its maximum extent. At that time, huge quantities of water were locked into the Antarctic and Greenland ice sheets as well as into smaller ice sheets, such as the Laurentide and Cordillera ice sheets, which covered most of present-day Canada. The net result was a much lower sea level that dropped by as much as 150 metres. By 14,000 years ago, the climate had warmed, causing two major natural events. First, as the ice sheets began to melt, the ocean waters rose, covering the land bridge between Siberia and North America. Around the same time, an ice-free corridor appeared between the Laurentide and Cordillera ice sheets. This corridor made it possible for the Old World hunters to migrate southward along the eastern slope of the Rocky Mountains. Unfortunately, no archaeological find confirms this ancient passageway across Beringia because it is deep beneath the Bering Sea, Bering Strait, and Arctic Ocean.

Vignette 3.2   The Clovis Theory: The Original Paradigm

Since the 1930s, the Clovis theory has served as the central paradigm for archaeologists studying the peopling of the Americas. According to the Clovis theory, some 12,000 years ago Old World hunters from Alaska and Yukon found an ice-free corridor through the vast ice sheet to the unglaciated areas of the New World. While Paleo-Indians left no evidence of their passage through the ice-free corridor, this migration route remains a key element of the Clovis theory. Armed with spears that had a distinctive fluted projectile point, these Paleo-Indians quickly spread across North and South America. Archaeological evidence clearly demonstrates that Paleo-Indians were hunting the woolly mammoth and other big-game animals in the Great Plains and the Southwest of what is today the United States as early as 11,500 years ago. At the same time, primitive hunters had reached the southern tip of South America. The empirical support for the Clovis theory is based on archaeological finds of flaked-stone projectile points. These distinctive spear points were first unearthed in 1932 at a New Mexico mammoth kill site near Clovis, New Mexico. Since then, the discovery of human remains at five sites dating more than 12,000 years ago have challenged the Clovis theory and the inland migration route. For example, recent archaeological findings, especially the Monte Verde site in Chile, which apparently dates to 12,500 years ago, suggest that Old World hunters reached South America well before the Clovis people. Added to that archaeological evidence, in 2014, the discovery of a skeleton in a submerged underground cave in Mexico's Yucatan Peninsula dates 12,000 to 13,000 years ago. Does this mean that the original paradigm is no longer valid?

Sources: Adapted from Thomas (1999); Bonnichsen and Turnmire (1999); Balter (2014).

The arrival of Old World hunters from Asia to North America took two distinct migration routes. The first movement of peoples went south to the more temperate climates of North America through the ice-free corridor. They were the hunters of large terrestrial game. The second movement saw these ancient hunters likely working their way south along the unglaciated islands off the British Columbia coast to reach what is now the United States. The other possibility is that they followed the game along the ice-free corridor (Vignette 3.2). Their main source of food remained the woolly mammoth and other big-game animals. Around 10,000 years ago, the woolly mammoth became extinct and its hunters, now described as Paleo-Indians, had to adjust to new circumstances. At the same time, the Laurentide and Cordillera ice sheets had retreated, allowing the people to begin the process of occupying these new hunting grounds in the Canadian North. The archaeological record for this period is very sketchy, leaving no clear picture of the pattern of occupancy by various Paleo-Indian groups until contact. Arid regions, such as the US Southwest, parts of the Pacific coast of South America, and the Arctic, have provided more archaeological evidence of human occupancy than is the case for more humid regions.

In this warmer Holocene climate, the open water along the Arctic coast attracted a second wave of Old World hunters, the Paleo-Eskimos. Some 9,000 years ago, they crossed the Bering Strait from Siberia to Alaska by primitive sea craft. By this time, ice sheets had retreated from the Canadian Arctic coast, allowing the sea-based hunting economy of the Paleo-Eskimos to flourish. These early migrants are named Denbigh after an important archaeological site in Alaska. Their culture, the Arctic Small Tool Tradition, was based on the use of flint to shape bone and ivory into harpoons and other tools. As sea hunters, they occupied Arctic coastal lowlands and eventually reached the coasts of Labrador and Newfoundland. Over time, the culture of these people either evolved or was replaced by a new wave of Paleo-Eskimo migrants—the Dorset—who in turn were replaced by the Thule. While the archaeological story is incomplete, two major cultural changes have been identified. The first one saw the Denbigh culture evolve into or replaced by the Dorset culture. The Dorset lived in semi-permanent houses built of snow and turf, heating them with soapstone oil lamps. Some archaeologists believe that the Thule, with their superior technology and weaponry, overwhelmed the Dorset and maybe even the Viking colony on Greenland (Thompson, 2010).

Around CE 900, the Thule culture appeared in coastal Alaska. Here, the Thule hunters had developed the technology and skills necessary to hunt the large bowhead whales. They migrated eastward, perhaps driven by population pressures or encouraged by a warming of the Arctic climate that resulted in an increase in open water in the Arctic Ocean. In any case, the Thule spread across the Canadian Arctic using their innovative transportation system of skin boats (kayak and umiak) and dogsleds to harvest seal, walrus, caribou, and the bowhead whale. By CE 1000, the Thule reached Greenland, where they probably encountered the Vikings.

After CE 1300, the climate began to cool, making ice cover more extensive. The climate continued to cool and the period between 1450 and 1850 is referred to as the Little Ice Age. With the more extensive ice cover, sea areas suitable for the bowhead whale became more limited. Unable to secure food from their favourite source, the bowhead whale, the Thule had to shift their hunting strategy and, therefore, became more mobile and formed smaller groups to hunt the seal and caribou. About this time, Europeans, such as the English explorer Martin Frobisher, came into contact with these people. Their descendants are the present-day **Inuit**. Ironically, the Little Ice Age coincided with the era of European exploration in the Arctic, and the colder temperatures and more extensive ice cover doubtless made such exploration more difficult, and likely was instrumental in the ill-fated Franklin expedition of 1845–7 (Vignette 3.5).

Contact

The New World was not, as Europeans believed, *terra nullius*—empty lands. The Eurocentric perspective regarded lands without permanently occupied land, such as signs of agriculture and settlements, as empty, though not necessarily uninhabited. Yet Aboriginal peoples inhabited these lands. Dickason (2009: 10) portrays this pre-contact world:

> When Europeans arrived, the whole of the New World was populated not only in all its different landscape and with varying degrees of density, but also with a rich cultural kaleidoscope of something like 2,000 or more different societies.

Exactly how many lived in Canada, let alone northern Canada, in these early days is unknown.[3] However, given the challenging environment for hunting peoples, the North must have had a small population spread over a vast territory. One estimate suggests that some 50,000 Indians and Inuit occupied the Subarctic and Arctic areas of Canada at the time of contact (Crowe, 1974: 20), while Mooney (1928) and Kroeber (1939) estimated a slightly higher figure of 56,000, with 16,000 Inuit in the Arctic and 40,000 Indians in the Subarctic. Such numbers show a population density of 130 km² per person, indicating that the hunting economy could only support a relatively small population due to the limited availability of game and its seasonal nature. In fact, the Arctic and Subarctic, as might be expected, had the lowest population density of all the Aboriginal cultural regions in North America, suggesting lower carrying capacity, which is a measure of the capacity of the land/sea to support people.

In 1857, the first comprehensive estimate of the Indian and Inuit population of British North America was made by Sir George Simpson for the Special Committee of the British House of Commons on the affairs of the Hudson's Bay Company (*Census of Canada*, 1876, vol. 4: ix–xiv). Simpson's figure of nearly 139,000 for British North America was based on reports made by local HBC officials on the number of Indians trading at their posts. Approximately 61,000 Indians were reported as trading at posts in the Subarctic. Added to these numbers was Simpson's guess that 4,000 Eskimos occupied the Arctic. In 1881, the census indicated about 108,000 Indians, Métis, and Inuit residing in the provinces and territories of Canada (Table 3.1). Nearly half (46 per cent) lived in the North-West Territories, which, in 1881, included Alberta and Saskatchewan. With 24 per cent of the Aboriginal population found in British Columbia, this province held first place in terms of density. Within the province, the greatest numbers fished and hunted along the narrow Pacific coast.

Table 3.1 Indian Population, 1881

	Indian Population	Total Population	Indian Percentage of Total Population
Prince Edward Island	281	108,891	0.3
Nova Scotia	2,125	440,572	0.5
New Brunswick	1,401	321,233	0.4
Quebec*	7,515	1,359,027	0.6
Ontario*	15,325	1,923,228	0.8
Manitoba*	6,767	65,954	10.3
British Columbia	25,661	49,459	51.9
North-West Territories	49,472	56,446	87.6
Canada	108,547	4,324,810	2.5

*In 1881, the geographic size of Quebec, Ontario, and Manitoba was smaller than today. As well, the North-West Territories included lands in present-day Alberta, Saskatchewan, much of Manitoba, Ontario, and Quebec as well as the three territories. Its capital city was Regina. After 1881, Manitoba, Ontario, and Quebec expanded their geographic size by obtaining land from the North-West Territories. In 1905, the provinces of Alberta and Saskatchewan were formed and the North-West Territories became the Northwest Territories.

Source: Statistics Canada, *Census of Canada*, 1880–81.

Where did Aboriginal peoples live? The geography of Aboriginal people at the time of contact is based on the diaries and accounts of early explorers, fur traders, and missionaries. The broad pattern involves a threefold geographic division, with the Algonquin Indians living in the eastern Subarctic while the Athapaskan Indians occupied the western Subarctic. To the north lived the Inuit. This same broad geographic settlement pattern exists today.

Determining the geographic location of individual **tribal groups** is much more problematic, but these peoples likely had occupied the territory shown in Figure 3.1 for some time, perhaps centuries. The geographic stability of the pre-contact hunting societies can only be guessed at because archaeological evidence is so limited. Only in the Arctic is there archaeological evidence and oral history documenting the forced replacement of one group by another—the Dorset by the Thule, who were the ancestors of the Inuit.

These small but highly mobile Indian and Inuit hunting groups had become finely attuned to harvesting the particular food resources in their geographic territory.

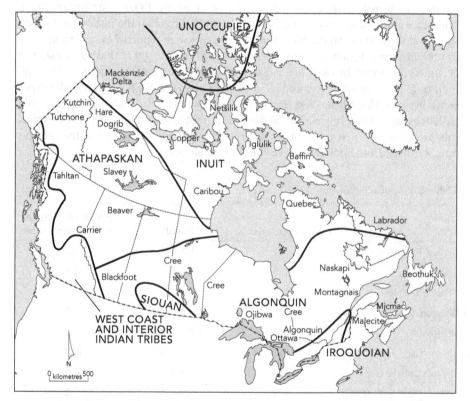

Figure 3.1   Geographic Areas Occupied by Aboriginal Peoples at Contact

Anthropologists and ethnographers have reconstructed the geographic areas of Aboriginal peoples at the time of contact. Yet, it is important to note that, prior to contact with Europeans, Aboriginal peoples were not always fixed in one geographic area. For instance, the Dorset had successfully occupied the Arctic, but by around CE 1000 the Thule had replaced them. On the other hand, tribes along the BC coast where fish and game were plentiful were much more sedentary.

Source: After Harris (1987: Plate 11).

Within that territory, hunters followed a seasonal rhythm of movement as they sought wild game. Through this seasonal movement, they exerted a form of political control over the territory. The main source of food was big game. For many, caribou was the preferred food; whale blubber and seal meat were common food sources for the Inuit. These hunting people moved according to the seasonal migrations of wild animals. Such a migratory hunting strategy is a form of sustainable resource use because the necessity of moving from one area to another limits the pressure on local resources. Limited weapon technology and cultural taboos also reduced the possibilities of over-hunting or overfishing.

The pre-contact hunting economy had many attractions but it also contained an element of risk. An unforeseen reduction in game, such as a shift in the annual migration route of caribou, could expose hunters and their dependants to severe shortages of food and perhaps starvation. As Ray (1984: 1) observed, "opinions are divided whether hunting, fishing, and gathering societies generally faced a problem of chronic starvation—the more traditional viewpoint, or whether they were the original affluent societies." The key to resolving this puzzle is reliability of food sources and the size of the population. Some geographic areas had a more readily available source of food than other areas. The west coast of British Columbia, for example, had abundant fish stocks, which permitted relatively dense and sedentary pre-contact Indian populations (Harris, 1997: 30). Caribou hunters, who depended on the seasonal migration patterns of caribou, had to deal with a more problematic food source. Over the long run, archaeological evidence suggests that the population in each cultural region attained a size that was in balance with its food sources, but short-term food shortages could occur.

The first Europeans to set foot on North American soil were the Vikings. During the tenth century, the Vikings regularly travelled from Greenland to Ellesmere and Baffin islands to trade with either the Dorset or Thule peoples. They also attempted to establish settlements along the coasts of Labrador and Newfoundland. The first and only Norse site found in Canada is located at L'Anse aux Meadows on the northern tip of Newfoundland's Great Northern Peninsula.

Some 500 years later, the Viking discoveries were all but forgotten, leaving Europe's geographic knowledge of the New World scanty at best. European explorers sought to reach the rich spice lands of Cathay by sailing across the Atlantic Ocean but reached North America. Reports of the new land's great riches, based on information obtained from local Indians, may have been misinterpreted or embellished in hopes of securing financial support for later voyages. In 1536, Jacques Cartier returned to France with a report of a golden "Kingdom of Saguenay" (Trudel, 1988: 368). This report was based on information supplied by Indians and supported by the Iroquois chief Donnacona, whom Cartier had captured and taken back to France. Cartier thought that he had discovered diamonds and gold from this mythical kingdom. These riches, upon careful examination in France, turned out to be quartz and iron pyrites—"fool's gold."

In a similar vein, Martin Frobisher, under orders from Queen Elizabeth I, set sail from England to find the Northwest Passage to the Orient. Instead, he discovered Baffin Island and the promise of great mineral wealth—gold. Like the Spaniards, the English sought gold and silver from this New World. Would Frobisher's discovery give England a source of gold to rival Spain's New World gold mines? Over 1,200 tonnes of rock from Baffin Island were transported to England, but this rock was found to contain not an ounce of gold! Nevertheless, Frobisher's two sea voyages in 1576 and 1577 mark the beginning of British exploration of Arctic Canada. As with most early voyages, contact with the local peoples was established and trade often followed. However, contact between these two very different peoples did not always end on a friendly note. For instance, after landing on the shore of Baffin Island in 1576, five of Frobisher's crew were captured by local Inuit during a skirmish and were never seen again (Cooke and Holland, 1978: 22–3). The following year, Frobisher lured an Inuk in his kayak to come near his ship. Captured and taken to England, the Inuk's skills at handling his kayak while hunting the royal swans were observed by the Queen (Crowe, 1974: 65). After a few years, he died of natural causes in England.

During the remainder of the sixteenth century and in the early seventeenth century, many explorers, including Davis, Hudson, and Button, continued to search for a route to the Orient. In the course of this search they explored the waters of Baffin Bay and Hudson Bay. These explorers failed to find a route to the Orient but the "casual" trade with Indians for furs marked the beginnings of a highly profitable enterprise and led to the formation of an English fur-trading company—the Hudson's Bay Company—in 1670.[4]

The Fur Trade

The story of the fur trade holds a special place in Canada's history. While many books and articles have discussed the fur trade, none have equalled the historic timeline, the staple thesis, and continental approach found in Harold Innis's *The Fur Trade*. Published in 1930, Innis's book is still acknowledged as one of the foundations of our understanding of the fur trade within the context of Canada's development and the place of Aboriginal peoples in that development. One of Canada's recognized scholars on this subject, Arthur Ray (2001: vi), summarized Innis's contribution:

> Innis argued that the resulting scramble for the beaver and other furs that aboriginal traders and hunters supplied set in motion a circular and cumulative process that propelled the industry across Canada and drew a succession of Native economies into the European economic orbit. The prime beaver country of North America encompassed the subarctic boreal forest extending across much of Canada from the Labrador coast to the Alaskan border.

In this historical context, the commodity that set the economic agenda for Canada's North was the beaver and other fur-bearing animals. The spatial extent of the fur trade economy depended on the habitat found in the boreal forest for these fur-bearing animals. But why was the fur trade always in search of new lands? This relatively rapid march across the boreal forest from east to west was necessitated by the overexploitation of the fur resource in one area, forcing a shift to another area. Another contributing factor to overexploitation was the competition between fur-trading companies that compelled trappers to no longer practise sustainable harvesting.

The trading of furs for European goods began in the fifteenth century with the Basque fishers who traded with Indians along the shores of Newfoundland. From these humble beginnings, the fur trade became the centrepiece of New France's economy. In 1627, the King of France granted the Company of New France a fur monopoly. As early as the sixteenth century, French fur traders, including the coureurs de bois located along the St Lawrence River, recognized that the best furs came from Quebec's Subarctic, where the long, cold winter resulted in long-haired fur pelts. Geography provided easy access to these northern Indians by means of southward-flowing rivers, such as the St Maurice and Saguenay.[5] The Saguenay River provided access to the Rupert River. From here, Indians brought their furs south to trading posts in New France. As well, French fur traders travelled north via the Saguenay, traversing Lake Misstassini and paddling along the Rupert River to its mouth at James Bay. With no competition from the British, New France enjoyed almost a hundred years of fur trading in the James Bay area. This monopoly ended in 1670 when the Hudson's Bay Company employed a sea route to reach the James Bay area. For over a century, the French and English competed for the highly prized fur pelts of the James Bay Cree (Figure 3.3). Even when French Canada fell to the British in 1763, two centres, London and Montreal, continued to compete for northern furs (Figure 3.5). Montreal fur traders of the North West Company, now a blending of French-Canadian and Scottish traders, established posts on the Saskatchewan River in order to intercept Indians on their way to trade at the Bay posts. After decades of fierce and sometimes bloody competition, the two companies merged in 1821 (Figure 3.6). Over the next 100 years or so, the Hudson's Bay Company dominated the Canadian fur economy.

The Hudson's Bay Company

Like all colonial powers, Britain and France looked to their colonies for riches. Wealth often took the form of gold and silver, but in the case of Canada's North, wealth took the form of beaver pelts. The fur trade drew Canada's North and its Indian inhabitants into the European economy, where the demand for furs was dependent on the European fashion industry (Vignette 3.3). The political implications of the fur economy are

---

Vignette 3.3   Beaver Hats: The Necessary Attire of Gentlemen

For several centuries, the fur trade was based on the beaver skin, highly valued in Europe for hats. This demand, driven by fashion, was both the strength and weakness of the fur trade. The beaver hat was a sign of upper-class status. As shown in Figure 3.2, the beaver hat had a number of styles, reflecting occupation/status in life. Back in Canada's trading posts, prices of beaver and other furs were subject to sharp fluctuations—often caused by variation in the demand for these pelts—and these price changes could greatly alter the number of furs required for European goods. The beaver was sought not for its entire pelt but for fur-wool, the layer of soft, curly hair growing next to the skin, which had to be separated from the pelt and from the layer of longer and stiffer guard hairs. This fur-wool was then felted for cloth or hats. The beaver hat went out of fashion in the 1850s and was replaced by top hats made of silk and other materials.

Source: Adapted from Wolfe (1982: 159–60).

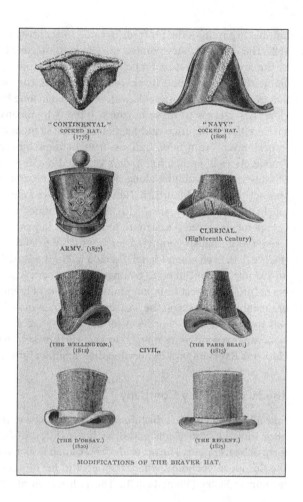

Figure 3.2 Beaver Hats

Source: Martin (1892: 125). © Public Domain.
Library and Archives Canada, nlc-710.

far-reaching and take the form of a core/periphery relationship within a mercantile economic system. In this historical period, the core nations were Britain and France while the North served as the fur hinterland. The implications of this core/periphery relationship were twofold. First, with a relatively small number of fur traders and fur-trading posts, Britain and France gained control over this vast land and its inhabitants by effectively involving the much larger Indian population as trappers. Second, the fur trade allowed the Indian tribes to maintain their hunting economy. At the same time, by trapping and exchanging furs for metal products, cloth, and weapons as well as other goods, Indian families improved the material aspect of their lives. As time progressed, however, the balance between hunting and trapping tipped heavily towards the fur trade and a form of dependency evolved, including the supply of food and clothing to destitute Indian families by the Hudson's Bay Company and, after 1870, by the federal government (Fumoleau, 2004: 21–3).

For over 300 years, the HBC dominated the Canadian fur trade. This British company was the most powerful economic, social, and political force in the Canadian North. Within the British mercantile system the HBC was an instrument of its foreign policy and a means of acquiring wealth, just as was the East India Company that was formed 70 years earlier. The role of these companies was twofold: to function as commercial enterprises and to exert British sovereignty.

In the seventeenth century, the best beaver pelts came from boreal forest lands along James Bay. This productive fur area provided New France with much of its wealth. In its wars with France, Britain would have liked nothing better than to weaken the economy of New France and, at the same time, engage in the lucrative fur trade. The unanswered question was how? The answer came from two French traders, Medard des Groseilliers and Pierre Radisson. They approached the British and convinced Prince Rupert that a great deal of money could be made by trading with Indians for furs along the Hudson Bay coast. These northern Indians currently traded their furs with the French, who were based on the St Lawrence. Groseilliers and Radisson knew that ships anchored near important northern rivers flowing into James Bay and Hudson Bay would soon attract local Indians and trade would ensue. Prince Rupert persuaded his cousin, Charles II of England, and some merchants and nobles to back the venture. In 1670, the King granted wide powers to the "Governor and Company of Adventurers" of the Hudson's Bay Company, and Rupert became the company's first governor. These powers included exclusive trading rights to all the lands whose rivers drain into the Hudson Bay. These lands were named Rupert's Land.

At first, the company sent ships to trade with the Indians who gathered at the mouths of rivers draining into Hudson Bay. Once the trade was completed, the ships returned to England. The Hudson's Bay Company soon established permanent trading posts, encouraging Indians living in the interior of the country to trade there. By locating at the mouths of rivers draining western Canada, the HBC extended its control over lands far into the interior, thereby increasing the number of Indians involved in trapping and, of course, maximizing the number of furs received. The HBC was a great commercial success from the start. It was also a political success, for its Bay forts provided an important foothold for British control over these lands.

The harsh climate around Hudson Bay prevented the fur traders from supplementing their food supply with agricultural products. They turned to local Indians to supply them with game and with other useful products of Aboriginal technology, such as snowshoes and canoes. These Indians who maintained permanent camps near the HBC forts became known as the Homeguard. Indian middlemen became traders who exchanged European goods for furs with inland Indians. This arrangement provided great profits to the middlemen and allowed them to control the inland Indians. Often, the middlemen Indians would not even permit other Indians to travel to the coastal trading posts.

The trading posts along the coasts of Hudson Bay and James Bay all followed the same pattern, being established at the estuaries of the major rivers draining the interior of the forested lands of what are now Manitoba, Saskatchewan, Quebec, and Ontario. The most important of these trading posts was York Factory. Situated at the mouth of the Hayes River, it offered the easiest river access to the rich fur lands of the interior. This post, because of its access to the river network leading into the interior of the Subarctic, soon began drawing furs from Indians as far distant as the Mackenzie River Basin (Zaslow, 1984: 6). By the early eighteenth century, then, the HBC's fur hinterland extended into the "unclaimed" territories indicated in Figure 3.3. A similar case

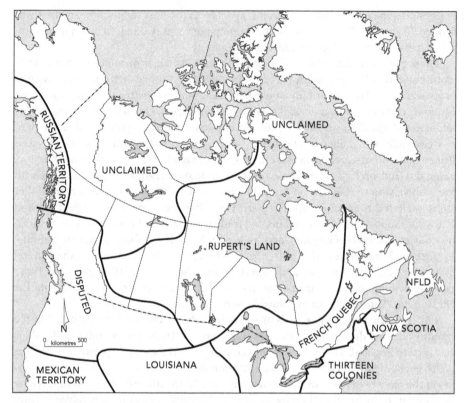

Figure 3.3 European Spheres of Influence, 1760

The struggle to gain control over lands in North America is part of the much larger story of European coloniza-
tion of the world. By 1760, most of North America had been claimed by Britain, France, Russia, and Spain. The
last unclaimed land lay in the northwest part of Canada and Alaska. In the coming century, British and Russian
fur traders would collide, with the Russians penetrating inland from the Pacific and the British moving inland
using the Mackenzie River system.

may be made for the Russian fur trade zone because Indians in the Yukon River Basin
may have made their way to the Pacific coast trading posts.

The Time of Competition

While the British had outflanked New France with their fur-trading posts on James and
Hudson bays, French traders countered this effort by penetrating deep inland to cut
across the river routes leading to York Factory (Figure 3.6). The stranglehold of the
Hudson's Bay Company on the fur trade in the western interior was broken when
French traders established inland posts on the Saskatchewan River. By 1750, French
competition had caused a sharp decline in the number of pelts reaching York Factory.
After the fall of New France in 1763, Montreal became the strategic headquarters for a
number of fur traders, first known as the "Pedlars" and, in 1784, as the North West
Company. In spite of a long supply route from Montreal to the western interior of
Canada, these Montreal fur traders were determined to intercept Indians on their way

---

Vignette 3.4   Geographic Advantages of Fort Chipewyan

In 1788, the North West Company established a fort on the northwestern shore of Lake Athabasca in present-day northeast Alberta. It was named Fort Chipewyan after the Indians who hunted in the area. Fort Chipewyan was an ideal place for the furthermost post on a fur-trading route. Not only was the fort located in prime beaver country where the pelts were of the highest quality, but its location allowed the northern brigade to leave Fort Chipewyan at breakup, reach Grand Portage, and return to Fort Chipewyan with a new supply of trade goods before freeze-up. From the east, the Montreal brigade brought a new supply of trade goods to Grand Portage and returned to Montreal with fur bales. Another geographic advantage of Fort Chipewyan's location was its ready access to the Athabasca River, which breaks up in the spring a month before the lake. This gave the western fur brigade an extra month for their canoe trips to rendezvous with their eastern counterparts.

Source: Donaldson (1989: 10).

---

to trade furs at the HBC posts on the shores of Hudson Bay and thus to obtain the highly prized furs from Canada's Northwest. The strategy of the Montreal traders was simple—reduce the distance Indian fur trappers had to travel to reach a trading post. The North West Company built a number of fur posts and connected them by its two main fur brigades. The Montreal brigade would bring supplies from Montreal to Grand Portage (later Fort William), which served as the supply depot where trade goods were shipped for later distribution to inland posts; the northern brigade originated at Fort Chipewyan, bringing fur bales to Grand Portage that would be taken to Montreal by the Montreal brigade (Vignette 3.4).

The strategy of permanent inland posts, however, had one serious problem—provisioning. The North West Company solved this problem by supplying its inland posts with pemmican, a food made by Plains Indians, and by having its **Métis** employees and Indian trappers supply the post with country food. Pemmican was light and nutritious and, most importantly, would not spoil. It consisted of dried meat (usually buffalo) pounded into a coarse powder and then mixed with melted fat and possibly dried Saskatoon berries. In this way, the fur trade of the late eighteenth century became dependent on food supplies from another natural region, the grasslands of the Canadian prairies. This arrangement indirectly involved the buffalo-hunting Plains Indians and Métis in the fur trade.

The success of the Nor'Westers' inland trading posts forced the Hudson's Bay Company to meet the competition by moving inland. In 1774, the English company established its first inland post at Cumberland House on the Saskatchewan River. This action caused the Montreal traders to move further west, and in 1776 Louis Primeau established a post on the Churchill River near the present-day settlement of Île-á-la-Crosse (Cooke and Holland, 1978: 95). In 1787, the remaining individual traders either joined forces with the now powerful North West Company or left the country. Competition between the North West Company and the HBC intensified, leading to the strategy of "matching" trading posts, i.e., if one company established a post in a new area, the other company built one nearby.

Figure 3.4 Governor George Simpson on a Tour of Inspection in Rupert's Land, Nineteenth Century

Following the merger of the two great fur-trading companies in 1821, Governor George Simpson of the Hudson's Bay Company went on a tour of inspection of trading posts. His objective was to make the new company more efficient, which meant reducing the number of posts and workers. Governor Simpson had his own, narrow-beam, eight-metre "express canoe" for travelling. He travelled with a crew of Iroquois voyageurs and a personal Scottish piper.

Source: HBC's 1926 calendar from a painting by L.L. Fitzgerald (from a photo of a painting by Cyrus C. Cuneo).

For the Aboriginal peoples, there were several consequences of this competition. First of all, the new posts allowed inland Indians direct access to traders. Second, it weakened the power base of the Cree, who lost their role as middlemen. The role of middlemen passed to the Chipewyans, who traded with the northern Indians, namely, the Yellowknife and Dogrib Indians. Finally, the competition between trading companies drew local Indians more fully into the fur trade and so made them more dependent on trade goods.

Amalgamation: A Time of Monopoly

The next phase in the fur trade saw a return to a monopolistic situation. With profits down and frequent bitter struggles between rival fur traders, the British Colonial Office wanted the North West Company and the Hudson's Bay Company to settle their feuding. In 1821, a parliamentary Act of the British government attempted to placate both parties by devising a coalition. The British government authorized that the name, charter, and privileges of the HBC would be assigned to the new firm but that its fur traders would include those of both the North West Company and the Hudson's Bay Company.

Under the direction of George Simpson, the "new" Hudson's Bay Company began a rationalization of its trading posts. Posts were abandoned, staff reduced, and prices increased. The North West Company's southern route was discontinued, leaving the Bay route via Hudson Strait to London. Most furs were shipped by York boats and canoes to York Factory and then loaded on sailing ships for England. Profits soon rose,

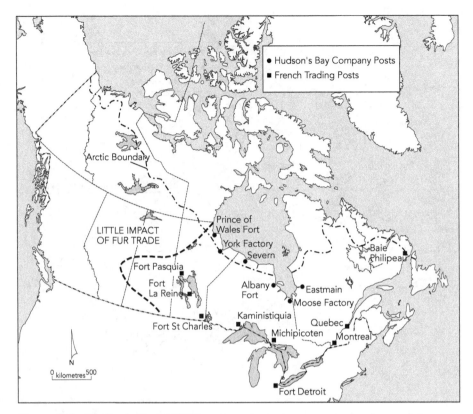

Figure 3.5   Fur Trade Posts, 1760

British and French fur-trading posts marked the limits of their colonial empires. In what is now Quebec, the conflict between France and England was over beaver pelts—the main form of wealth in the North. In the late seventeenth century, the British, through the Hudson's Bay Company, developed a sea route through Hudson Strait and the company established trading posts at the mouths of rivers flowing into James Bay and Hudson Bay, thus allowing inland Cree Indians easy access to their trading posts. The key trading posts were Moose Factory (1673), Albany Fort (1679), and Severn House (1680). This strategy effectively redirected the trade in beaver pelts from posts in New France to Hudson's Bay trading posts, thus damaging the main economic activity in New France and thereby weakening New France's economy. The building of York Factory (1684) allowed the Hudson's Bay Company to tap into the vast beaver lands, stretching westward to the Rocky Mountains.

reaching unimagined levels. The company controlled not only Rupert's Land but also the North-Western Territory that extended further west to British Columbia and north to the Arctic Ocean.

While his main objective was to increase HBC profits, Simpson realized that such profits could only be achieved if the Indian way of life also "prospered," i.e., if Indians devoted most of their time to trapping. He supported the concept of supplying Indians with food when country food—fish and game—was in short supply. Short supply was a product of the fur trade, largely because the demand for more game to meet the needs of fur traders resulted in overexploitation around fur posts and because trappers, faced with depleted game stocks in the more accessible areas, had less time for hunting in their traditional hunting grounds. Another factor was an advance in hunting technology from bows and arrows to muskets and then rifles, thus tipping the balance in favour of the hunter.

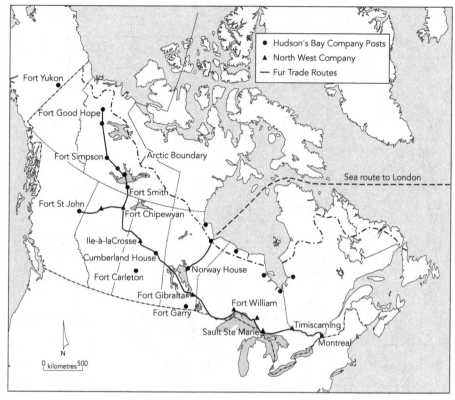

Figure 3.6 Fur Trade Posts and Main River Routes, 1820

Just before the Hudson's Bay Company and the North West Company amalgamated, each had its own river routes to supply their trading posts and to ship fur pelts. Neither company had an interest in the Arctic because Europeans valued beaver and other pelts found in Canada's Subarctic. Fashions were changing, however, and by the mid-nineteenth century the beaver hat was no longer popular in Europe (see Vignette 3.3).

A new challenge to the Bay route appeared in the mid-nineteenth century. A combination of steamboat and rail service to New York first came into play in 1859 when the steamboat, the *Anson Northrup*, left the upper reaches of the Red River in the United States and arrived at Fort Garry (Winnipeg). Within a decade, the fur trade shifted from the Bay route to the US steamboat/ox trail route to St Paul, Minnesota. Fur brigades now followed a southern route to Fort Garry. From there, the fur bales were loaded on steamboats bound for St Paul and then the bales were transferred to rail cars bound for New York. The last stage in the journey to London, England, was by transatlantic steamer. Later, when the Canadian Pacific Railway was completed, the furs were shipped along this Canadian railway to Montreal and then on to London.

Exploration and the Fur Trade

The fur trade had always been tied to exploration, as the traders continued to search for new trapping areas. Fur traders were eager to explore new lands and to make

Figure 3.7 York Factory, Hudson Bay, 1925

Founded in 1684, York Factory at the mouth of the Hayes River was the central base of operations for the Hudson's Bay Company for over 200 years. Indians brought fur pelts great distances by canoe to trade for British goods. The company shipped its fur pelts to London, England, through Hudson Strait, and York Factory received trade goods and other supplies in the same manner. Until the fall of New France, York Factory played a critical role in the French–English struggle on Hudson Bay for control of the fur trade. For the next century, the post had a major role in the expansion of the fur trade into the interior of western Canada and as the major administrative, transshipment, and manufacturing centre for the Bay's operations in Canada's northwest country.

After a rail connection from New York to St Paul on the Mississippi River was completed in 1857, its days were numbered. By the late nineteenth century, the HBC began to ship fur pelts by riverboat up the Red River, by rail to New York, and then by ship to London. From that point on, York Factory's role in transatlantic shipping of fur pelts declined. In 1931, York Factory ceased to ship pelts to England. From 1931 to 1957, its final role was to serve the local Indian population. In 1957, this Hudson's Bay post closed, and in 1968 it became a National Historic Site under Parks Canada.

Source: Canada. Dept. of Interior/Library and Archives Canada/e004665616

contact with distant Indians who might supply them with furs. As always, the push into new lands was driven by the need to reach the Indian hunters before a rival fur trader.

Fur traders occasionally undertook exploration for other reasons. For example, Samuel Hearne's remarkable overland journey from Prince of Wales Fort at the mouth of the Churchill River to the Arctic Ocean in 1771–2 was motivated by the desire to substantiate reports of a "rich" copper deposit along the Arctic coast. In 1768, northern Indians had brought several pieces of copper to the Hudson's Bay trading post of Fort Churchill (Morton, 1973: 291). The governor of the post, Moses Norton, sent Samuel Hearne in search of the deposit. In his first two attempts, Hearne failed. Crossing the Barrens was no easy feat, even with Indian guides and company servants. Only by travelling with the Chipewyan guide Matonabbee and his family was Hearne able to reach the mouth of the Coppermine River in 1771 and examine the copper deposit (Hearne,

1958; Rich, 1967: 298). The success of this amazing journey was due to Chipewyan knowledge of the land and animals. Hearne not only reached the shores of the Arctic Ocean but he also travelled west to Slave River. Near Great Slave Lake, Matonabbee traded European goods for furs trapped by Dogribs, which he would later exchange for more trade goods at Fort Churchill (Morton, 1973: 298).

Hearne made a number of important observations about the Chipewyans. For the first time, Europeans had some inkling of the enormous extent of travel by Indians living in the Subarctic. Matonabbee and his group ranged a vast hunting territory, extending from Hudson Bay to the Arctic Ocean to Great Slave Lake. These forested and tundra lands, interconnected by rivers and lakes, were not the exclusive hunting grounds of the Chipewyans, nor were they marked by fixed boundaries. For example, during the summer, some of these Indians moved into the Barrens, which also served as the hunting territory of the Inuit. But in these sparsely populated lands, the chance of contact was slim and direct contact could easily be avoided. Yet, at this particular time in history, the Chipewyans were the dominant group on the Barrens, partly because of their leader and partly because they were armed with muskets and other European weapons. As middlemen in the fur trade, they were able to terrorize the Inuit and exploit neighbouring Indian tribes such as the Dogribs and still profit from trade at Fort Churchill.

With the end of the Napoleonic Wars in 1815, the Royal British Navy undertook a series of Arctic expeditions to solve the enigma of the Northwest Passage and to present a challenge to its navy. Sir John Franklin commanded three of these expeditions. The first two were overland journeys designed to survey the Arctic coast while his last one was the famous but tragic sea voyage. His first overland journey, the Coppermine expedition (1819–21), focused on recording the coastline east of the Coppermine River. Franklin underestimated the perils of Arctic exploration; he was ill prepared and promises of support from the two competing fur-trading companies often failed to materialize. At that time, only two explorers had reached the Arctic Coast—Samuel Hearne at the mouth of the Coppermine River in 1771 and Alexander Mackenzie at the Mackenzie River Delta in 1789. Between these two points lay hundreds of kilometres of uncharted coastline. Franklin's first journey ended badly as little coastline was surveyed; all suffered terribly from hunger and cold; and 10 members of the expedition died from starvation and two others by gunfire. Much of the blame for the failed expedition lay at Franklin's feet. His second overland expedition (1825–7) was clearly a success, largely because Franklin recognized the rigours of Arctic exploration and was much better prepared and better supported by the Hudson's Bay Company. Again, his task was to map the unknown Arctic coast east and west of the mouth of the Mackenzie River. His party reached the Mackenzie Delta and then split, with Franklin heading west and a group led by John Richardson, his second in command, heading east to the Coppermine River. Their accomplishments were significant as Franklin and Richardson charted over 1,600 km of the Arctic coast, thus suggesting a southern route to the Northwest Passage. Franklin's final expedition set sail from England in 1845 in search of the elusive Northwest Passage by proceeding south from Barrow Strait through Peel Sound. Unfortunately, ice conditions were particularly bad and his two ships, the *Erebus* and the *Terror*, were trapped in ice just north of King William Island. All perished (Vignette 3.5). In 2008, the Canadian government began a series of expeditions to find the sunken ships. A discussion of these expeditions is found in Chapter 8.

Vignette 3.5   The Franklin Expedition to Find the Northwest Passage

Sir John Franklin's final and tragic expedition to find the Northwest Passage began in 1845, with two well-equipped ships, the *Erebus* and the *Terror*, and with a total crew of 129. Earlier explorers, it seemed, had set a course for his success. Back in 1819–21, William Edward Parry had discovered the entrance to the Northwest Passage at Lancaster Sound. Parry sailed west to Barrow Strait and then Viscount Melville Sound—but a wall of thick ice blocked further travel to M'Clure Strait and the Beaufort Sea. Parry set up camp at Winter Harbour on Melville Island, demonstrating that British ships and their crew could winter in the Arctic, thus expanding the exploration time. In 1839, an overland Hudson's Bay Company expedition led by Peter Dease and Thomas Simpson charted the remaining unknown Arctic coast from Coppermine to the mouth of the Great Fish River, leaving only the mystery of the north–south sea route of some 500 km from Barrow Strait to the Arctic coast at Great Fish River. With the shifting Arctic ice pack a constant threat to these wooden sailing ships, Arctic exploration was a very risky business. Since the first half of the nineteenth century was still in the grip of the Little Ice Age, sailing in Arctic waters was much more hazardous than in the twenty-first century, when an ice-free Northwest Passage during the late summer became a reality.

In the summer of 1845, Franklin's two ships had reached Baffin Bay and soon entered Lancaster Sound. Unfortunately, by the late summer of 1845, the severity of ice floes made sailing hazardous and the expedition was forced to seek a winter site at Beechey Island just off the southwest corner of Devon Island. Over the next two summers, the two ships slowly made their way south through the ice-choked Peel Sound. In June 1847, Franklin had died, leaving Captain Crozier in command. During the following winter, the ships were lodged in thick ice just north of King William Island. By then, the hope of reaching the Beaufort Sea or sailing back to England was gone. On 22 April 1848, the *Terror* and *Erebus* were abandoned. Captain Crozier, who knew about the Great Fish River, attempted to lead the 105 remaining sailors across King William Island to that river first discovered by George Back some 15 years earlier. Most perished on King William Island; a few reached the mainland at Starvation Cove on the north tip of Adelaide Peninsula. While the mouth of the Great Fish River was only less than 100 km further south, travelling upstream to Fort Reliance would have been impossible. Back in England, the organization for the greatest land and sea search in Arctic history began: the ensuing search for Franklin and his men resulted in one of the greatest naval rescue efforts in the nineteenth century. Over the course of nearly 12 years much of Canada's Arctic coastline was explored and mapped by upward of 30 separate expeditions mounted to search the Arctic for Franklin and his men (see, e.g., Berton, 1988). The first news of the fate of the doomed Franklin expedition came in 1853 when an overland Hudson's Bay Company search party headed by John Rae reported that local Inuit had seen a ship and remains of sailors on King William Island (Woodman, 1991). Six years later, an expedition headed by Leopold McClintock and William Hobson, and sponsored by Lady Franklin, was successful in finding the remains of part of the Franklin party near King William Island as well as the only written account of the disastrous expedition.

Source: Cooke and Holland (1978: 212–16).

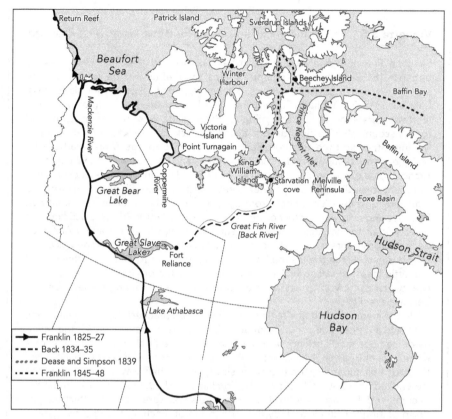

Figure 3.8   Search for the Northwest Passage: Land and Sea Expeditions, 1825 to 1848

Several overland expeditions were designed to chart the Arctic coast, thereby making the search for the Northwest Passage more assured of success. These expeditions mapped the Arctic coast from the Mackenzie River to the Great Fish River. Franklin's first (1825–27), recorded the coastline from the Mackenzie River to the Coppermine River. A decade later an expedition (1834–5) led by George Back travelled along the Great Fish River to its mouth at Chantrey Bay, just south of King William Island. Then, in 1839, Peter Dease and Thomas Simpson undertook the mapping of the coast from the Coppermine to just beyond the mouth of the Great Fish River. In 1819, the sea expeditions began with the remarkable accomplishment of William Edward Parry, who discovered the entrance to the Northwest Passage (Lancaster Sound) and then sailed westward to Melville Island. Franklin, on his fatal last Arctic adventure, planned to find the southern link from Barrow Strait to waters near Great Fish River and then proceed in a westerly direction to the Beaufort Sea.

Source: Adapted from National Maritime Museum, Greenwich, London at http://www.rmg.co.uk/file/9431

Commerce Comes to the Arctic

For many centuries, the fur economy centred on the boreal forest took the form of a partnership between the fur trader and the Indian trappers. The trapping economy began first in New France, then spread to the Provincial Norths and much later to the Northwest Territories. The Inuit, for better or worse, did not become part of the fur trade until the early years of the twentieth century—the Arctic habitat for fur-bearing animals like the beaver was not suitable. However, white fox did thrive in the tundra and when their pelts caught the eye of the fur garment industry, the Hudson's Bay and other fur traders entered the Arctic.

Figure 3.9 HMS *Investigator*

The HMS *Investigator* was one of many British naval vessels searching for the two missing ships of the Franklin expedition. In 1853, the ship was abandoned in thick ice in M'Clure Strait. This strait poses the most difficult ice conditions along the northern route of the Northwest Passage. In 1969, the SS *Manhattan*, an American oil tanker refitted to test the viability of transporting oil from Alaska through the northern leg of the Northwest Passage to the east coast of North America, was unable to penetrate the ice in M'Clure Strait. For the first time in recorded history, M'Clure Strait was ice-free for a short time in 2007. Parks Canada archaeologists in July 2010 found the ship on the sea bottom off Banks Island at a depth of 11 metres.

Source: Lieut. S. Gurney Cresswell, *A Series of Eight Sketches in Colour, of the Voyage of the H.M.S. Investigator During the Discovery of the North-West Passage* (London: Day and Son, and Ackermann and Co., 1854). Library and Archives Canada, Acc. No. R9266-757.

Whaling drew the Inuit into another form of the commercial economy. English, Scottish, and American ships were drawn to the Arctic waters in search of another form of wealth—whales. In the eastern Arctic, Baffin Bay and Hudson Bay were popular whale-hunting areas; in the western Arctic, whaling ships followed the Alaskan coast to the mouth of the Mackenzie River. These whalers helped find the elusive Northwest Passage.

Whaling began in Davis Strait in the seventeenth century. Later, whaling ships ventured further north and even into Lancaster Sound. Dutch, German, English, and Scottish whaling ships plied the Arctic waters in summer voyages. Between 1820 and 1840, whaling reached its greatest intensity with over a hundred ships involved. By 1860, American ships appeared in Hudson Bay, and several decades later the Americans were whaling in the Beaufort Sea. Wintering over by whaling ships was most common in the more inaccessible areas, such as Hudson Bay and the Beaufort Sea. By the time a sailing ship reached Hudson Bay or the Beaufort Sea, sea ice had begun to form, making whaling impossible. These ships, spending the winter months frozen in a sheltered

Table 3.2 Mackenzie Eskimo* Population, 1826–1930

Year	Population Estimate	Source
1826	2,000	Franklin (1828: 68–228)
1850	2,500	Usher (1971a: 169–71)
1865	2,000	Petitot (1876a: x)
1905	250	RCMP (1906: 129)
1910	130	RCMP (1911: 151)
1930	10	Jenness (1964: 14)

*In 1977, the term "Inuit" replaced "Eskimo" in official Canadian usage because the latter was considered pejorative.
Source: Smith (1984: 349). http://anthropology.si.edu/handbook.htm

harbour, would get an early start in the spring, leaving time to return to their home ports before the beginning of the second winter.

The impact of the whalers on the Inuit was greatest in places where wintering over was common. While the population of whalers was not great (perhaps 500 men annually wintered in the Beaufort Sea and 200 in Hudson Bay), this contact drew the Inuit into a new economic system. Engaged in the whaling industry as pilots, crewmen, seamstresses, and hunters, Inuit hunters, now armed with rifles, greatly increased their take of caribou in order to supply the whalers with game. In exchange for these services, the Inuit obtained trade goods. Contact with the whalers had its downside, however, counterbalancing economic gains. Excessive drinking sometimes led to violence, but by far the most negative impact of contact with Europeans and Americans was exposure to new diseases. Epidemics of smallpox and other contagious diseases swept through contact Inuit communities, reducing their populations. Table 3.2 shows the estimated population decline of one group of Inuit over the course of a century. Even so, the sudden disappearance of the whalers, upon whom many Inuit had come to depend, was a blow. The whaling industry saw its products lose favour with southern consumers as new products came into the marketplace. First, petroleum replaced whale oil in the late nineteenth century as a fuel for lighting. Then, a few decades later, the market for baleen collapsed when steel products were substituted. By World War I, commercial whaling in the Arctic had ceased.

Trapping in the Arctic

Until the beginning of the twentieth century, fur trading was confined to the Subarctic. When European fur buyers decided that the Arctic white fox pelt had commercial value, the Hudson's Bay Company quickly spread its operations into the Arctic and soon had a series of Arctic trading posts, including Wolstenholme (1909) on the Ungava coast, Chesterfield (1912) on the Hudson Bay coast, Aklavik (1912) near the mouth of the Mackenzie River, and Padlei (1926) in the Barren Lands. As with the Indian tribes, the fur economy changed the traditional Inuit ways. In the Repulse Bay area, Inuit camps became sites of winter trapping, while sealing at the ice edge began to replace breathing-hole sealing as the main winter activity (Damas, 1968: 159).

Figure 3.10   A Beluga or White Whale, Herschel Islanders, and RCMP Officer, 1924

In 1924, Herschel Island had an HBC trading post and an RCMP detachment. In that same year, Herschel Island witnessed the first Arctic court case and subsequent hanging of two Inuit from the Central Arctic, who were convicted of killing several people, including an RCMP officer. In the 1920s, the Inuvialuit along the Beaufort Sea coast and at Herschel Island commonly harvested beluga whales. The island also served as Yukon's most northerly RCMP detachment, which was originally opened to monitor American whalers and thus exert Canadian sovereignty. Whaling had a huge impact on the Inuvialuit. The economic and social interaction between the Inuvialuit and the whalers, most of whom wintered over, was considerable. In 1895–6, for example, 12 whalers with 1,000 to 1,200 crew wintered at the mouth of the Mackenzie River and Herschel Island.

Source: Yukon Archives, Ernest Pasley fonds #9238.

Impact on First Nations

The fur trade, by introducing European goods, institutions, and values to the Aboriginal people, caused many changes to the traditional Aboriginal way of life and rearranged the geographic distribution of many Indian tribes. Traditional Aboriginal material culture was greatly diminished by substituting European goods for those produced from the local environment. Indian women found an iron needle much easier than a bone needle for sewing skins. Similarly, iron kettles and copper cooking pots replaced the traditional wooden and bone ones, making domestic life easier, and wool Hudson's Bay blankets became a popular trade item among the Native peoples themselves. European firearms, iron knives, and axes were in great demand, replacing traditional weapons such as bows and arrows. Such an advance in weaponry often tipped the balance of power between those Indian tribes with muskets and those with bows and arrows. The fur trade had implications, therefore, for tribal relations, control of territory, and access to fur-trading posts.

In a similar fashion, Europeans challenged Aboriginal customs and spiritual beliefs. Most spiritual changes occurred when missionaries introduced Christianity to the North. These missionaries accelerated the process of cultural change. By the middle of the nineteenth century, Anglican and Roman Catholic missions had been established throughout the North. Usually, Anglican and Roman Catholic priests located their missions adjacent to the fur posts. Their objective was simple—to convert the Indian peoples to their version of Christianity. It was not long before the shamans/ medicine men no longer held a central position in Aboriginal society. In the nineteenth century the introduction of boarding-school "European-style" education by the two churches was an important step towards assimilation and the erosion of traditional Aboriginal beliefs and languages. Until the post-World War II years, the federal government encouraged the Anglican, Roman Catholic, and other Christian missions to educate Aboriginal children. The languages of instruction were English and French and the children were not allowed to speak their own language. Regardless of the "good intentions" of the missionaries, their efforts further weakened the confidence of Aboriginal peoples in their own cultures, causing them to be more dependent on but at the same time not part of Western culture.

The fur trade drew the Indians and, much later, the Inuit into a new pattern of social and economic relations. Yet the degree of involvement of Aboriginal people varied over time and space. While the exchange of furs for trade goods provided the Indians and Inuit with access to European technology, the main impact was on those Aboriginal people involved in direct trading with Europeans. Others, more distant from the fur traders, were only marginally involved. Relations between trappers and traders were not constant over time. Whenever trapping and/or hunting was unable to satisfy the needs of Aboriginal people, they turned to their fur trader for help. Food security was always an issue. In this way, a rather balanced, mutually advantageous relation in which each partner was more or less equal and certainly in need of the other changed into an unbalanced dependency, with Native people trapped into bartering for Western tools and weapons now essential for hunting, fishing, and trapping. Goods were also needed to run the household. This dependency culminated in an Aboriginal welfare society, which Ray (1984) has argued was not a sudden event associated with the decline of the fur trade but rather was rooted in the early fur trade and the role of the Hudson's Bay Company:

> The Hudson's Bay Company was partly responsible for limiting the ability of Indians to adjust to the new economic circumstances at the beginning of [the twentieth] century. Debt-ridden, repeatedly blocked from alternate economic opportunities, and accustomed to various forms of relief for over two centuries, Indians became so evidently demoralized in the twentieth century, but the groundwork for this was laid in the more distant past. (Ray, 1984: 17)

The geographic expression of dependency on the fur trader and its consequences for Indian economies is illustrated in an 1820 map constructed by Heidenreich and Galois (Harris, 1987: Plate 69). They classify the Indian economies into seven types, depending on their involvement in the fur trade. The disruption to local Indian economies was magnified as the level of involvement in the fur trade increased. The more disrupted Indian economies were so entangled in the fur economy that the pressure on wildlife, but especially the beaver, often resulted in an ecological collapse. The pressure to overhunt and trap came from the demands of the fur trader for country food and from the need for more trade goods.

Vignette 3.6   Food Security for Hunting Families

Living on the land had many attractions. Game was not always readily available and food shortages could occur. For that reason, food security was a serious and some-times deadly weakness in the hunting economy. Not surprisingly, then, rations were sought from fur traders to tide them over. Foodstuffs were obtained in other ways, such as helping unload supply HBC supply ships or gaining temporary work at con-struction sites. From the hunter's perspective, work was simply a different form of obtaining goods and foodstuffs from Europeans. Such arrangements complemented hunting and increased food security.

The Caribou Inuit (Ahiamiut) provide an example of the risk of depending on game for food. This small group of around 120 lived in the heart of the Barren Grounds where the Beverley Herd of caribou spends the summer. These animals then congre-gate to migrate to the boreal forest for the winter. While the highly mobile Caribou Inuit knew well the migration routes of the caribou, they depended heavily on them for their food. When hunting went well, all had an ample supply of food. Yet, this small group was particularly vulnerable when the caribou took an alternative migration route. In the 1950s, the Ahiamiut fell on particularly hard times and starvation had already claimed lives before the outside world knew of their plight. By 1957, Ottawa had resettled the survivors in coastal settlements where store food was readily avail-able. One such relocation centre was Eskimo Point (renamed Arviat, meaning "place of the bowhead whale," in 1989). For the Ahiamiut, the overall effect was devastat-ing—first the months of hunger, then relocation that meant a loss of their traditional hunting life. Professor Williamson (1974: 90) reported that starvation had reduced the population of the Ahiamiut from about 120 in 1950 to about 60 in 1959.

The fur trade brought new technology to Indian communities but it also drew them into a different economic system and cultural world. Concepts of private and public property were foreign concepts. In 1870, the transfer of Hudson's Bay Company lands to Canada sent shock waves through the Métis communities in the Red River Valley who feared the loss of their lands and way of life. While the Hudson's Bay Company sup-ported the fur economy and discouraged agricultural settlements, Ottawa was deter-mined to settle the West. To prepare the land for agricultural settlement, Ottawa began to survey the land in order to determine ownership. Land surveys sparked two Métis rebellions, and led to treaties and the assignment of Prairie Indians to reserves where they became marginalized on their own land. Fortunately for the Woodland Indians, the forested lands of the Subarctic were not suitable for agriculture and therefore the Crown lands remained available for hunting and trapping. While landownership had changed with the transfer of Rupert's Land and the North-Western Territory to Canada,[6] the daily lives of the Native peoples remained firmly tied to the fur trade.

Resource Development Begins

Until the 1950s, Ottawa did little to integrate the North and its Aboriginal peoples into Canada's economy (Rea, 1968). The main reasons were that the federal government had its hands full with nation-building in southern Canada and dealing with two world

---

Vignette 3.7   Ottawa's Hands Are Full

In the decades immediately following Confederation, little attention was paid to "developing" the North; rather, it was a matter of holding on to this vast territory with a minimum of effort and cost. This strategy was applied to the Arctic, which was considered by Ottawa as the most remote and barren lands in Canada. In the nineteenth century, land was valued in terms of its potential for agriculture and those lands were in southern Canada.

Canada's first Prime Minister, Sir John A. Macdonald, was concerned that American homesteaders might migrate northward into the "unoccupied" fertile Canadian prairies, thereby leading to their annexation to the United States. For this reason, his primary concern was to exert political control by building a railway from Ontario to the Pacific Ocean and then to settle these newly acquired fertile lands. The railway was completed in 1885, and the settling of the prairies continued into the early part of the twentieth century. During this time, most of Rupert's Land and all of the North-Western Territory and the Arctic Archipelago were left to the fur economy. Threats to Canadian sovereignty in the Arctic came from Otto Sverdrup's 1898–1902 explorations of Ellesmere Island and surrounding islands. Sverdrup discovered three islands and then claimed them for Norway. After learning of American expeditions to the Arctic Ocean and fearing the discovery of more islands, the Borden government sponsored the Arctic Expedition led by Vilhjalmur Stefansson. This controversial expedition saw the sinking of one ship, the loss of lives, and the discovery of five major islands, including Mackenzie King and Borden islands.

---

wars. Until World War I, the national priority was the settlement of western Canada. The Canadian Pacific Railway was designed to transport wheat and other raw materials to ocean ports and thus created a commercial basis for the West. As the railway crossed the Canadian Shield in Ontario and Quebec, a commercial route appeared for the region's minerals and timber to find a way to markets.[7]

As for investing in the North, cash-strapped Ottawa was happy to wait for the private sector to undertake resource development in the North and for the private sector to build the necessary transportation network to move their products to market. The only exception was the Canadian Arctic Expedition, 1913 to 1918, which allowed Ottawa to strengthen its claim to the Arctic Archipelago. As for the northern Indians, Métis, and Inuit, the federal government was content to allow them to remain in the fur economy, thus keeping its education and health services expenses for them as low as possible.[8] In regard to the Inuit, since no treaty was struck with them, the federal government did not consider them a federal responsibility. However, a 1939 Supreme Court ruling compelled Ottawa to accept the Inuit as a federal responsibility.

The Klondike Gold Rush

The Klondike Gold Rush represents a classical example of a boom/bust mining venture. The rush began in 1896. Two Tagish men, Skookum Jim Mason and his nephew Dawson Charlie, and George Carmack, an American prospector married to Skookum Jim's sister Kate, made the world's most famous gold strike on Rabbit Creek (later

Figure 3.11　Panning for Gold on Bonanza Creek

The Klondike gold rush began when an American prospector and his two Tagish companions discovered gold nuggets at Rabbit Creek, a small tributary of the Klondike River in the Klondike district of the Yukon. The creek was soon renamed Bonanza Creek.

Source: SKINNER/BLM

renamed Bonanza Creek), a small tributary of the Klondike River. Prospectors panned for gold—a very simple method of gold mining requiring little capital investment or skill. Word spread quickly of the discovery within the small Yukon population of approximately 5,000. Many rushed to the Klondike district to stake claims, but it was a year before the outside world learned of this find.

As the first on the scene, local prospectors quickly staked their claims and panned great quantities of placer gold from the river gravel (Figure 3.11). By the next spring, much of the easily accessible gold had been recovered and, as soon as the ice broke on the Yukon River, the successful miners left to enjoy their newly found wealth in Seattle and San Francisco. This journey was long—by sternwheeler from Dawson to St Michael, at the mouth of the Yukon River, and then by ocean steamer to Seattle and San Francisco. Upon landing in these cities, the motley crew of prospectors received tumultuous receptions from the huge crowds on the wharves. Gold fever struck fast, causing many to try their hand. Many thousands flocked to the Klondike in hope of discovering gold, but already the most accessible placer gold was gone. Yet, the people kept coming and the population exploded. In July 1896 Dawson City didn't yet exist. One year later it had a population of 5,000 and the year after that, 30,000. Yukon was suddenly transformed from a fur-trapping economy to a resource economy. In 1898, news of a gold discovery at Nome, Alaska, caused many to leave the Klondike for Alaska. One critical reason for this out-migration was that the days of panning and digging for gold on small individual claims was over, being replaced by **hydraulic mining**. Such mining was very harmful to the environment because it involved extracting the river gravel from the frozen ground with powerful steam-powered hoses. Another method was dredging, as monster dredging machines slowly scoured creek beds for the gold, leaving in their wake long ridges of the discarded gravel. The impact on Dawson City, the "Paris of the North," was immediate, with its population soon falling from 30,000 to below 5,000.

The Yukon River served as a natural water route and privately operated steamers took miners and freight from Whitehorse to Dawson City. Many miners rafted down the river with their supplies. From Dawson City, Bonanza Creek, the site of the first strike, was just 17.7 kilometres.

The transportation route did have a bottleneck—getting from Skagway, Alaska, to Whitehorse. Miners from San Francisco and Seattle landed at Skagway and then faced the difficult Chilkoot Trail. Originally an Aboriginal trail for trade between the coastal Indians and those whose territory was inland, the steep and rugged trail, with a summit at 939 metres, made this trip extremely challenging. As the miners entered Canada at the summit of the Chilkoot Pass or the even more treacherous White Pass, near the Bennett Lake

Figure 3.12   S.S. *Yukoner* Heading Upstream through Five Finger Rapids

Sternwheelers plied the 800-km river journey from Bennett Lake on the BC–Yukon border to Dawson City, carrying those gold seekers who could afford the fare. Many other prospectors made the journey on hastily built river rafts.

Source: MacBride Musuem of Yukon History Collection, Image 4062.

headwaters of the Yukon River in far northern British Columbia, they had to ensure the officers at the North West Mounted Police post that they had sufficient supplies to survive the winter. A ton of supplies was required of each miner, which meant many back-breaking trips to the summit, even with Native packers to help with the loads. By 1900, the White Pass and Yukon Railway provided a much easier means of reaching Whitehorse. Financed by British investors, the railway ran from Skagway to Whitehorse, but by then the Yukon gold rush was over. The railway and steamships were no longer catering to individual miners but had turned to transporting freight for the mining companies in the Dawson area.

The Klondike gold rush forced Ottawa's hand to take a more active administrative role in the North. In 1896, close to a thousand prospectors were spread out along the Yukon River and its tributaries. Almost all were Americans. Many had worked placer gold claims in California and British Columbia. But with tens of thousands more arriving from the United States, the issue of Canadian sovereignty arose. Prior to the gold rush, Indians and a handful of white traders and prospectors had inhabited Yukon. With so few white people, Ottawa had sensed no need to send its officials there, but now the federal government quickly dispatched detachments of the North West Mounted Police to Yukon to "show the flag" and to enforce Canadian laws and regulations. Fortunately, a small number of NWMP officers arrived in the Yukon in 1894 because of the disputed boundary with Alaska. More officers were added in 1898. The Mounties issued licences, collected taxes, and kept law and order. In this way, Canadian sovereignty was demonstrated.

The local Indigenous peoples had their first exposure to resource development and, for the most part, this experience was not a good one (Vignette 3.8). According to Crowe (1974: 121), the coming of so many white men shattered the world of the Tagish, Tutchone, and other Indian tribes in Yukon. These Indians lost control of their traditionally occupied lands and some became involved in the gold economy. Their participation varied from prospecting to packing supplies from the coast over the Chilkoot Pass (Crowe, 1974: 122). Others found a place as wage earners, working as deckhands on the riverboats or even as carpenters in Dawson. Hunters sold game to the miners.

The Northern Vision

Until the late 1950s, Ottawa paid little attention to the North (Vignette 3.7). The Northern Vision, articulated in a political rally by John Diefenbaker, called for the development of the North's resources with the federal government investing in roads and other forms of infrastructure necessary to stimulate private investment. His speech hit a cord with the public, and the Progressive Conservatives won the 1957 federal election.

Beyond the political rhetoric, what was the economic basis of this Northern Vision? The answer lies in the rising demand for raw materials and commodities by industry in the United States. The massive iron ore project in northern Quebec provides an example. After World War II, the American iron and steel industry had exhausted its domestic supplies of iron ore and had to import iron ore from foreign sources. At that time, the rich iron deposits in the **Labrador Trough** attracted industry attention. The drawback was the remote location. To overcome this remoteness, several American and Canadian steel companies formed the Iron Ore Company of Canada and provided it with the funds to build a railway, a townsite, and a number of open-pit mines. At that time, the federal government played no role in environmental

Vignette 3.8 The Klondike: Aboriginal Society in Turmoil

As Ken Coates (1988: 73) has written, "the Yukon Indians have faced the industrial frontier for over a century. Their situation provides a useful longitudinal study of Native reaction to the forces of cyclical mining development, seasonal industrial activity, and white encroachment on traditional hunting territories." Change began with fur trading but the discovery of gold quickened its pace. A trickle of prospectors had been arriving in Yukon for three decades before the incredible Klondike gold rush at the end of the nineteenth century, when, "[b]y 1898, . . . the gold-seekers were desperate to cross the [Chilkoot P]ass, [and] the Indian packers could make $100 a day and more, carrying loads of up to 200 pounds across the mountains" (Crowe, 1974: 122). In the twentieth century, Yukon Native peoples experienced the encroachment of World War II highway and pipeline developments, and later saw lead/zinc and other mines open and close, pipeline proposals, oil exploration off the Arctic coast, and an ongoing influx of newcomers to their lands. Indians took advantage of wage employment, but for the most part they remained on the land as hunters and trappers. This mixed economy took hold with traditional hunting and trapping activities at its core but supplemented by seasonal wage employment.

Figure 3.13 North West Mounted Police at Pleasant Camp on the Dalton Trail, about 1898

Mining prospectors are central to any understanding of the actual or mythical "frontier" in Canada and the United States. They represented desperation, rough living, lawlessness, and sometimes sudden and unimaginable wealth and debauchery. The North West Mounted Police in Yukon during the gold rush maintained a semblance of order on this frontier.

Source: © McCord Museum

Figure 3.14   Resource Town of Schefferville, 1964

The Iron Ore Company of Canada created this resource town in 1954. Located in the forested Canadian Shield, the town was named after Bishop Lionel Scheffer, who in 1954 served as the Vicar Apostolic of Labrador. At that time, Innu resettled at Schefferville, and in 1968 a new Matimekosh reserve was established for them.

Source: George Cottreau

matters or in ensuring the well-being of local Innu and Naskapi tribes. The 1950s were the tail end of a long period of a laissez-faire approach by Ottawa, leaving resource development to private firms. It would be another two decades before the Canadian public had thoughts about alternative approaches to northern development and considered the broader issues of the environmental and social impacts of resource projects.

The grandest expression of a resource megaproject took place in the Labrador Trough located in the rugged and rocky Canadian Shield of northern Quebec. Undertaken by the Iron Ore Company of Canada on behalf of the American steel mills, the investment ranked as one of the largest capital investments of the 1950s. First, a resource town, Schefferville, Quebec, was built from scratch in 1954 to serve as the hub of this huge operation. Schefferville is only two km from the Labrador border on the north shore of Knob Lake. The townsite on the shores of Knob Lake is shown in Figure 3.14.

The development of open-pit iron ore mines was the next order of business. Mining operations take place within 50 km of the town of Schefferville. The iron ore reserves, estimated at 250 million tonnes, represented one of the largest in the world. Then, a railway was built to Sept-Îles, where the company in 1950 built the port's first industrial dock for loading ore onto ships. Today, Sept-Îles remains the most important port for the shipment of iron ore in North America.

The Iron Ore Company included major US steel companies such as National Republic, Armco, Youngstown, and Wheeling-Pittsburgh. The purpose of this privately funded project was to supply much-needed iron ore, especially to United States steel

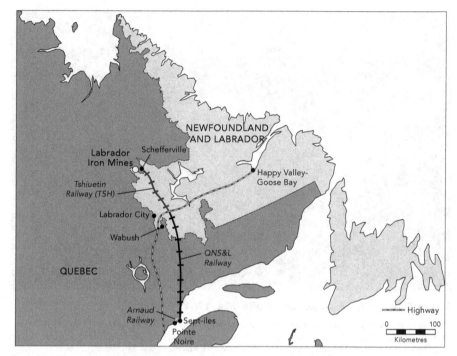

Figure 3.15 Map of Quebec's Iron Mines, Railway, and River Port

While iron ore is still mined and shipped by rail to Sept-Îles, much has changed since the Iron Ore Company of Canada placed its mark on northern Quebec. One change is the involvement of local Innu First Nations. For example, the Tshiuetin Rail Transportation Inc. railway is owned and operated by a consortium of these First Nations.

plants. The principal shipping routes were (1) up the St Lawrence Seaway to Cleveland, Ohio, by lake carriers; (2) along the Atlantic seaboard by ocean carriers; and (3) across the Atlantic Ocean to Rotterdam (Gern, 1990: Appendix A). The mines and the resource town, in turn, needed power. In 1954, the development of the hydroelectric complex at Churchill Falls on the Churchill River began. While Churchill Falls was situated in Labrador, the iron ore mines were in Quebec near the mining town of Schefferville. Labrador's second hydro-generating station was built in the early 1960s to supply power to iron ore mines in Labrador City and Wabush. Watson (1964: 467–8) describes the transformation of a northern region from "undeveloped to developed" and the link of such development to the American steel industry:

> A railway was built from the little fishing village of Sept-Îles—now a thriving port—up over the high, formidable, scarp-like edge of the faulted Shield, across extremely rugged terrain, deeply eaten into by the rejuvenated sharply entrenched rivers, a distance of 360 miles to the Knob Lake iron field. Here the town of Schefferville soon grew up, a major outpost in the wilderness. The making of the port, the building of the railway, and the installation of nearby

hydroelectric dams and works, were all major operations, but with American backing they were soon completed, and in 1954 the production of ore started, and over 2 million tons of ore were shipped out to the iron-hungry cities of the St Lawrence and Great Lakes lowland.

Watson fails to mention the social impacts of this project on the local Aboriginal people—the Naskapi and Montagnais (Innu)—or the environmental consequences of open-pit mining in an area underlain by permafrost. Unlike the James Bay Hydroelectric Project, no one raised the question of Aboriginal title to these yet unceded lands or the matter of environmental damage.[9] As Aboriginal leaders began to comprehend the effect on their people of the massive developments taking place in the North, their voices, along with those of environmentalists, were raised, challenging two major development projects proposed in the 1970s—the James Bay Project and the Mackenzie Valley Gas Pipeline Project.

With the North gaining value as a source of raw materials and commodities for the US market, the question was: "How can the handicap of geography be overcome?" Ottawa preferred the private capital model used to develop the iron ore resources of northern Quebec. But few companies had the resources to undertake such investments. Private industry pressured the federal government to provide the infrastructure. As noted earlier, Diefenbaker's northern development concept, the "Northern Vision," called for the opening of the northland by building transportation routes and communication lines, thereby linking northern resources to southern markets. In 1958, under Prime Minister Diefenbaker, a series of programs was put into place, including "Roads to Resources." The Roads to Resources program provided funds for new roads leading to potentially valuable natural resources in the northern reaches of the provinces. The federal share of these monies was determined by a formula of sharing costs with the provincial governments. The territorial counterpart to this program was the Development Road Program, with the federal government paying for all the construction costs.

During the next decade, the Northern Vision held Canada's attention. In 1969, Richard Rohmer, a lawyer, author, Air Force general, and political adviser, presented a more complex version of public involvement in northern development. His Mid-Canada Development Corridor proposal called for the building of a northern railway across the mid-north, a transportation corridor meant to stimulate settlement and development similar in effect to the building of the CPR across the Canadian West in the 1880s. In many ways, this grand scheme was an elaboration of Diefenbaker's Northern Vision. According to Rohmer, investment and planning would come from the federal government, ensuring Canadian control and ownership. Rohmer envisioned this massive federal undertaking as a means to strengthen both Canada's economy and its national purpose. He saw the Mid-Canada Corridor concept as a counterweight to the American ownership of Canadian natural resources.

Canadians, particularly non-Aboriginal northerners, responded warmly to Rohmer's concept of a developed North. Federal officials, on the other hand, were cool to his idea of building a railway across the Subarctic, partly because of the potential drain to the treasury but mostly because Ottawa considered it "unsound." In the West, a handful of provincial officials reacted suspiciously, fearing that this transportation system would serve the interests of central Canada rather than those of the western provinces. The mixed response to this project indicated that while Canadians wished to see the North developed, there were differences of opinion as to how the goal was to be accomplished and who should pay for it.

Concerns and interests in the land and resource development were now on the table, forcing governments to address their concerns but not so much their interest in the land and resource development. Decisions by the Supreme Court of Canada over the next 40 years, as well as proposed and already started megaprojects, pushed Aboriginal interests into a legal framework, and by 2014 the cry from Aboriginal leaders was for "sharing the resources," or, more strongly, "what does such resource development do for or to the generations to come?"—but that is a story for later. In the next chapter, population provides the focus. The North's demography was affected by two events. One was relocation of Aboriginal peoples to settlements. The other was the resource boom of the 1950s and 1960s when thousands of Canadians moved to northern resource towns and government administrative centres.

Challenge Questions

1. During the last ice age, ocean levels dropped, exposing more of BC's coast. Explain why the ocean levels dropped and why this drop in sea level might have allowed Old World hunters to island-hop south of the Cordillera Ice Sheet?
2. Why did early explorers take a few Aboriginal people back to Europe and what often happened to them?
3. Beginning in the seventeenth century, how did the Hudson's Bay Company hurt the fur economy of New France?
4. Why did the Hudson's Bay Company and private fur traders take until the twentieth century to establish posts in the Arctic?
5. Why would Sir John Franklin have had a better chance of making it through the Northwest Passage with today's ice conditions?
6. Prime Minister Diefenbaker's Northern Vision focused on extending Canada's transportation system northward, while Richard Rohmer proposed an east–west railway line in the Canadian Subarctic. What were the advantages and disadvantages of Diefenbaker's Northern Vision? What geographical and political problems were inherent in Rohmer's proposal? Which proposal came to reality?
7. In terms of capital investment, what are the key differences between the Klondike gold rush and the iron ore project in northern Quebec?
8. Of all the images of Canada's North, why have two visions—homeland and hinterland—governed Canadian thought and writings over the past century and longer?

Notes

1. One of Canada's outstanding historical geographers, Arthur Ray, stated in the 2001 edition of Innis's book that "*The Fur Trade* is arguably the most definitive economic history and geography of the country ever produced" (Ray, 2001: v). Canada, from Innis's perspective, was a product of its geography with each region containing a staple that would trigger its economic future.

2. Who the first North Americans were remains a puzzling question for archaeologists. A growing number of archaeologists suspect that early humans arrived in North America well before 13,000 years ago. Their suspicions are partly based on the Bluefish Caves site, where stone tools and bones from several prehistoric animals, including mammoth, antelope, bison, and large cat, and dated from 25,000 to 40,000 years old, were found. Unfortunately, the site does not provide geological support that the stone tools can be associated with the smashed bones, i.e., there is a possibility that the stone tools found their way to this site at a more recent time. Other support for the earlier peopling of the Americas comes from the tools dated as 12,500 years old found at Monte Verde in Chile. The Chile find gives support to the arrival of humans some 25,000 to 30,000 years ago. Ancient stone tools found near Grimshaw, Alberta, may have been left at this site some 20,000 years ago—before the site was covered by the Cordillera Ice Sheet (Chlachula and Leslie, 1998). The distinguished archaeologist J.V. Wright supports the notion of humans arriving before the last ice advance. Wright (1999: 18) wrote that:

> Archaeological, physical anthropological, ... linguistic, ... and genetic evidence ... has been used to argue for three major migrations from Asia into the Western Hemisphere. While controversy exists regarding the specific timing of these events, one sequence has been suggested as follows: the first migration between 30,000 and 15,000 years ago involved the Paleo-Indians whose descendants would constitute the vast majority of native peoples of the Western Hemisphere; a second migration between 15,000 and 10,000 [years ago] gave rise to the members of the Eyak-Athabascan linguistic family and probably some of the current linguistic isolates such as Haida and Tlingit; and a third migration between 9,000 and 6,000 years ago resulted in the historic members of the Eskimo-Aleut. . . .

3. From the time of contact to the late nineteenth century, population estimates were based on accounts of explorers, fur traders, and government officials. The Royal Commission on Aboriginal Peoples suggests a pre-contact population of 500,000. Heidenreich and Galois (Harris, 1987: Plate 69) argue that, in the early seventeenth century, there were about 250,000 Aboriginal people living in what is now known as Canada, suggesting a drop of 50 per cent. By the early 1820s, the Aboriginal population in Canada was estimated at 175,000, indicating a substantial decline from both the pre-contact and the early seventeenth-century figures. The Aboriginal population continued to fall well into the nineteenth century. With the formation of Canada, regular census-taking began. In 1871, the first census recorded just over 102,000 Aboriginal people, which may mark the low point in their population size.

   Some tribes, such as the Beothuks, the Mackenzie Delta Inuit, the Sadlermiut, and the Yellowknives, disappeared and their lands were occupied by Europeans or by other tribes. The popular view is that the Beothuks were victims of intentional genocide who were forced to move from the coast by the English and French settlers. Along the coast, food was more plentiful than in the interior of the island of Newfoundland, where food sources were more restricted, caribou being the main food source. According to Pastore (1992), the Beothuks did not want to live near English and French settlers, so they decided to remain in the interior. Prior to the arrival of Europeans, these hunters and gatherers had complex travel and settlement patterns that mixed inland and coastal locations into a successful nomadic lifestyle. Their life depended on being in the right place at the right time to obtain food from the sea and land. This geographic arrangement was interrupted by the arrival of Europeans who occupied coastal sites.

4. The Hudson's Bay Company's royal charter gave the company "sole rights to trade and commerce" in Canada. Two trade goods, guns and blankets, were of special importance. Indians preferred blankets of exceptionally pure, bright colours. Each blanket was graded with a point system corresponding to weight and size. The number of points represented the number of beaver skins required for obtaining a blanket.

5. Transportation is indispensable for resource exports, whether fur pelts or gold ingots. In the days of the fur trade, rivers provided natural highways. Later, resource development depended on expensively constructed roads and railways. Provinces were able to incorporate their northern hinterlands into their economies by building roads and railways. The British Columbia government provides one example. In the early post-World War II period, BC extended its provincial highway system to the Peace River district to provide the rest of the province with a link to that region and to the Alaska Highway and to give Peace River settlers their long-awaited direct outlet to the Pacific coast. The John Hart Highway, built from Prince George through the Pine Pass to a junction with the Peace River district's road system after 1945, pioneered the route to be followed by the Pacific Great Eastern Railway.

6. The Canadian North came into being as a political entity in 1870 when Rupert's Land and the North-Western Territory were transferred from Britain to Ottawa. The HBC agreed to surrender Rupert's Land to the British Crown, which transferred it to Canada. Canada paid £300,000 in compensation to the HBC and allowed the company to keep its 120 trading posts. Canada also agreed that the company could claim one-twentieth of the land in the Canadian Prairies. Just what riches these northern lands contained was unknown. A decade later, the British government transferred to Canada the rest of its Arctic possessions, the Arctic Archipelago (even though all the islands had not yet been discovered). By 1880, the geographic extent of Canada had been realized, though Newfoundland was to join Confederation much later.

    Still, Canada began to wonder what other riches it had inherited. In 1888, a Senate Committee on the Resources of the Great Mackenzie Basin investigated this question. Later, the Senate published a "highly enthusiastic report on the potential for agriculture, fisheries, forestry, mining, and petroleum, setting the precedent for the optimistic and promotional tone that has continued to this day [1970s] to pervade government pronouncements on northern resources" (Rowley, 1978: 79). While the Senate report sparked considerable interest, the national priority was to settle the prairies. Little commercial activity took place in the Subarctic and northern life continued to revolve around the fur trade and the Hudson's Bay Company.

7. Resource development also took place in the late nineteenth century along the southern fringe of the North in Ontario and Quebec, forest products and minerals being the main attractions. The building of the two national railways provided access to these northern forests and minerals. In the case of forest resources, the American timber supplies in New England were no longer adequate to meet growing US demands. Soon, Americans looked to their northern neighbour and the vast forests in Ontario and Quebec. In a classic core/periphery relationship, the Subarctic forests attracted capital from the giant newspaper firms in major American cities; in turn, the Canadian forests supplied these firms with pulp that was processed into paper at plants near American metropolitan centres.

8. Canadian Indians, such as the Quebec Cree, fall under the Indian Act (1876). By 1930, most First Nations had signed treaties. The main exceptions were in British Columbia, Quebec, Yukon, and Newfoundland. Until 1939, Ottawa did not acknowledge a responsibility towards the Inuit, but in that year a Supreme Court decision determined that the Inuit of northern Quebec were a federal rather than provincial responsibility.

9. Modern treaties began in 1975 with the signing of the James Bay and Northern Quebec Agreement. In exchange for land, cash, and a form of regional government, the Quebec Cree and Inuit allowed the James Bay Project, which was already well underway, to proceed. Recognizing the need for land claim agreements in other areas not covered by treaty, the federal government established a more orderly process for settling land claims known as comprehensive land claim negotiations. In 1984, the first comprehensive land claim agreement, the Inuvialuit Final Agreement, came into force.

The basis of Aboriginal title was first expressed legally in the **Royal Proclamation of 1763**. Why did King George III make such a proclamation? The Seven Years War between the British and French was a struggle for military supremacy in North America. Each European nation allied itself with Indians living on lands that it controlled or claimed. The British allied themselves with the Iroquois. At the end of this war, George III issued the Royal Proclamation, declaring that lands west of the Appalachians were to remain Indian lands. The British needed the support of the Iroquois to control these lands, which formerly fell under French control. The problem facing Britain was that British subjects in the American colonies wanted to settle the area, but under the Proclamation they were not to cross a "line" following the Appalachian Divide from Maine to Georgia and then to the St Marys River in Florida. Efforts to prevent settlers from moving west proved futile, however, and just before the American Revolution as many as 100,000 colonists may have been living west of this imaginary line (Hilliard, 1987: 149). After the American Revolution, the newly formed United States of America declared the lands extending from the Appalachians to the Mississippi River open for settlement. Thousands of Americans poured over the divide. Within several decades, virtually all of the land east of the Mississippi had passed into American ownership and the original inhabitants were exterminated, assimilated, living on reservations, or had been forced to migrate westward (Hilliard, 1987: 163).

References and Selected Reading

Balter, Michael. 2014. "Bones from a Watery 'Black Hole' Confirm First American Origins." *Science* (16 May): 608–81.

Berger, Thomas R. 1977. *Northern Frontier, Northern Homeland: The Report of the Mackenzie Valley Pipeline Inquiry*, 2 vols. Ottawa: Minister of Supply and Services.

Berton, Pierre. 1988. *The Arctic Grail: The Quest for the North West Passage and the North Pole, 1818–1909*. Toronto: McClelland & Stewart.

Bone, Robert M. 2002. "Colonialism to Post-Colonialism in Canada's Western Interior: The Case of the Lac La Ronge Indian Band." *Historical Geography* 30: 59–73.

Bonnichsen, Robson, and Karen Turnmire. 1999. "An Introduction to the Peopling of the Americas." In Robson Bonnichsen and Karen Turnmire, eds, *Ice Age People of North America: Environments, Origins, and Adaptations*, 1–27. Corvallis: Oregon State University Press.

Buxton, William J., ed. 2013. *Harold Innis and the North: Appraisals and Contestations*. Montreal and Kingston: McGill-Queen's University Press.

Canada. 1884. *Census of Canada, 1880–81*. Ottawa: Department of Agriculture.

Canada, Royal Commission on Aboriginal Peoples. 1996. *Report of the Royal Commission on Aboriginal Peoples: Looking Forward, Looking Back*, vol. 1. Ottawa: Minister of Supply and Services Canada.

Chlachula, Jiri, and Louise Leslie. 1998. "Preglacial Archaelogical Evidence at Grimshaw, the Peace River Area, Alberta." *Canadian Journal of Earth Sciences* 35, 8: 871–84.

Cinq-Mars, Jacques, and Richard E. Morlan. 1999. "Bluefish Caves and Old Crow Basin: A New Rapport." In Bonnichsen and Turnmire, eds, *Ice Age People of North America*, 200–12.

Coates, Kenneth. 1985. *Canada's Colonies: A History of the Yukon and Northwest Territories*. Toronto: Lorimer.

————. 1988. "On the Outside in Their Homeland: Native People and the Evolution of the Yukon Economy." *Northern Review* 1: 73–89.

Cooke, Alan, and Clive Holland. 1978. *The Exploration of Northern Canada, 500 to 1920: A Chronology.* Toronto: Arctic History Press.

Crowe, Keith J. 1974. *A History of Original Peoples of Northern Canada.* Montreal and Kingston: McGill-Queen's University Press.

Damas, David. 1968. "The Eskimo." In C.S. Beals, ed., *Science, History and Hudson Bay,* vol. 1. Ottawa: Queen's Printer.

————. 2002. *Arctic Migrants/Arctic Villagers: The Transformation of Inuit Settlement in the Central Arctic.* Montreal and Kingston: McGill-Queen's University Press.

Denevan, William M., ed. 1976. *The Native Population of the Americas in 1492.* Madison: University of Wisconsin Press.

————. 1992. "The Pristine Myth: The Landscape of the Americas in 1492." *Annals, Association of American Geographers* 82, 3: 369–85.

Dickason, Olive Patricia, with David T. McNab. 2009. *Canada's First Nations: A History of Founding Peoples from Earliest Times,* 4th edn. Toronto: Oxford University Press.

Donaldson, Yarmey. 1989. "Alberta's First Fort." *Western People,* 8 June, 10.

Driver, Harold E. 1961. *Indians of North America.* Chicago: University of Chicago Press.

Duffy, Patrick. 1981. *Norman Wells Oilfield Development and Pipeline Project: Report of the Environmental Assessment Panel.* Ottawa: Federal Environmental Assessment Review Office.

Dussault, René, and George Erasmus. 1994. *The High Arctic Relocation: A Report on the 1953–55 Relocation.* Ottawa: Canadian Government Publishing.

Feit, Harvey A. 1981. "Negotiating Recognition of Aboriginal Rights: History, Strategies and Reactions to the James Bay and Northern Quebec Agreement." *Canadian Journal of Anthropology* 2: 159–72.

Fumoleau, René. 2004. *As Long As This Land Shall Last: A History of Treaty 8 and Treaty 11, 1870–1939.* Calgary: University of Calgary Press.

Gern, Richard. 1990. *Cain's Legacy: The Building of Iron Ore Company of Canada.* Sept-Îles, Que.: Iron Ore Company of Canada.

Harris, R. Cole. 1987. *Historical Atlas of Canada, vol. 1, From the Beginning to 1800.* Toronto: University of Toronto Press.

————. 1997. *The Resettlement of British Columbia: Essays on Colonialism and Geographical Change.* Vancouver: University of British Columbia Press.

Hearne, Samuel. 1958. *A Journey from Prince of Wales's Fort in Hudson's Bay to the Northern Ocean: 1769–1772.* Toronto: Macmillan.

Hicks, Jack, and Graham White. 2000. "Nunavut: Inuit Self-Determination through a Land Claim and Public Government?" In Jens Dahl, Jack Hicks, and Peter Jull, eds, *Nunavut: Inuit Regain Control of Their Lands and Their Lives,* 30–115. IWGIA Document No. 102. Copenhagen: Centraltrykkeriet Skive A/S.

Hilliard, Sam B. 1987. "A Robust New Nation, 1783–1820." In Robert D. Mitchell and Paul A. Groves, eds, *North America: The Historical Geography of a Changing Continent, 149–71.* Totowa, NJ: Rowman & Littlefield.

Innis, Harold. 1930. *The Fur Trade in Canada.* Toronto: University of Toronto Press.

Kroeber, Alfred L. 1939. *Cultural and Natural Areas of Native North America.* Berkeley: University of California Press.

Lux, Maureen K. 2001. *Medicine That Walks: Disease, Medicine and Canadian Plains Aboriginal People, 1880–1940.* Toronto: University of Toronto Press.

McGhee, Robert. 1996. *Ancient People of the Arctic.* Vancouver: University of British Columbia Press.

McMillan, Alan D. 1995. *Aboriginal Peoples and Cultures of Canada: An Anthropological Overview.* Vancouver: Douglas & McIntyre.

Marcus, Alan R. 1991. "Out in the Cold: Canada's Experimental Inuit Relocation to Grise Fiord and Resolute Bay." *Polar Record* 27, 163: 285–96.

————. 1995. *Relocating Eden: The Image and Politics of Inuit Exile in the Canadian Arctic.* Dartmouth, NH: University Press of New England.

Martin, Horace T. 1892. *Castorologia: or, The History and Traditions of the Canadian Beaver.* Montreal: Drysdale, 1892.

Miller, J.R. 1989. *Skyscrapers Hide the Heavens: A History of Indian–White Relations in Canada.* Toronto: University of Toronto Press.

Mooney, James. 1928. *The Aboriginal Population of America North of Mexico.* Washington: Smithsonian Institution.

Morton, Arthur S. 1973. *A History of the Canadian West to 1870–71.* Toronto: University of Toronto Press.

Newhouse, David R. 2001. "Modern Aboriginal Economies: Capitalism with a Red Face." *Journal of Aboriginal Economic Development* 1, 2: 55–61.

Nuttall, Mark. 2000. "Indigenous Peoples, Self-Determination and the Arctic Environment." In Mark Nuttall and Terry V. Callaghan, eds, *The Arctic: Environment, People, Policy.* Singapore: Harwood Academic Publishers.

Page, Robert. 1986. *Northern Development: The Canadian Dilemma.* Toronto: McClelland & Stewart.

Pastore, Ralph T. 1992. *Shanawdithit's People: The Archaeology of the Beothuks.* St John's: Breakwater Books.

Ray, Arthur J. 1974. *Indians in the Fur Trade: Their Role as Trappers, Hunters, and Middlemen in the Lands Southwest of Hudson Bay, 1660–1870.* Toronto: University of Toronto Press.

———. 1976. "Diffusion of Diseases in the Western Interior of Canada." *Geographical Review* 66: 139–57.

———. 1984. "Periodic Shortages, Aboriginal Welfare, and the Hudson's Bay Company, 1670–1930." In Shepard Krech III, ed., *The Subarctic Fur Trade: Aboriginal Social and Economic Adaptations,* 1–20. Vancouver: University of British Columbia Press.

———. 1990. *The Canadian Fur Trade in the Industrial Age.* Toronto: University of Toronto Press.

———. 2001. "Introduction." In Harold Innis, *The Fur Trade in Canada: An Introduction to Canadian Economic History,* v–xix. Toronto: University of Toronto Press.

Rea, K.J. 1968. *The Political Economy of the Canadian North: An Interpretation of the Course of Development in the Northern Territories of Canada to the Early 1960s.* Toronto: University of Toronto Press.

Rich, E.E. 1967. *History of the Hudson's Bay Company, 1670–1870,* vol. 2. London: Hudson's Bay Record Society.

Rowley, Graham. 1978. "Canada: The Slow Retreat of 'the North.'" In Terence Armstrong, George Rogers, and Graham Rowley, eds, *The Circumpolar North,* 71–123. London: Methuen.

———. 1998. *Cold Comfort: My Love Affair with the Arctic.* Montreal and Kingston: McGill-Queen's University Press.

Saku, James C., and Robert M. Bone. 2000. "Looking for Solutions in the Canadian North: Modern Treaties as a New Strategy." *Canadian Geographer* 44, 3: 259–70.

Scott, Colin H., ed. 2001. *Aboriginal Autonomy and Development in Northern Quebec and Labrador.* Vancouver: University of British Columbia Press.

Smith, Derek G. 1984. "Mackenzie Delta Eskimos." In William C. Sturtevant, ed., *Handbook of North American Indians, vol. 5, Arctic.* Washington: Smithsonian Institution.

Tester, F.J., and Peter Kulchyski. 1994. *Tammarniit (Mistakes): Inuit Relocation in the Eastern Arctic 1939–63.* Vancouver: University of British Columbia Press.

Thomas, David Hurst. 1999. *Exploring Ancient Native America: An Archaeological Guide.* New York: Routledge.

Thompson, Niobe. 2010. "Inuit Odyssey." *The Nature of Things,* CBC-TV. 12 Aug. www.cbc.ca/documentaries/natureofthings/2009/inuitodyssey/.

Trudel, Marcel. 1988. "Jacques Cartier." In James H. Marsh, ed., *The Canadian Encyclopedia,* vol. 1, 368. Edmonton: Hurtig.

Watson, J.W. 1964. *North America: Its Countries and Regions.* London: Longmans.

Williamson, Robert G. 1974. *Eskimo Underground: Socio-cultural Change in the Canadian Central Arctic.* Uppsala: Institutionen for Allman.

Wolfe, Eric R. 1982. *Europe and the People without History.* Berkeley: University of California Press.

Woodman, David. 1991. *Unravelling the Franklin Mystery: Inuit Testimony.* Montreal and Kingston: McGill-Queen's University Press.

Wright, J.V. 1995. *A History of the Native People of Canada, vol. 1, 10,000–1,000 B.C.* Mercury Series, Archaeological Survey of Canada, Paper 152. Ottawa: Museum of Civilization.

———. 1999. *A History of the Native People of Canada, vol. 2, 1000 B.C.–A.D. 500.* Mercury Series, Archaeological Survey of Canada, Paper 152. Ottawa: Museum of Civilization.

Zaslow, Morris. 1984. *The Northwest Territories, 1905–1980.* Canadian Historical Association Historical Booklet No. 38. Ottawa: Canadian Historical Society.

———. 1988. *The Northward Expansion of Canada, 1914–1967.* Toronto: McClelland & Stewart.

- 4 -

Population Size and Its Geographic Expression

The thinly populated North lies well beyond Canada's **ecumene**, with fewer than 5 per cent of Canada's population—less than 1.5 million people. A growing percentage of these northerners are Aboriginal people. In fact, Aboriginal Canadians form nearly one-quarter of the North's population and make up over half of those living in the Territorial North. Such a remarkable demographic rebound has economic, political, and social consequences; as well, this rebound has unleashed a determination on the part of First Peoples and some in government positions to complete its consequences (Saul, 2014).

History has shaped the northern population and its geographic expression. In the mid-twentieth century, two events contributed to a dramatic increase in the size of the North's population. The first was the resource boom that began in the 1950s and petered out by the end of the 1980s. In response to the economic opportunities created by that boom, thousands of southern workers moved to the North, often settling in newly created company towns such as Pine Point, NWT, and Tumbler Ridge, BC. About the same time, the Aboriginal population began to expand, resulting in what demographers call a **population explosion.** In these Native settlements, store food and medical services were readily available, allowing the Aboriginal **birth rate** to increase sharply while the **death rate** declined abruptly.

Within the northern population, other changes have taken place. The two most important are:

- Aboriginal peoples have increased their proportion of the northern population and now form nearly 24 per cent (Figure 4.2 and Table 4.2). Within the Arctic, the figure is closer to 85 per cent.
- While the vast majority (94 per cent) reside in the Provincial North, its population size has diminished since 1981 while that of the Territorial North has grown (Figure 4.1).

In 2014, Canada had a population of 36 million. From a geographer's perspective, Canada's population falls into four zones with zones 3 and 4 found in the North (Figure 4.1). Zone 3 lies largely in the Subarctic and contains over 90 per cent of the North's population. The largest northern cities are located in zone 3. Zone 4, on the other hand, is sparsely populated, accounting for less than 10 per cent of the North's

Table 4.1 Population of the Canadian North, 1981, 2001, and 2011

Region	1981	% of Total Northern Population	2001	% of Total Northern Population	2011	% of Total Northern Population	% Change 1981– 2011
Territorial North	68,894	4.7	92,779	6.3	107,275	7.2	55.7
Provincial North	1,402,289	95.3	1,376,755	93.7	1,382,456	92.8	–1.4
Total Canadian North	1,471,883	100.0	1,469,534	100.0	1,489,766	100.0	1.2

Sources: Statistics Canada (1987, 2002, 2007a, 2013a).

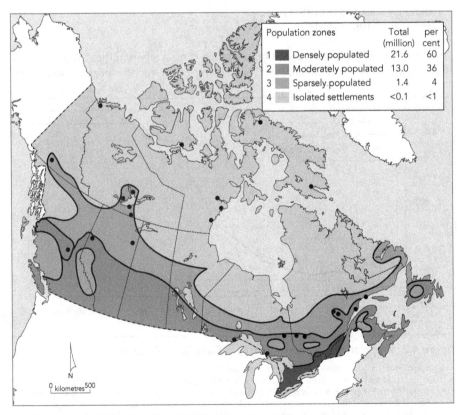

Figure 4.1 Canada's Population Zones and Northern Centres

The largest centres in the Provincial and Territorial Norths are indicated on the map. Their names and population sizes are found in Tables 4.7 and 4.8.

population. The largest cities are located in a belt along the southern border of the Provincial North or, in a few cases, just south of this border. The names and populations of these 24 centres are found in Tables 4.7 and 4.8.

Population Change Since 1867

Since Confederation in 1867, the northern population has increased from 60,000 to just under 1.5 million today. However, the **growth rate** in the North has varied over this period and has been near zero for the last 30 years (Figure 4.2 and Table 4.1). Change in the size and character of the northern population has been brought about by the interplay of two demographic variables: **fertility** and migration. While census divisions along the North's southern edge sometimes extend south of permafrost, Figure 4.2 shows how the North's population has grown and levelled off over time.[1] These demographic changes take the shape of an S-curve, suggesting that the North's **carrying capacity** was reached at the end of phase 2. Within the total population curve, the proportion of Aboriginal population has increased significantly in phase 3. The three phases of population change are:

- slow but steady increase (1871 to 1941)
- rapid increase (1941 to 1981)
- no change (1981 to 2015).

Phase 1. In 1871, the first post-Confederation census reported that the population of the North was approximately 60,000 (Figure 4.2). Over the next 20 years, the

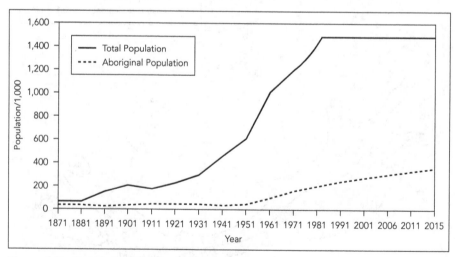

Figure 4.2   Population Change in Northern Canada, 1871–2015

This graph provides a generalized view of population change in the Canadian North over time. The sharp increase from 1951 to 1981 reflects migration into the North by southern Canadians, largely in response to job opportunities in the resource industry. From 1965 to the present, most population increase is attributed to the **natural increase** in the Aboriginal population. According to Statistics Canada, in 1986 Aboriginal people formed 14.3 per cent of the total population; by 2011, this figure had jumped to nearly 24 per cent (Table 4.2 and Appendix IV). Assuming that the demographic forces at play remain the same in the coming years, the total population is likely to remain just under 1.5 million while the proportion of the Aboriginal population could jump from 14 per cent in 1986 to 30 per cent by the 2021 census.

Source: Estimates based on Census of Canada, various years.

settlement of the southern fringe of the boreal forest by settlers, loggers, and miners accounted for its population growth. The Aboriginal population changed little in size or distribution. By 1901 the North's population had reached 100,000, and by 1931 it had more than doubled to 250,000. Almost all population increase took place in the northern areas of Ontario and Quebec. The Aboriginal population may have recovered from the sharp decline due to the flu epidemic that swept across the North in the 1920s but Aboriginal people continued to live on the land until the 1950s relocation efforts of the federal government. At that time, the people transitioned very swiftly from hunters to village dwellers. One outcome was a sharp increase in fertility rates with the more sedentary lifestyle, better access to medical services, and a more stable food supply. This marked the start of the Aboriginal population explosion. Another outcome was the so-called "lost generation" of adults who found the cultural transition too difficult.

Phase 2. During the 40 years from 1941 to 1981, population in Canada's North jumped from 350,000 to nearly 1.5 million. Almost all this population increase took place in the Subarctic. The main factor accounting for the enormous increase was a resource boom that began in the late 1940s, resulting in a massive **in-migration** of southern Canadians. Other factors were (1) a high **rate of natural increase** among the Aboriginal population; and (2) a policy change in Ottawa that began in 1957 with Prime Minister John Diefenbaker's "Northern Vision."[2] By promoting development in the North, both the federal and provincial governments increased public spending in the North, including the relocation of government offices and their staff to the North.

Phase 3. Since 1981, the North's resource boom has slowed and its population size has remained relatively stable at just under 1.5 million. Migration was the key demographic factor contributing to this population stabilization. Over this period, out-migration has exceeded in-migration—which is just the reverse of the situation in phase 2. Even though the **population increase** of Aboriginal northerners was high, the North's population remained just under 1.5 million (Table 4.1). The main point is clear. As Canada entered the second decade of the twenty-first century, the North's population remained stalled at just under 1.5 million. Yet, the Territorial North continues to grow, as do a few areas of the Provincial North, especially northern Alberta where oil sands projects have continued to attract workers from across Canada. One of the fastest-growing regions is Nunavut, where the population nearly reached 32,000 by 2011 (Vignette 4.1). On the other hand, much of the Provincial North's economy is based on a declining forest industry, which results in out-migration. The worst-hit areas remain northern Ontario and Quebec, due to the drop in demand for softwood lumber and pulp from the United States.

Can We Expect a Declining Northern Population?

Since demographic trends associated with phase 3 could continue well into the future, a slow but steady downward slide in the North's total population and an increase in the proportion of Aboriginal population are a strong possibility. The failure of resource megaprojects, such as the Mackenzie Gas Project, to take hold plus the slowdown of older mining operations, such as diamond mining in the Northwest Territories, have dampened population growth in the Territorial North while the continued decline of the forest industry keeps the population of the Provincial North below its peak of 1.4 million in 1981 (Table 4.1). Of course, much will depend on the health of the northern resource economy. Another resource boom could see a surge in people moving to the North. However, times have changed since the last resource boom—companies no longer create resource towns. Instead, firms fly in and fly out experienced southern workers. As a result, after the project is completed, no population footprint remains. The lead/zinc mines at Nanisivik on Baffin Island and Polaris on Cornwallis Island provide two examples. Then, too, the weak state of the forest industry remains a drag on the Provincial North and could cause more out-migration.[3]

Coupled with these factors, the many small Native communities offer few economic opportunities and are troubled by very high unemployment rates, a situation that may cause greater numbers of younger Aboriginal adults to try their luck in southern cities. In fact, this population movement has already begun, with a growing number of Aboriginal people living in urban areas. For example, in 2006 nearly 20 per cent of Canada's Inuit population did not live in their four land claim regions (western Arctic, Nunavut, far northern Quebec, and northern Labrador) and most of these Inuit resided in southern cities, but by 2011 this had climbed to just under 27 per cent (Bone, 2006; Statistics Canada, 2014a: Figure 2).

Population Density

Population density provides a rough indicator of the capacity of a region to support a given population. Since the physical geography of the North has a very low carrying capacity, i.e., the ability of the land to support a population, it helps explain why the North's population is so low. This argument is even more persuasive for the Arctic but this natural region continues to see a rapidly growing Inuit population. This situation raises a serious socio-political question: Does the Arctic face a modern version of the Malthusian population trap? The answer to this question is explored below.

In 1971, Milton Freeman expressed concern about the imbalance between Inuit population size and their resource base. Freeman was particularly concerned that "a rapid rate of population growth generally prevents any successful attempt to remedy the prevailing unfortunate economic situation." In 1988, Colin Irwin repeated this concern. Since then, the Inuit population has continued to increase at a greater rate than the economy, and a fear has arisen of an Arctic version of the **Malthusian population trap** (Vignette 4.1). In classical Malthusian terms, this trap eventually results in severe food shortage and starvation until a balance between population size and food supply returns. In Nunavut, a modern version of the Malthusian trap leads not to starvation and a reduction in numbers, but to a greater numbers and a dependency on the federal government through transfer payments. In 2014–15, support through Territorial Formula Financing (equivalent to equalization payments to provinces) for the three governments is: Nunavut, $1,456 million; Northwest Territories, $1,209 million; and Yukon, $851 million. Beyond these federal transfers, other funds are sent to the territories and provinces, and, on a per capita basis, they reveal the degree of dependency of each jurisdiction on Ottawa: Nunavut will have received $40,080 per person; Northwest Territories, $29,136 per person; Yukon, $24,310. Prince Edward Island, at $3,735 per person, represents the highest figure for Canada's 10 provinces while Alberta, Saskatchewan, and British Columbia, at $1,258, will have received the lowest amount per person (Department of Finance, 2013).

Population Distribution

The uneven distribution of the northern population can be examined from two geographic perspectives, but both reveal a similar spatial pattern—a sharp north/south division with most people living in the southern half of the North.

The first perspective views northern population in terms of the two natural regions. The Arctic, with a population of approximately 60,000, contains only a small fraction (4 per cent) of the North's total population. The Subarctic, on the other hand, is home to over 1.4 million people or nearly 96 per cent of the northern population. The chief explanation for this difference lies in their respective resource bases and their historic development. The Arctic, home for the Inuit and their Thule ancestors for over a thousand years, provided a most challenging environment for a hunting economy and, not surprisingly, its carrying capacity under that economic system could only support a small number of people, perhaps less than half the size of the current population. The most recent figures for the Inuit population reveal that, by 2011, the Inuit population was almost 60,000, but only 44,000 (73 per cent) lived in the Arctic (Table 4.6). The majority reside in four Inuit land claim regions known collectively as **Nunangat** but individually as Nunatsiavut (northern Labrador), Nunavik (northern Quebec), Nunavut, and Inuvialuit (Northwest Territories) while the rest live mainly in southern cities.

Until the twenty-first century, few Inuit relocated to southern cities. Now a clear trend indicates that a small but growing stream of Inuit migrants is a new fact of demographic life. For instance, the 2011 National Household Survey (Statistics Canada, 2014a) reported that nearly 30 per cent of Inuit reside in Edmonton, Montreal, Ottawa, Yellowknife, and St. John's. Unlike the Subarctic, few southern Canadians have resettled in the Arctic. The principal reason is the weak economy, which offers few business/job opportunities. Other factors are an unattractive climate for most Canadians, limited urban amenities and public services, and, except for the Dempster Highway from Dawson, Yukon, to Inuvik, NWT, the absence of a highway system. On the other hand, the vast majority of people residing in the Subarctic are Canadians who relocated from southern Canada to the Subarctic largely because of economic opportunities related to the resource economy and the public service sector.

The second perspective looks at population in the two political norths. The first point is that the two norths have opposite population trends over the 30-year period (1981–2011) with the Provincial North's share of population dropping from 95.3 per cent to 92.8 per cent (Table 4.2). In 2011, for instance, 107,265 people or 7.2 per cent of

Table 4.2 Percentage of Population of the Territorial and Provincial Norths, 1981, 2001, 2006, and 2011 (% Aboriginal population in brackets)

Political Region	1981	2001	2006	2011
Territorial North	4.7 (47%)	6.3 (51%)	6.9 (52%)	7.2 (53%)
Provincial North	95.3 (12%)*	93.7 (20%)	93.1 (21%)	92.8 (22%)

*Ethnic mobility offers an explanation for the low Aboriginal percentage in 1981. Ethnic mobility refers to the phenomenon of more Aboriginal people, especially Métis, recording their ethnic identity as Aboriginal in the last three censuses.

Sources: Statistics Canada (1987, 2002, 2007a, 2013a).

Table 4.3 Northern Population by Territorial and Provincial Norths, 2001 and 2011

Province/Territory	2001	2011	% Change
Yukon	28,674	33,900	18.2
Northwest Territories	37,360	41,465	11.0
Nunavut	26,745	31,910	15.3
Territorial North	**92,779**	**107,310**	**15.7**
British Columbia	198,289	197,747	−0.1
Alberta	100,479	129,020	28.4
Saskatchewan	32,029	36,557	14.1
Manitoba	66,622	70,906	6.4
Western Canada	**397,419**	**434,230**	**9.3**
Ontario	416,475	400,656	−3.8
Quebec	515,126	504,100	−2.1
Newfoundland/Labrador	47,735	43,515	−8.8
Eastern Canada	**979,336**	**948,271**	**−3.2**
Provincial North	**1,376,755**	**1,382,456**	**0.0**
Canadian North	**1,469,534**	**1,489,766**	**0.1**

Source: Appendix II.

the North's population lived in the three territories while 1,489,766 or 92.8 per cent inhabited the northern areas of provinces. Also, from 1981 to 2011, the Territorial North saw its population increase by nearly 16 per cent while the Provincial North experienced virtually no increase. The second point is that northern Ontario and Quebec account for the bulk (65.5 per cent) of the population of the Provincial North, but these same areas (northern Ontario and Quebec) suffered a substantial population decline of 26,000, dropping from 931,000 in 2001 to 905,000 in 2011 (Table 4.3). The Labrador population dropped even more—by almost 9 per cent. In sharp contrast, western Canada exhibited considerable growth due to resource developments such as those taking place in the Alberta oil sands. From 2001 to 2011, the population of northern Alberta increased by just under 30 per cent, while the populations in northern Saskatchewan and Manitoba grew by 14.1 and 6.4 per cent, respectively.

The Number and Distribution of Aboriginal Peoples

The number of Aboriginal people in the Canadian North is substantial—and, unlike the non-Aboriginal population, growing. For instance, in 1986, nearly 210,000 Aboriginal Canadians lived in northern Canada, constituting 14 per cent of the northern population (Table 4.4). By 2011, their numbers had risen to 357,000, forming 24 per cent of the North's population. Since their rate of natural increase has been around 1.5 per cent per year and is likely to continue, the Aboriginal population in the North will expand, possibly reaching 400,000 by 2016.[4]

Table 4.4 Aboriginal Population in Canada's North, 1986 and 2011

Region	1986	Aboriginal % of Total Population	2011	Aboriginal % of Total Population
Territorial North	35,525	46.9	56,235	52.4
Provincial North	173,755	12.2	300,435*	21.7
North	209,280	14.3	356,670	23.9
Canada	737,035	2.8	1,400,685	4.3

*In the 2011 census, more people self-identified as having Aboriginal identity than in previous censuses. By the next census, this figure is expected to increase far beyond the natural rate of increase because of the 2013 ruling of the Federal Court of Appeals that Métis and **non-status Indians** are "Indians" under the Constitution Act, 1867. In 2014, the matter went to the Supreme Court of Canada, which as of May 2015 had not yet ruled on this matter.

Source: Appendix IV.

Within the Territorial North, Aboriginal peoples have formed the majority for the last decade and a half. In 1986, they made up only 47 per cent of the total territorial population (Table 4.4), but by 2001 their share was 51.7 per cent. By 2011 this had increased to 52.4 per cent (Table 4.5). This trend seems destined to continue, with the Aboriginal proportion of the Territorial North's population likely to reach 53 and possibly 54 per cent by the time of the next census.

Equally significant, all three territories witnessed an increase in their Aboriginal populations. From 1986 to 2011, the Aboriginal share of Yukon's total population increased from 21 per cent to 23 per cent; the Northwest Territories saw a similar trend, from 49 per cent to 52 per cent, while Nunavut went from 85 to 86 per cent from 2001 to 2011. Over this 10-year period, the total Aboriginal population jumped from 48,000 to just over 56,000. This pattern of population increase is particularly high among the Inuit population, whose annual natural rate of increase has exceeded 2 per cent (Vignette 4.1). Such rates are normally found in developing countries.

Two demographic factors explain the increasing proportion of Aboriginal people within the total population of the North: (1) a decline in southern workers and their families moving to the North; and (2) the still high but slowly declining rate of natural increase among Aboriginal populations. These two factors explain the increasing proportion of the Aboriginal population but also indicate a slowing of their natural rate of increase. Take, for example, the Territorial North, where Aboriginal people formed 51.7

Table 4.5 Aboriginal Populations for Canada's Three Territories, 2001 and 2011

Territory	2001 Aboriginal Population	2001 Aboriginal % of Total Population	2011 Aboriginal Population	2011 Aboriginal % of Total Population
Yukon	6,540	22.8	7,705	23.1
NWT	18,730	50.1	21,160	51.9
Nunavut	22,720	84.9	27,360	86.3
Territorial North	47,990	51.7	56,225	52.4

Sources: Statistics Canada (2004, 2013a).

Vignette 4.1 Population Explosion in Nunavut

While the rate of natural increase has diminished in most areas of the North, Nunavut's high fertility rate continues to fuel its rapidly expanding population. Of Canada's three Aboriginal peoples, the Inuit have the highest fertility rate—more than double the national average. From 1996 to 2011, Nunavut's population increased from 24,730 to 31,906, an increase of nearly 30 per cent (Statistics Canada, 2013a). By 2031, Statistics Canada estimates that Nunavut's population will increase another 15,000 or so, reaching somewhere between 47,000 to 49,000—and almost all of that growth is expected to come from the natural increase of the Inuit population (Statistics Canada, 2013f: Table 7).

---

per cent of the total population in 2001 compared to 52.4 per cent in 2011 (Table 4.5). Current indications suggest that these factors will remain at play for the short run. However, the Aboriginal population is moving through its version of the **epidemiological transition**, which includes a reduction in the **fertility rate**, a constant **crude death rate**, a falling **infant mortality rate**, and an increase in **life expectancy**. The implication is that, if Aboriginal fertility rates continue to decline and eventually match the national average, the territorial population growth will slow, perhaps stop, and even decline. The Inuit are the last Aboriginal peoples to enter the epidemiological transition and their fertility rate remains very high, but a downturn is expected—at least according to this theory.

The Inuit Population

An examination of the Inuit population provides an opportunity to see more deeply into shifts in Aboriginal population found in northern Canada. After the relocation program took hold, life in settlements resulted in high annual rates of natural increase, which propelled the Inuit population from just over 4,000 in 1961 to 27,000 in 2011. As noted in Vignette 4.1, a Statistics Canada report (2013f) sees a continuation of population growth over the next 20 years. The average annual growth rate of the Aboriginal population would range between 1.1 and 2.2 per cent from 2006 to 2031. In comparison, the growth rate of the non-Aboriginal population would average 1.0 per cent annually. This report goes on to state that the Inuit population, augmented solely by natural increase, will grow at an average annual rate of between 1.3 per cent and 2.5 per cent from 2006 to 2031. Throughout this period, the Inuit rate of natural increase would remain the highest of all Aboriginal identity groups, regardless of the scenario considered.

The higher fertility rates and therefore higher annual growth rates of the Inuit population, compared to those of Canadian **Indians** and Métis, is likely due to a cultural lag, that is, the Inuit participation in settlement life and exposure to Canadian society and its value system have been shorter than is the case for **status Indians** and Métis populations. The assumption is that as this cultural lag dissipates, Inuit exposure to Canadian values, such as family size, the acute housing shortage in

Nunavut, higher education levels for young Inuit women, less need for larger families to survive on the land, and other socio-economic factors unique to Arctic Canada will combine to lower both fertility and growth rates—at least to the level of other Aboriginal peoples.

The Inuit are the fastest-growing Aboriginal population and their numerical growth is well ahead of the national figure. Statistics Canada has recorded Inuit populations by its regions for 2006 and 2011 and these data provide additional insights into the spatial characteristics of this population. Within Nunangat (Arctic), the Inuit population gained nearly 4,000 persons. Most of this increase occurred in Nunavut and Nunavik (Arctic Quebec). Undoubtedly, the most surprising change was the drop in Nunangat's population and the increase in the number living in other parts of Canada, especially in southern Canadian cities. In 2011, nearly 27 per cent of the Inuit population lived outside of their traditional Arctic homeland, up from 22 per cent in 2006 (Table 4.6). Does this 5 per cent jump indicate a trend? Perhaps, but confidence in such a trend would increase if similar results are recorded in the next census. Does this mean that these16,000 Inuit consider themselves permanent residents of southern Canada? Probably not, because most began as temporary residents who came south for education, employment, training, medical attention, and other services not available in Nunangat. Some have returned home, but with prospects back home limited, some with the required education and skills have opted to join the middle-class urban society; others, usually those without much education or skills, have fallen through the cracks and live on the streets (Kishigami, 1999, 2015).

This pattern of migration and urban adjustments/failure is not unique to the Inuit but represents a common pattern across Canada (see Vignette 4.2). While geography (distance and the cost to return home) is a lesser constraint for those living in the Subarctic, the attraction of cities is powerful for different reasons. In sum, an urban

Table 4.6 Size and Growth of the Inuit Population, Canada and Regions, 2006 and 2011*

Regions	2006	2011	% Change 2006–11	Absolute Change	2006 % of Total Inuit Population	2011 % of Total Inuit Population
Nunatsiavut	2,160	2,325	7.6	165	4.3	3.9
Inuvialuit	3,115	3,310	6.3	195	6.2	5.7
Nunavik	9,565	10,750	12.4	1,185	19.0	18.1
Nunavut	24,635	27,070	9.9	2,435	48.8	45.5
Total Nunangat	39,475	43,460	10.1	3,985	78.2	73.1
Outside Nunangat	11,005	15,985	45.3	4,980	21.8	26.9
Canada	50,485	59,445	17.8	8,960	100.0	100.0

*Rounding of population counts may result in differences in sums.

Note: NHS supplied the data for 2011. Since its methodology differs somewhat from the data collected by Statistics Canada in 2006, the 2011 Inuit population figures may be less accurate than those for 2006.

Sources: Bone (2012: Table 4.3); Statistics Canada (2013f).

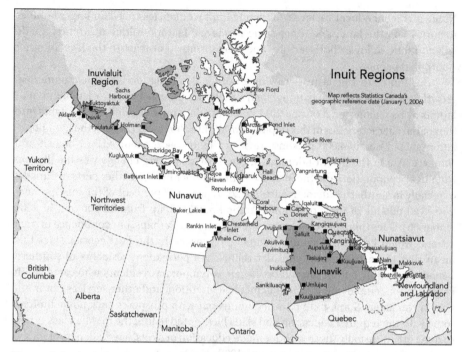

Figure 4.3 Inuit Settlements in Nunangat, 2011

Only Inuvik, a town created under the Diefenbaker government, has a majority of non-Inuit residents.

Source: 2006 Census of Canada. Produced by the Geography Division, Statistics Canada, 2006, at: geodepot.statcan.ca/ Diss2006/Maps/ThematicMaps/OMA_CSD_Maps/InuitRegionsAboriginal_Reference_ec.pdf.

Aboriginal middle class has emerged in Canada's major cities; at the same time, others, less well equipped for urban life, form an underclass.

Migration

Migration into the North is no longer the dominating demographic factor. Well-paying jobs and opportunities for quick advancement were attractive forces. During the rapid expansion of the resource industry from 1950 to 1980, the North had these qualities and in-migration was large enough and persistent enough to accelerate population growth, alter the demographic structure of the population, and affect its ethnic composition. Viewed in terms of the **push–pull migration theory**, attractive economic opportunities pulled southerners into the North. However, since the 1980s, the reverse has happened. The North's economy slowed and workers and their families returned to southern towns and cities where "place" and jobs were powerful magnets: in many instances, even for those with northern jobs, environmental factors and living conditions in the North "pushed" them back to southern Canada. Provincial norths have suffered the most and recorded a net loss. In fact, for the last decade, the northern areas of Ontario, Quebec, and Newfoundland and Labrador have lost people, especially young adults, because of limited economic opportunities.

The decline of the forest industry, especially in its logging sector and mills, has meant a reduced labour force. As a result, many families left for southern Canada. Many mills in single-industry towns have closed because of the lack of demand from the United States, and the volume of logging has greatly decreased, causing layoffs in both sectors. On the other hand, Alberta and, to a lesser degree, British Columbia and the Northwest Territories are gaining population largely because of the expanding oil and gas industry. Moving to the North has its attractions for those in the public sector due to high wages, cost-of-living allowances, and subsidized housing. As well, the prospect of rapid advancement (often a result of workers replacing those who have opted to return south) is another attraction. These extra benefits have helped southern workers to overcome their concerns about the cold climate and lack of amenities in the North.

Northern migrants, however, often have been temporary residents. Many young people go north to make money for five years or so and then return home with a "nest egg" for a new start (Petrovich, 1990). Some southern transplants, on the other hand, do become enamoured with the North and make it their permanent home. In resource towns and regional centres along the southern fringe and in the three territorial capital cities, more and more newcomers are laying down roots. Countering that process are (1) workers residing in the south who commute by air to northern work sites (Herkes, Mooney, and Smith, 2013) and (2) the issue of sustainability in the non-renewable resource sector. Non-renewable resource projects often have a relatively short lifespan. In the last two decades, the mine closures such as those at Faro in Yukon and Leaf Rapids in Manitoba caused many long-term residents to lament the loss of their community and to complain about uprooting their families to relocate elsewhere—perhaps a social reason why resource companies today often fly their workers in and out rather than developing resource towns. Regional loyalties are readily acknowledged as a key geographic theme. For southern migrants, such loyalties are more easily formed in Subarctic than the Arctic. First, the climate in the Subarctic is not so extreme. Second, the Subarctic contains a number of urban centres, especially resource towns, where community design and structure, as well as the pace and pattern of life, are similar to that found in southern Canada. Third, the Arctic, with 86 per cent of the population Inuit, represents a different cultural and urban setting from that familiar to southern Canadians, and many urban amenities are not available.

Nevertheless, many southern Canadians are economic migrants attracted by high-paying jobs. For these transplanted Canadians, the North is a frontier for economic opportunity but not a place of permanent settlement. As newcomers, they remain highly mobile, and if their employment ends or an attractive job opens up elsewhere they are likely to move. Few stay longer than five years. To a large degree, the attraction of **place** in southern Canada is difficult to overcome (Halseth, 2010). Based on records from 1995 to 2008 for Yukon, the median length of residency for out-migrants (a resident who moved from Yukon in 2008) was three years, while for non-migrants (a resident who remained in Yukon in 2008) the length of residency was seven years (Yukon Bureau of Statistics, 2010: 8). This tendency to move back home so quickly is reinforced by a desire to be closer to family and friends, to enjoy a wider range of urban amenities, and to return to a more temperate climate. For those who stay longer than three years and then leave Yukon, two other factors often motivate them to return to the south:

- A high degree of job uncertainty common to all resource industries, whether mining, forestry, or oil exploration. During times of expansion there is an in-migration of workers, but during economic contraction the same workers may seek employment outside the North.
- A tendency for young couples to leave the North when their children reach school age. The general feeling is that schools in the North do not provide the same level of education as schools in the south.

Natural Increase

The rate of natural increase of a population consists of differences between births and deaths for a given time period divided by its total population and expressed as a percentage. Within the North, its two populations, Aboriginal and non-Aboriginal, have different rates of natural increase, with a much higher Aboriginal rate. For example, in 2011, Nunavut's birth rate was the highest in Canada (Appendix III). However, except for the Inuit rate of natural increase, the Aboriginal rate has declined and may soon match the national figure (Statistics Canada, 2013b). By 2011, the annual rates for the three territories had declined to 0.67 for Yukon, 1.13 for the Northwest Territories, and 2.0 for Nunavut (Appendix III). Does this northern decline in natural increase over the last decade represent a long-term trend? In the author's opinion, the answer is "yes." But how do we explain this apparent trend? The **demographic transition theory** focuses on Western experiences with economic and social changes and therefore does not address the forces at play within the Aboriginal population (Vignette 4.3). Nevertheless, with Aboriginal northerners now living in settlements, the forces of change affecting the

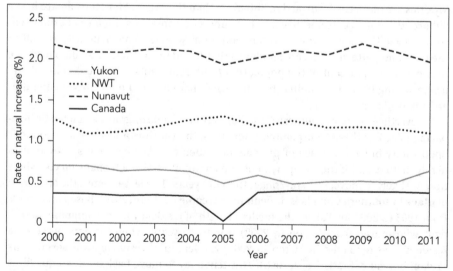

Figure 4.4 Rate of Natural Increase for Canada and the Territories, 2000 to 2011

Source: Appendix III.

Vignette 4.2   A Growing Phenomenon: Aboriginal Population in Saskatchewan Cities

Those Aboriginal people who relocated from the land to northern trading posts had assured food supplies and accessible medical assistance. Consequently, their populations increased rapidly due to a high rate of natural increase and the virtual absence of out-migration. Even today their settlements continue to increase, although their rate of natural increase has slowed and out-migration has become a factor. Most migrants have moved to towns and cities located in Canada's ecumene. Saskatchewan's north reflects this demographic pattern. From 2001 to 2011, this Subarctic region, with close to 90 per cent of its residents Aboriginal by identity, witnessed a population increase from 32,029 to 36,557 (Appendix II; Statistics Canada, 2013a). Over the same 10-year period, the number and percentage of the Aboriginal populations in the province's two northern gateway cities, Prince Albert and Meadow Lake, jumped dramatically. For Prince Albert, the figures for 2001 and 2011 were 10,185 (30.5 per cent) and 13,930 (40.6 per cent); for Meadow Lake: 1,535 (34.1 per cent) and 2,180 (44.4 per cent) (Statistics Canada, 2013a, 2013b). Further south, the Aboriginal population in Saskatoon, the largest city in Saskatchewan, reached 21,335 (9.8 per cent) in 2011 (Statistics Canada, 2013c). Of the many attractions of southern cities, three factors are especially significant: access to urban amenities, including shopping, post-secondary education, and advanced medical services; availability of employment and housing; and a critical mass of established Aboriginal residents who assist newcomers. These factors alone ensure that the flow of northern newcomers to Saskatchewan urban areas and other Canadian cities will continue. For Meadow Lake and Prince Albert, the question is: Will Aboriginal residents form a majority of the city populations by the next census in 2016?

Sources: Appendix II; Statistics Canada (2013a, 2013g, 2013h, 2013i, 2013j).

general Canadian society also reach those in the Aboriginal community. The mass media, especially television, are powerful cultural transmitters and levellers.

Demographic Structure

The demographic structure of a population—its age and sex composition—is measured in age cohorts, usually based on five-year intervals. Since the population processes of fertility, mortality, and migration shape a population over time, the demographic structure provides a profile of the age and sex of that population for a particular instant in time, as well as clues regarding its population future and probable implications for its economy and society.

Demographers pay particular attention to the age composition of a given population. A youthful population has a high proportion of its members under the age of 15. Most developing countries fit into that demographic age picture while industrial countries do not. The 2011 census revealed that 16.8 per cent of Canada's population is under 15 years of age, but in the Canadian North those under 15 comprise just under 20 per cent (Statistics Canada, 2014b). Taking the Territorial North as an example of the range of "youthful" populations, Yukon has 17.3 per cent under 15, the Northwest Territories 21.7 per cent, and Nunavut 32.7per cent (Statistics Canada, 2014b). While an older population exists in most areas of the Provincial North but especially in northern Ontario and Quebec, the rest of the Provincial North has a much younger population. For example, in Saskatchewan's North, as measured by Census Division 18, 32.7 per cent of its population was under the age of 15, with most of these young northerners of Cree, Dene, and Métis ancestry (Statistics Canada, 2013d). What these age figures demonstrate is that Canada has a dual population in terms of demography, and that duality signifies a parallel duality in economic well-being. The statistics also indicate that the Aboriginal birth rate and, therefore, natural rate of increase are declining (Vignette 4.3).

Vignette 4.3 Dependency Ratio and Resource-Sharing

The **dependency ratio** serves as an indicator of the proportion of non-working-age members of a population compared to those of working age. A high ratio means those of working age—and the overall economy—face a greater burden in supporting the population. Demographers define the productive members of society by those of working age, that is, between 15 and 64 years of age. Those outside of this age **cohort** are considered either too young or too old to work. Nunavut, for example, has a heavy economic burden compared to Canada because Nunavut, as of 2011, had an extremely youthful population, with 32.7 per cent of its population under the age of 15. In comparison, Canada as a whole had only 16.8 per cent in that age bracket. While this difference is softened when the proportion of senior citizens is added into the mix, the implications of a **young population** are considerable. First, a young population requires more investment in social infrastructure and social programs, ranging from daycare facilities to additional classroom space for kindergarten and primary students. Second, as this young population bulge moves through the life cycle, public investment must be directed to new high schools and post-secondary facilities and then to job training and job creation programs.

Transfer payments to Nunavut from Ottawa help, but resource development has so far provided little revenue. Unlike other provinces and territories, Nunavut is the only territory without an agreement for the transfer of responsibility for mining, oil, and gas exploration and development—and most importantly, sharing of resource royalties. Ottawa still collects resource royalties in Nunavut. The additional income stream to Nunavut would help balance its budget. However, the Premier of Nunavut, Paul Okalik, held out little hope: "Such a deal would see the GN [Government of Nunavut] getting a share of [non-]renewable resource royalties. So far, however, there has been little or no response from the federal government" (Dolphin, 2014).

Urban Population

Economic reality encourages residents of the North to live in small urban centres. Even so, the cost of living in these centres, especially in the Arctic, is much higher than it is in southern Canada. For example, without an agricultural base, foodstuffs must be imported at a high cost.[5] These centres have the following characteristics:

- The largest cities are found on the southern edge of the Provincial North (Table 4.7). In fact, in 2011, the Territorial North had only one city, Whitehorse, Yukon, with a population over 20,000. Whitehorse's population was just over 23,000 (Table 4.8).
- Most Arctic centres have less than 1,000 residents. Only one centre—Iqaluit—has more than 6,500 inhabitants. In spite of their small size, almost all Arctic communities, unlike most Subarctic ones, have recorded high rates of population increase from 2001 to 2011. The population of Iqaluit, for example, jumped from 5,200 in 2001 to almost 6,700 by 2011—an increase of nearly 28 per cent (Table 4.8).

The southern edge of the North does not correspond perfectly with the boundaries of census division used by Statistics Canada to define the North. Several large cities lie along this edge or just to its south. Many have either lost population or gained very little over the last decade (2001 to 2011). The four largest of these cities are Saguenay

Table 4.7 Top 12 Provincial North Urban Centres by Population Size, 2001–11

Urban Centre	2001	2006	2011	% Change 2001–11
Saguenay, Quebec	154,938	156,305	157,790	1.8
Thunder Bay, Ontario	121,986	122,907	121,596	0.0
Prince George, BC	85,035	83,225	84,232	−0.1
Sault Ste Marie, Ontario	78,908	80,098	79,800	1.1
Wood Buffalo (Fort McMurray)*	41,466	52,643	65,565	58.1
Grande Prairie, Alberta	36,983	47,107	55,032	48.8
Timmins, Ontario	43,686	42,997	43,165	−1.2
Rouyn–Noranda, Quebec	39,621	40,650	41,798	5.5
Alma, Quebec	32,930	31,864	33,018	0.3
Val-d'Or, Quebec	32,423	32,288	33,265	2.6
Baie-Comeau, Quebec	30,401	29,674	28,789	−5.3
Sept-Îles, Quebec	27,623	27,827	28,487	03.1
Total	699,492	747,585	773,858	1.1
Provincial North	1,376,755	1,368,204	1,382,456	0.0
Percentage of Provincial North	50.8	54.6	56.0	

*In 1995 the province of Alberta created an amalgamated Regional Municipality of Wood Buffalo, which includes Fort McMurray, several Aboriginal settlements, and the oil sands camps. Statistics Canada, for 2011, listed the Wood Buffalo population (which includes Fort McMurray) as 65,565 and the Fort McMurray population as 61,374. The Regional Municipality, for the purpose of gaining more funding for infrastructure and services, has conducted its own population counts. This "census" includes non-permanent residents, i.e., those who commute to oil sands and related employment and who work in the municipality at least 30 days a year, as well as homeless people and those in small Aboriginal communities. The Wood Buffalo population, according to the Regional Municipality, reached 125,032 at the end of June 2015.

Sources: Statistics Canada (2007a, 2014c); Regional Municipality of Wood Buffalo (2015).

Table 4.8 Top 12 Territorial Urban Centres by Population Size, 2001–11

Urban Centre	2001	2006	2011	% Change 2001–11
Whitehorse, Yukon	19,058	20,481	23,276	22.1
Yellowknife, NWT	16,541	18,700	19,234	16.3
Iqaluit, Nunavut	5,236	6,184	6,699	27.9
Hay River, NWT	3,510	3,648	3,606	2.7
Inuvik, NWT	2,894	3,484	3,484	20.4
Fort Smith, NWT	2,185	2,364	2,093	–4.2
Rankin Inlet, Nunavut	2,177	2,358	2,266	4.1
Arviat, Nunavut	1,899	2,060	2,318	22.1
Behchokò (Rae-Edzo), NWT	1,552	1,894	1,926	24.8
Baker Lake, Nunavut	1,507	1,728	1,872	24.2
Cambridge Bay	1,309	1,477	1,608	22.8
Pond Inlet	1,220	1,315,	1,549	27.0
Total	59,088	65,683	69,931	18.4
Territorial North	92,779	101,310	107,310	15.7
Percentage of Territorial Population	63.7	64.8	65.2	

Sources: Statistics Canada (2007a, 2014c).

(formerly Chicoutimi–Jonquière) at 157,800, up 1.8 per cent from 2001; Thunder Bay, 121,596, down slightly from 2001; Prince George, 84,232, down 0.1 per cent from 2001; and Sault Ste Marie, 79,800, up by 1.1 per cent from 2001 (Table 4.7). However, the 12 largest centres in the Territorial North have all gained population and thus bucked this trend found in the Provincial North (Table 4.8).

Two cities in northern Alberta—Grande Prairie and Wood Buffalo (a regional municipality, with the vast majority of its population in Fort McMurray)—are growing at very rapid rates and thus provide an exception. The population of Grande Prairie, with its diversified economy based on agriculture, forestry, and natural gas, jumped from 37,000 in 2001 to 55,000 in 2011, an increase of 48 per cent. Fort McMurray, a single-industry town in the heart of the oil sands developments, grew from less than 42,500 in 2001 to over 65,000 in 2011. In general, the weak economy of the Provincial North (with the exception of Alberta) lies behind the population decline of cities. On the other hand, major towns and cities in the Territorial North continue to expand because of the high rate of natural increase among Aboriginal residents.

Classification of Urban Centres

Economic and cultural elements allow for the division of northern population centres into three categories: regional centres, resource towns, and Native settlements.[6] *Regional centres* range from small service centres like The Pas in northern Manitoba to capital cities like Yellowknife in the Northwest Territories. In all instances, the capital cities are the major service centres in the territories. These regional centres are the source of certain goods and services for those living within their catchment areas. Central place theory seeks to explain the relative size and geographic spacing of urban centres as a function of shopping behaviour. Its central argument is that basic or lower-order goods and services are found in all urban centres while specialized or higher-order ones are found only in larger urban centres. Successively larger centres offer a greater variety of goods and services and so command a broader market territory. For example, Yellowknife provides a wider range of goods and services than do smaller centres, such as the nearby village of Behchokò (Rae-Edzo). Residents of Behchokò have to travel to Yellowknife to obtain a higher order of good or service—e.g., shopping for a truck—not available in their home community. For a firm to offer a set of goods at a particular place, it must sell enough to meet operating costs and, presumably, to turn a profit. The concept that a minimum level of demand is necessary to allow a firm to stay in business is called the threshold. The same argument can be made for public services, i.e., hospitals and high schools are located in regional centres while nursing stations and primary schools are found in smaller centres. Fort Simpson and Wrigley provide a simple example of the urban hierarchy. Fort Simpson, with a population in 2011 of 1,238, is the regional centre in the Deh Cho region of the Northwest Territories. Fort Simpson has both a hospital and a high school (Statistics Canada, 2014c). Wrigley, with only 133 residents in 2011, has neither a hospital nor a high school. Students from Wrigley who wish to attend high school must travel nearly 200 km or move to Fort Simpson. Similarly, residents of Wrigley requiring medical care beyond what the local nurse can offer have to go to the Fort Simpson hospital (Statistics Canada, 2014c).

Resource towns or single-industry towns are products of the resource economy. Many were created during the resource boom from 1950 to 1980. They accounted for much of the population increase during those 30 years. Fort McMurray, on the other hand, is an exception today. This resource town has seen most of its population growth occur in the last 20 years. Given the enormous size of the oil sands, the Fort McMurray population may continue to increase rapidly. For example, its population grew from 35,213 in 1996 to 41,466 in 2001, an increase of 17.8 per cent (Statistics Canada, 2007a). Over the next five years, the population reached 52,643, gaining nearly 24 per cent over the 2001 figure (Statistics Canada, 2007a). This rate of increase continued to 2011, when the Fort McMurray population reached 65,565 (Statistics Canada, 2014b, 2014c). With several huge oil sands construction projects underway plus the expansion of existing heavy oil plants, the demand for workers exceeds the supply, driving wages higher and higher. High wages and steady employment attract many from across Canada and beyond, especially from economically depressed areas troubled by high rates of unemployment.[7] Not surprisingly, Newfoundland has supplied so many workers that Newfoundlanders make up around one-third of Fort McMurray's population (Mahoney, 2002: A6; Storey, 2009; Keough, 2013).

Resource towns are normally based on a non-renewable resource, such as coal, nickel, or uranium. They are vulnerable to the vicissitudes of global prices for their natural resources. Tumbler Ridge, for instance, has spanned two global business cycles and thus lived, died, and was reborn as a centre for coal production in the northeastern area of BC's Cordillera.[8]

Vignette 4.4 Dominance of Capital Cities in the Territories

Whitehorse is the capital of Yukon, and is also the transportation, business, and service centre for the territory. In 2011, its population was just over 23,000, making it the largest city in the three territories. In the same year, Whitehorse made up nearly 69 per cent of the territory's population of 33,900. This modern city is located in the Yukon River Valley and along the surrounding terraces. In the past, the Yukon River represented a major transportation link between Whitehorse and Dawson. Since 1947, the Alaska Highway not only connected the urban centres of Watson Lake, Whitehorse, and Dawson, but also has provided a link with Alaska and the provinces of British Columbia and Alberta.

Yellowknife is the capital of the Northwest Territories. Located on the north shore of Great Slave Lake, Yellowknife was originally a gold-mining town but its two mines, Con and Giant, closed in 2003 and 2004, respectively. With the opening of several diamond mines in the Territorial North, Yellowknife has added diamond processing to its business functions. In 1967, Yellowknife was selected as the capital of the Northwest Territories. Its population in 2011 was just over 19,000, making it the second largest city in the three territories. In 2011, Yellowknife formed 46 per cent of the population of the Northwest Territories (41,465).

Iqaluit is the third-largest city in the territories. Located on Baffin Island, Iqaluit had a population of 6,699 in 2011. Its rapid growth (almost 28 per cent from 2001 to 2011) is largely due to its role as the political capital of Nunavut. Unlike the other two capital cities in the Territorial North, Iqaluit formed only 21 per cent of Nunavut's 2011 population of 31,910.

Resource towns based on a non-renewable resource were in vogue in the resource boom that began in the 1950s. Gradually, these communities lost their raison d'être and were replaced by an air-commuting system that brings workers to and from the mines. By the twenty-first century, the construction of a resource town is a rare event and its approval is much more complex than before. All of this constitutes another phase in the life cycle of a project, coming after the company's decision to proceed (Table 4.9). Accordingly, the mining industry has several hurdles to pass before obtaining permission from government. This step, now the second one, occurs before ground is broken and it includes submitting an environmental impact statement, participating in an environmental assessment with public hearings, and possibly concluding an impact and benefits agreement with one or more local Aboriginal groups.

The lifespan of an ore body and hence the resource town is limited, often less than 30 years. These towns have gone through rapid population growth, population stability, and population collapse. Pine Point in the Northwest Territories, Uranium City in northern Saskatchewan, Leaf Rapids in northern Manitoba, and Schefferville in northern Quebec provide examples of mining towns that have gone through such a 30-year cycle of population expansion and decline. All lost their primary mining function, causing the miners and their families to move. By 1993, Pine Point was abandoned, while Schefferville had a population of 213 in 2011 (Statistics Canada, 2014c) and Uranium City had shrunk to less than 50 inhabitants, most of whom were either Métis or Indians who lived in the area prior to the mining developments. Leaf Rapids faced a

Table 4.9 Classic Population Life-Cycle Model for Resource Towns*

Phase	Population Characteristics	Associated Events
1	Uninhabited site	Company announces plans to build a resource town.
2	Sudden increase in population size	Workers and their families arrive in recently completed company town.
3	Population size reaches peak and then remains stable	Resource production attains its maximum and the need for additional workers ceases.
4	Sharp decline in population size	Company decides to close operations. Workers and their families depart.
5	Return to an uninhabited site	Mine closed and its buildings and housing demolished.

*Note that in today's world, a second phase would involve completing certain environmental matters before the federal or provincial governments would grant approval.

Source: Bone (1998: 250). Reprinted by permission of Cahier de Geographie du Québec.

similar fate with the closure of its mine in 2002. In 2001 its population totalled 1,309, but by 2011 it had only 453 residents (Statistics Canada, 2014c).

The mining town of Pine Point illustrates the full resource town cycle common in the height of the resource boom that began in the 1950s (Bone, 1998: 252–3). In the late 1950s, Cominco decided to develop a lead/zinc mine just south of Great Slave Lake. The problem of access to outside markets was solved by the construction of the Great Slave Lake Railway by the Canadian National Railway. With considerable help from Ottawa, construction of the new town of Pine Point and its lead/zinc mine began in 1962. Three years later, the mine was in full production. From 1962 to 1976, this resource town witnessed extremely rapid growth, reaching nearly 2,000 inhabitants by 1976. By that time, the company had extracted most of the more accessible and higher-quality ore and the cost of mining began to increase. In 1983, the mine ceased production and most work was focused on milling the ore for shipment to a smelter at Trail, British Columbia. From 1976 to 1986, the size of the community began to decrease, dropping to 1,500 by 1986. With the final export of processed ore in 1988, the mill was closed and preparations were underway to turn the townsite back to nature. All buildings were destroyed or removed. By 1991, the town of Pine Point had disappeared from the landscape. As one former Pine Point resident observed:

Don't think people realize how hard it is to leave a town where all three children were born. . . . Seems unreal to see your home town slowly disappear. It's too bad that some other industry couldn't have been brought in to keep the town alive. . . . It's a shame to see buildings (arenas, school, etc.) abandoned when so many other settlements have nothing of the sort and could use the facilities. (Kendall, 1992: 134)

Native settlements in the North are a product of the post-colonial era in Canada. Early settlements coalesced around fur-trading posts, when trapping and fur trading

Figure 4.5   Whitehorse, Yukon

Located in the boreal forest, the city of Whitehorse lies along the river terrace of the deeply entrenched Yukon River. Whitehorse is the largest city in the Territorial North.

Source: City of Whitehorse. Reprinted by permission.

were important economic pursuits. More recently, many settlements were created in the 1950s when northern Aboriginal peoples were relocated by the federal government to facilitate the delivery of state services such as health care and education. Unfortunately, these settlements based on the trapping economy have too small an economic base to support a larger population. This problem remains at the root of Aboriginal dependency and poverty in northern communities.

Many settlements are isolated and have fewer than 1,000 inhabitants. Road access from the south is limited. For some, only air transportation is available. Even with a modest supply of game, the cost of living is extremely high because store food and other goods are transported by truck or air from southern cities. In the territories, the federal government, directly or indirectly, contributes most funds for public services and housing. In the Provincial North, the Métis receive support from provincial governments. In spite of these shortcomings, Native settlements continue to grow because they represent a portion of their traditional homeland. In 2007, the Cree of Kashechewan, a remote First Nation community of about 1,900 people on the shores of the Albany River in northern Ontario, suffered a serious flood and their water and sewage systems were compromised. The question then surfaced: Should they relocate to a more secure site? One option was to move to Timmins where urban amenities and services were readily available. Such relocation, however, would remove them from their traditional homeland. Not surprisingly, they chose to remain in their flood-prone community (CBC News, 2007). Culture nearly always trumps economic reality, although when serious flooding inundated the community for the fourth straight year in April 2015 and everyone had to be evacuated yet again, the local chief remarked that the time had come for the community to relocate to higher ground (Canadian Press, 2015).

Figure 4.6   Fort McMurray, Alberta

This aerial view of Fort McMurray (popularly referred to as Fort Mac) is from the north looking south. Fort McMurray lies at the confluence of the Athabasca and Clearwater rivers. The Grant MacEwan Bridge over the Athabasca River connects the downtown with the new Thickwood subdivision (in the foreground). Fort McMurray's 2011 population exceeded 65,500.

Source: Photo by Gord McKenna

The operating cost of Native settlements is high because of their remote location (high cost of delivering public services) and small size (lack of economies of scale). With a negligible local tax base, local councils depend on transfer payments from territorial and provincial governments. Given rising costs of delivering existing services and the costs of building additional facilities, a major question facing governments is whether they can properly support these communities. Faced with increasing demand, the government of Nunavut has wondered if it should stop funding "weak" communities and concentrate resources on "healthy" ones. This issue boiled down to: "Should we close Clyde River?" As shown in Vignette 4.5, there is no easy answer to such a question. Thus far, Clyde River is still a functioning community, and its population grew to 934 by 2011, an increase of 14 per cent over its 2006 population.

**Vignette 4.5   Are "Have-Not" Native Settlements Doomed
or Cultural Survivors?**

A consultant's draft report on Clyde River in Nunavut suggested that the existence of this community might have to be "reviewed." According to the report, the problem in Clyde River is not unique and is found in many other Arctic communities with no economic activities other than hunting and public construction projects. Local tax revenues fall far short of the funds needed to provide local services. With high unemployment rates, a critical shortage of housing, and an insufficient tax base to pay for public services, increasing transfer payments from other levels of government are necessary to maintain a minimum quality of life in these communities. Unfortunately, the territorial government is hard-pressed to keep such funding at its current level and has little room to increase funding. Capital projects, such as sewer and water systems, provide temporary relief, but the cost of maintaining these systems becomes an additional burden on the community. A local Clyde River official, Mr Palluq, bitterly said of the consultant's report: "The guy writing this was getting $200 a day. And he says we should cut back employee benefits. Lots of these people have been working for the municipality for $12 an hour for 15 years on half-days." But cuts are coming and retiring hamlet employees may no longer receive a payment for unused sick leave. Clyde River's mayor, James Qillaq, unhappily said, "It seems like it's a slope and we're still sliding down."

Can Clyde River expand its economic base? In southern Canada, single-industry towns that have faced closure, notably Chemainus on Vancouver Island, have sometimes turned to the arts and tourism (Barnes and Hayter, 1992). In the case of Clyde River, this isolated Inuit community represents an ideal location for an Inuit culture and language centre; hence, such a centre, Piqqusilirivvik, was completed in 2010. Here, young Inuit from across Nunavut learn from Elders about culture, language, and the land/ice, thanks to grants from the federal and territorial governments (O'Neill, 2009).

Source: McKibbon (1999).

Labour Force

As a resource frontier, the economy of the Northwest Territories is heavily dependent on the resource industry, commodity prices, and global demand. Relatively wide fluctuations in its economy and employment situation are a painful but normal occurrence in resource frontiers. To that end, Employment and Development Canada (2013) reported that:

> The Northwest Territories (NWT) has continued to face difficulties in recent years, with employment declining in several key industries across most age groups, largely due to struggles in diamond mining. As a result of decreasing job opportunities, NWT saw a net outmigration of 2,100 people, mostly headed to Alberta, where opportunities are more widespread.

Resource frontiers such as the Canadian North have a particular pattern of industrial employment where the service or tertiary sector provides most employment. Such

Figure 4.7   Clyde River, Nunavut—Going, Going, Gone?

The challenge facing the Inuit community of Clyde River is typical of isolated Native communities. Clyde River, located on a fjord on Baffin Island, represents a cultural homeland for its residents, but the community lacks an economic base other than traditional seal hunting. Yet, its residents remain committed to the community, indicating the strong pull of place. Defying the odds, its population just keeps increasing—from 820 in 2006 to 934 in 2011.

Source: Dr. Ansgar Walk, used by Public License, Creative Commons Share Alike 2.5 Generic (CC BY-SA 2.5) at https://commons.wikimedia.org/wiki/File:Clyde_River_Community_1997-08-07.jpg

a structural pattern is due to the nature of their economies, which depend on the export of resources. In Canada's case, the state provides funds to ensure a degree of equality in terms of public services. This public policy is a cornerstone of Canada's social contract. In the Canadian North, geography and jurisdictional complexity exacerbate the need for a large public sector. Geography, with its combination of a vast space and few people, makes the delivery of public services expensive. In Nunavut, most residents live in very small communities where administration, education, and health services operate without the benefit of economies of scale found in larger centres. The principle of access to public services regardless of the size of the community is necessary but expensive. For this reason, the ratio of government employees to residents is much higher than in the rest of Canada. Another reason is that administrative bodies representing different levels of government and Aboriginal organizations often exist in the same community. While each has its own mandate, this jurisdictional arrangement provides a series of hierarchical layers of government for relatively few people living in urban centres. In Nunavut, for example, Nunavut Tunngavik Inc. (NTI) is the Inuit organization responsible for the implementation of the Nunavut Land Claim Agreement and, according to the Clyde River Protocol that formally outlines the relationship between NTI and the government of Nunavut, that responsibility involves regular consultation with government officials (Légaré, 2000). NTI's goal is to ensure that Inuit interests are expressed in legislation. The Tungavik Federation of Nunavut (later renamed Nunavut Tunngavik Inc.), for example, took the lead in a harvester support program (Wenzel, 2009).

Because labour force statistics are compiled by provinces and territories, a single set of employment figures for the North is not available. To throw some light on this subject, however, the labour force for the Northwest Territories can serve as a proxy for the Territorial North and a number of areas in the Provincial North. Its labour force has been adjusted to fit into three standard categories: primary, secondary, and tertiary employment (Table 4.10). As a result of so few economic opportunities in the North, those of working age are often classified as "not in the workforce."[9] The data in Table 4.10 represent employed workers. A summary of that table reveals:

- On first glance, given the dominating role of the resource industry in terms of value of production, it would seem logical that primary activities would be the major employer. This is not the case. Table 4.10 demonstrates the relatively weak employment position of the primary sector, comprising only 7.3 per cent of the total employed workers compared to the tertiary sector at 83.8 per cent, with public administration leading the way at 23.7 per cent of the total employed labour force.
- With tertiary employment dominant, the manufacturing activities found in the secondary sector are virtually non-existent, a pattern of employment structure common among resource frontiers and developing countries. In 2012, employment in manufacturing amounted to less than 1 per cent of the total employed

Table 4.10 Employment by Economic Sectors, Northwest Territories, 2012

Economic Sector	Subsector	Number of Employees	Percentage
Primary		1,800	7.3
	Agriculture, forestry, fishing, hunting	100	0.1
	Mining, oil and gas	1,700	7.2
Secondary		2,000	8.9
	Construction	1,500	6.6
	Manufacturing	200	0.9
	Utilities	300	1.4
Tertiary		18,900	83.8
	Education	2,100	9.1
	Health care and social assistance	2,300	10.0
	Retail trade	2,300	10.3
	Public administration	5,400	23.7
	Transport and warehousing	1,700	7.7
	Accommodation and food services	1,100	4.7
	Other	4,000	18.3
Total		22,700	100.0

Source: Adapted from Employment and Development Canada (2013: Table 2).

workers. In 2009, manufacturing in Yellowknife took a hit as two subsidized diamond processing facilities—Laurelton Diamonds and Arslanian Cutting Works NWT Ltd—closed (Employment and Development Canada, 2013).

Summing Up

For the next decade, the North's population will likely remain just under 1.5 million. At the same time, its ethnic composition will reflect a stronger Aboriginal component.

Two factors—a falling rate of natural increase among the Aboriginal population and an increase in net out-migration (including Aboriginal migrants)—suggest that a population decline may occur. A third factor is that resource projects tend to use air-commuting to supply labour to their remote mines and mega-construction jobs rather than create resource towns. In the following chapter, the involvement of local Indigenous workers is examined in the context of past mining operations and the most recent megaproject in the Arctic, the St Mary iron mine project.

Northern reality is not always pretty. Far too many Aboriginal communities are trapped in an ever-worsening economic quagmire involving a limited local tax base, few employment opportunities, and growing dependency on higher levels of government. In addition, their populations are growing rapidly and that spells trouble for public services and housing. Clyde River, for instance, typifies the dilemma facing such communities. Funds necessary to supply modern services, ranging from health care to education, to provide economic support for the unemployed, and to operate and maintain public services such as housing are far beyond the local tax base. Financial support from the government of Nunavut and ultimately the federal government is essential. The fiscal solution is beyond the capacity of northern governments. The choice is simple: either provide more transfer payments to local governments or suffer the consequences of social fallout from such Aboriginal communities.

On a closing note, population is more than numbers. Aboriginal culture, Native settlements, and the future for the Aboriginal northern population are significant political issues that reflect the struggle between culture and economics on the landscape of northern Canada. This raises the question: *How important is human diversity?* The popular anthropologist, Wade Davis, sees it as our greatest legacy. Davis coined the term **ethnosphere** to capture the notion of a global social web of human life:

> You might think of this social web of life as an "ethnosphere," a term perhaps best defined as the sum total of all thoughts and intuitions, myths and beliefs, ideas and inspirations brought into being by the human imagination since the dawn of consciousness. (Davis, 2009: 2)

At the centre of this web are the many languages and cultures, which, unfortunately, have dwindled in our economically and socially globalized world. In the North, Aboriginal languages and cultures have already been lost and more are on the edge of extinction.

Finally, the challenge for the twenty-first century is to tackle northern poverty found in many Aboriginal communities. One answer lies in redistributing northern wealth in a more equitable way. Decisions by the Supreme Court of Canada have opened a Pandora's box over control of ancestral lands and resource development, and thus the concept of resource-sharing. The basic question is: Should the provincial, territorial, and federal governments share revenue/royalties from such developments with Aboriginal organizations? Sharing of the resources is a basic theme in the remaining chapters in this book.

Challenge Questions

1. Why has the population of the Territorial North increased over the last decade or so while that of the Provincial North has declined?
2. Why is migration such a powerful factor in slowing the population growth in the North?
3. Could a case be made that the relocation of Inuit into settlements was a key factor in the socio-political development of the Inuit, which led to the creation of Nunavut?
4. What is the difference between population increase and natural increase?
5. What are the economic implications of a youthful population compared to an older population?
6. Why is the North's labour force so heavily employed in the service or tertiary sector?
7. Suggest some reasons why the North's population is so concentrated in small urban places.
8. Why do so many resource towns follow a boom/bust life cycle?
9. With Nunavut unable to fund its language program, should Canadians take measures to ensure the well-being of Inuktitut and other endangered Aboriginal languages?
10. Clyde River's population continues to increase, but can the community survive in the long run? With a very small tax base, its survival depends on continued funding from the government of Nunavut. As its population increases, however, so should its funding from Iqaluit. Is this a viable proposition for the Nunavut government?

Notes

1. Within territories and provinces, the statistical units employed by Statistics Canada may change over time. Territories and provinces have the right to ask for internal boundary changes to meet their particular statistical needs. Within territories and provinces, the largest geographic level of a statistical unit is the census subdivision.
2. A huge immigration was associated with the resource boom of the 1950s and 1960s. Ottawa played a role in that boom. In 1957, Prime Minister John Diefenbaker announced his "Northern Vision." Abandoning its earlier laissez-faire policy, Ottawa began to aggressively promote development, particularly in the Provincial North: highways were built to resource sites, and new administrative and resource towns were created.
3. The out-migration of northerners varies across the North by ethnicity and by region. Southerners who settle in the North often relocate to southern Canada within five years. This high turnover accelerates in mining towns. Downsizing or even mine closing usually translates into most miners and their families leaving the North. Aboriginal residents are just showing interest in relocating to southern cities and towns. Others (usually the more educated ones) are gravitating to northern regional centres or, in the Territorial North, to the three capital cities (but not to resource towns). Here, they seek employment within the public sector.
4. The question is how we can attribute the decline in the rate of natural increase to the Aboriginal population. First of all, we assume that the northern non-Aboriginal

population has a similar **crude birth rate** found in the national population, i.e., around 0.4 per cent annually, while the Aboriginal rate of natural increase would be 4 to 6 times higher, i.e., in the range of 1.6 to 2.4 per cent annually. Since ethnicity is not recorded at the time of birth, to calculate such high rates among the Aboriginal population we assume there is a relationship between the percentage of Aboriginal population in the territories and the resultant rate of natural increase. For instance, slightly over half of the NWT population (51.9 per cent) indicated Aboriginal ancestry in the 2011 National Household Survey. Thus, its rate of natural increase for 2011 is composed of an estimated Aboriginal rate of 2.0 plus an estimated rate of 0.4 for the rest of the population, i.e., 2.0 + 0.4 = 2.4/2 or 1.2. Similarly for Nunavut, its 2.0 rate of natural increase is explained by the percentage of Aboriginal population (86.3 per cent), i.e., 0.86 (2.0) + 0.14 (0.4) = 1.78. A case for a higher crude birth rate for the Inuit is supported by the 2011 age distribution (Statistics Canada, 2013c). By taking the youngest **cohort** (0–4 years), a proxy for the crude birth rate is created. The percentage of those in this cohort for First Nations was 10.7, Métis 7.7, and Inuit 12.1. The estimated result for Nunavut of 1.95 for the rate of natural increase is close to the official figure of 2.0 [0.86 (2.2) + 0.14 (0.4) = 1.95]. Yukon provides the weakest case. Its rate of natural increase is 0.67 while our calculations reach only 0.49: 0.23 (2.0) + 0.77 (.04) = 0.49.

5. The cost of living and doing business in northern centres tends to increase with distance from southern cities that supply most foodstuffs and building materials. The highest cost of living and doing business is found in places where air transportation is the only means of reaching southern cities. Also, the costs involved in constructing and maintaining buildings are well above similar construction and maintenance costs in southern cities because of the long, cold winter and the presence of permafrost.

6. In the 1950s, Aboriginal Canadians were relocated to small settlements where they traditionally traded their furs. In this text, these tiny villages are classified as Native settlements. Within a generation, some moved to regional centres for a variety of reasons, including specialized health services and employment opportunities. Most migrants were drawn from the growing number of young, more educated Aboriginal Canadians. In regional centres, many are now employed by Aboriginal organizations and government agencies.

7. In terms of its potential labour force, the North has very few professional and skilled workers, especially among the Aboriginal population. For that reason, most skilled workers are hired in southern Canada and brought to northern worksites. Since these workers face much higher living costs and a lack of urban amenities found in southern centres, they will not move to the North unless employers (companies and governments) pay high wages and offer incentives. Often, these incentives include subsidized housing and an allowance to travel to southern Canada. The federal government provides several additional incentives, including a northern isolation allowance for its employees and a northern personal tax deduction for all northern employees and business people. Since northern communities vary widely in terms of their isolation, determining who should get these incentives is difficult. As well, the degree of isolation could change with a highway reaching a community. Northern isolation allowance payments provide cash compensation. Revenue Canada (now the Canada Customs and Revenue Agency) has used Hamelin's concept of nordicity to create two northern zones. Places in Zone A are more isolated than locations in Zone B, so that individuals in Zone A filing personal income tax receive more tax relief than those in Zone B. All places in Yukon, the Northwest Territories, and Nunavut fall into Zone A. Zone B corresponds to places in the Provincial North, including resource towns. Examples include Chibougamau in Quebec, Red Lake in Ontario, The Pas in Manitoba, La Ronge in Saskatchewan, Fort McMurray in Alberta, and Tumbler Ridge in British Columbia.

8. With huge coal reserves in the nearby mountains, Tumbler Ridge was created as a planned resource community in 1981. Situated in the eastern flank of the Rocky Mountains, the town's future seemed secure because the coal companies had long-term contracts to supply coal to Japanese steel mills. Yet, the global economy slid into a recession in the 1980s, causing the demand for coal to drop dramatically and appearing to spell an end to Tumbler Ridge (Halseth and Sullivan, 2002). China, however, needed coal, and by the early twenty-first century, coal mines in northeast BC were in full production, giving Tumbler Ridge a second chance. Tumbler Ridge's population changes reflect its ups and downs as a resource town. In 1986, Tumbler Ridge had 4,566 residents and the town reached a population peak in 1991 with 4,794 people. Over the next 10 years, the population declined, reaching a low of 1,851 people in 2001. It rebounded to 2,454 in 2006 due largely to the restart of coal mining (Statistics Canada, 2007g).

9. The potential labour force, that is, those over 15 years of age and under 65 years of age, has two different employment characteristics in much of the North and certainly in the Northwest Territories. On the one hand, non-Aboriginal workers have much higher employment rates than Aboriginal workers. On the other hand, Aboriginal men and women have two major barriers: (1) often they do not have the job experience or training necessary to qualify for jobs; (2) they are often not able to access the job market because they reside in places where few jobs are available. For example, in 2014, 7,400 persons, or 23 per cent of the NWT potential labour force, were designated as "not in the labour force," with the vast majority residing in isolated Aboriginal communities (NWT Bureau of Statistics, 2014: 20). In such Aboriginal communities, it is a vicious circle—little incentive and opportunity exist to complete high school, attend trade schools, and become part of the skilled labour force, and consequently the people remain trapped. In sum, the economic well-being of the territorial economies varies and so does the unemployment rate. In October 2013, this rate varied: 4.9 per cent in Yukon, 8.1 in Northwest Territories, and 15.6 in Nunavut (Statistics Canada, 2013e: Table 282-0100).

References and Selected Reading

Barnes, T., and R. Hayter. 1992. "'The Little Town That Did': Flexible Accumulation and the Community Response in Chemainus, British Columbia." *Regional Studies* 26: 647–67.

Bone, Robert M. 1998. "Resource Towns in the Mackenzie Basin." *Cahiers de Géographie du Québec* 42, 116: 249–56.

————. 2006. "Inuit Research Comes to the Fore." In Jerry P. White, Susan Wingert, Dan Beavon, and Paul Maxim, eds, *Aboriginal Policy Research: Moving Forward, Making a Difference*, vol. 3. Toronto: Thompson Educational Publishing.

————. 2012. *The Canadian North: Issues and Challenges*, 4th edn. Toronto: Oxford University Press.

Canadian Press. 2015. "Flooding Forces Full Evacuation of Kashechewan." *National Post*, 27 Apr. http://news.nationalpost.com/news/canada/flooding-forces-full-evacuation-of-kashechewan-more-than-1800-displaced-from-remote-james-bay-community.

CBC News. 2006. "Kashechewan: Water Crisis in Northern Ontario." 9 Nov. www.cbc.ca/news/background/aboriginals/kashechewan.html.

————. 2007. "Ottawa Nixes Relocation for Flood-Prone Kashechewan." 30 July. www.cbc.ca/canada/ottawa/story/2007/07/29/kashechewan-deal.html#skip300x250.

Davis, Wade. 2009. *The Wayfinders: Why Ancient Wisdom Matters in the Modern World*. Toronto: House of Anansi Press.

Department of Finance, Canada. 2013. *Federal Support to Provinces and Territories*. 17 Dec. http://www.fin.gc.ca/fedprov/mtp-eng.asp.

Dolphin, Myles. 2014. "Words of Praise for Okalik." *Northern News Service Online*. 28 Apr. http://www.nnsl.com/frames/newspapers/2014-04/apr28_14ok.html.

Employment and Development Canada. 2013. "Change in Employment by Industry in Northwest Territories between 2007 and 2012." *Labour Market Information.* Table 2.17 Oct. http://www.esdc. gc.ca/eng/jobs/lmi/publications/e-scan/nwt_nt_yt/ts-escan-201303.pdf.

Freeman, Milton M.R. 1971. "The Significance of Demographic Changes Occurring in the Canadian East Arctic." *Anthropologica* 13, 1 and 2: 215–37.

Halseth, Greg. 2010. "Understanding and Transforming a Staples-based Economy: Place-based Development in Northern British Columbia, Canada." In G. Halseth, S. Markey, and D. Bruce, eds, *The Next Rural Economies: Constructing Rural Place in Global Economies.* Wallingford, UK: CAB International.

————— and Lana Sullivan. 2002. *Building Community in an Instant Town: A Social Geography of Mackenzie and Tumbler Ridge, British Columbia.* Prince George: University of Northern British Columbia Press.

Herkes, J., J. Mooney, and H. Smith. 2013. *Exploring Residency in Yukon.* A report prepared by Ecofor Consultants for the Yukon government. http://economics.gov.yk.ca/Files/2013/ ResidencyReport.pdf.

Irwin, Colin. 1988. *Lords of the Arctic: Wards of the State.* Special edition of *Inungnut.* Rankin Inlet: Keewatin Inuit Association.

Kendall, Glen. 1992. "Mine Closures and Worker Adjustment: The Case of Pine Point." In Cecily Neil, Markku Tykklainen, and John Bradbury, eds, *Coping with Closure: An International Comparison of Mine Town Experiences,* ch. 6. London: Routledge.

Keough, S.B. 2013. "Examining the Cultural Imprint of Newfoundlanders in Fort McMurray, Alberta." *Focus on Geography* 56: 23–31. doi: 10.1111/foge.12008.

Kishigami, Nobuhiro. 1999. "Why Do Inuit Move to Montreal? A Research Note on Urban Inuit." *Études/ Inuit Studies* 23, 1 and 2: 221–7.

—————. 2015. "Low-income and Homeless Inuit in Montreal, Canada: Report of a 2012 Research." *Bulletin of the National Museum of Ethnology* 39, 4: 575–624.

Légaré, André. 2000. "La Nunavut Tunngavik Inc.: Un examen de ses activités et de sa structure adminis- trative." *Études/Inuit/Studies* 24, 1: 197–224.

McKibbon, Sean. 1999. "Report: Clyde River's Very Existence Is Questionable." *Nunatsiaq News.* http:// www.nunatsiaqonline.ca/archives/nunavut991030/nvt91029_01.html.

Mahoney, Jill. 2002. "Where the Jobs Are." *Globe and Mail,* 12 Mar., A6.

Martel, L., and Alain Bélanger. 1999. "An Analysis of the Change in Dependency-Free Life Expectancy in Canada between 1986 and 1996." In A. Bélanger, *Report on the Demographic Situation in Canada 1998–1999,* 164–86. Statistics Canada Catalogue no. 91-209. Ottawa: Minister of Industry.

Newman, James L., and Gordon E. Matzke. 1984. *Population: Patterns, Dynamics, and Prospects.* Englewood Cliffs, NJ: Prentice-Hall.

Northwest Territories (NWT) Bureau of Statistics. 2006. NWT Labour Supply Presentation, 5 July. http://www.statsnwt.ca/labour-income/labour-supply/.

—————. 2008. *Newstats: Labour Market Activities, 2006 Census.* http://www.statsnwt.ca/census/2006/ Labour%20Force_2006.pdf.

—————. 2014. *Statistics Quarterly* 36, 1 (Mar.). http://www.statsnwt.ca/publications/statistics- quarterly/sqmar2014.pdf.

O'Neil, Katherine. 2009. "Traditional Skills To Be Taught at Nunavut's New Cultural School." *Globalcampus,* 9 Aug. http://www.theglobeandmail.com/news/national/traditional-skills-to-be-taught-at- nunavuts-new-cultural-school/article12011.

Petrovich, Curt. 1990. "The Best Reason to Come North . . . Or Is It?" *Arctic Circle* 1, 1: 36–41.

Pleizier, Christina. 2006. Personal communication, 13 Apr., Indian and Northern Affairs Canada.

Regional Municipality of Wood Buffalo. 2015. "Census 2015: The Count Is On." http://www.rmwb.ca.

Saku, James C. 1999. "Aboriginal Census Data in Canada: A Research Note." *Journal of Native Studies* 19, 2: 365–79.

Saul, John Ralston. 2014. *The Comeback: How Aboriginals Are Reclaiming Power and Influence.* Toronto: Viking.

Statistics Canada. 2004. *Aboriginal Identity Population, 2001 Counts, for Canada, Provinces and Territories— 20% Sample Data.* www12.statcan.ca/English/census01/products/highlight/Aboriginal/Page.cfm? Lang=E&Geo=PR&View=1a&Table=1&StartRec=1&Sort=2&B1=Counts01&B2=Total.

————. 2007a. *Population and Dwelling Counts, for Canada, Provinces and Territories, Census Divisions and Census Subdivisions (Municipalities), 2006 and 2001 Censuses—100% data.* www12.statcan.ca/english/census06/data/popdwell/Filter.cfm?T=302&S=1&O=A.

————. 2007b. *Births and Birth Rate by Province and Territory.* www40.statcan.gc.ca/l01/cst01/demo04b.htm.

————. 2007c. *Deaths and Death Rate by Province and Territory.* www40.statcan.gc.ca/l01/cst01/demo07b.htm.

————. 2007d. *Portrait of the Canadian Population in 2006, by Age and Sex: Provincial/Territorial Populations by Age and Sex.* www12.statcan.ca/english/census06/analysis/agesex/ProvTerr7.cfm.

————. 2007e. *Age Groups and Sex for the Population of Canada, Provinces, Territories, Census Divisions and Census Subdivisions, 2006 Census, 100% Data.* www.statcan.ca/bsolc/english/bsolc?catno=97-551-XWE2006013.

————. 2007f. "Whitehorse City." *Community Profiles 2006.* www12.statcan.ca/english/census06/data/profiles/community/Details/Page.cfm?Lang=E&Geo1=CSD&Code1=6001009&Geo2=PR&Code2=60&Data=Count&SearchText=whitehorse&SearchType=Begins&SearchPR=60&B1=All&Custom.

————. 2007g. "Tumbler Ridge." *Community Profiles 2006.* www12.statcan.ca/english/census06/data/profiles/community/Search/SearchForm_Results.cfm?Lang=E.

————. 2008a. "Table 8: Size and Growth of the Inuit Population, Canada and Regions, 1996 and 2006." www12.statcan.ca/english/census06/analysis/aboriginal/tables.cfm.

————. 2008b. "Aboriginal Peoples: Highlight Tables, 2006 Census." www12.statcan.ca/english/census06/data/highlights/Aboriginal/index.cfm?Lang=E.

————. 2010a. "Canada's Population Estimates: Third Quarter 2010." *The Daily,* 22 Dec. www.statcan.gc.ca/daily-quotidien/101222/dq101222a-eng.htm.

————. 2010b. "Deaths." *The Daily,* 23 Feb. www.statcan.gc.ca/daily-quotidien/100223/dq100223a-eng.htm.

————. 2010c. "Quarterly Demographic Estimates—June to March 2010." *Canada's Population Estimates.* www.statcan.gc.ca/pub/91-002-x/91-002-x2010001-eng.htm.

————. 2010d. *Canada Yearbook:* Table 24.9, "Birth Rate, by Province and Territory, 2003/2004 to 2008/2009." www.statcan.gc.ca/pub/11-402-x/2010000/chap/pop/tbl/tbl09-eng.htm.

————. 2011. *Annual Demographic Estimates: Subprovincial Areas, 2005–2010.* 3 Feb. www.statcan.gc.ca/pub/91-214-x/2009000/tablelist-listetableaux3-eng.htm.

————. 2013a. "Population and Dwelling Counts, for Canada, Provinces and Territories, 2011 and 2006 Censuses." http://www12.statcan.gc.ca/census-recensement/2011/dp-pd/hlt-fst/pd-pl/Table-Tableau.cfm?LANG=Eng&T=101&S=50&O=A.

————. 2013b. "Factors of Demographic Growth, 1982/1983 to 2012/2013." *Demographic Estimates: Canada, Provinces and Territories.* 25 Nov. http://www.statcan.gc.ca/pub/91-215-x/2013002/part-partie1-eng.htm.

————. 2013c. "Age Distribution and Median Age for Selected Aboriginal Identity Categories, Canada, 2011." National Household Survey. Table 4. 24 Apr. http://www12.statcan.gc.ca/nhs-enm/2011/as-sa/99-011-x/2011001/tbl/tbl04-eng.cfm.

————. 2013d. *Population by Broad Age Groups and Sex, 2011 Counts for Both Sexes, for Canada and Census Divisions.* 19 July. http://www12.statcan.gc.ca/census-recensement/2011/dp-pd/hlt-fst/as-sa/Pages/highlight.cfm?TabID=1&Lang=E&Asc=1&PRCode=01&OrderBy=1&Sex=1&View=1&tableID=21&queryID=4.

————. 2013e. "Labour Force Characteristics, Unadjusted, by Territories (3 month moving average)." Catalogue no. 71-001-XIE. 8 Nov. http://www.statcan.gc.ca/tables-tableaux/sum-som/l01/cst01/lfss06-eng.htm.

————. 2013f. "Table 7: Population Counts and Proportion of Persons with an Aboriginal Identity by Province and Territory of Residence, Canada, 2006 and 2031, Four Projection Scenarios." *Population Projections by Aboriginal Identity in Canada, 2006 to 2031.* 13 May. http://www.statcan.gc.ca/pub/91-552-x/2011001/tbl/tbl07-eng.htm.

————. 2013g. "Table 2: Number and Distribution of the Population Reporting an Aboriginal Identity and Percentage of Aboriginal People in the Population, Canada, Provinces and Territories,

2011." National Household Survey. 24 Apr. http://www12.statcan.gc.ca/nhs-enm/2011/as-sa/ 99-011-x/2011001/tbl/tbl02-eng.cfm.

————. 2013h. Prince Albert, CY, Saskatchewan (Code 4715066) (table). *Aboriginal Population Profile, 2011*. National Household Survey. Statistics Canada Catalogue no. 99-011-X2011007. 13 Nov. Accessed 3 Sept. 2014. http://www12.statcan.gc.ca/nhs-enm/2011/dp-pd/aprof/index. cfm?Lang=E.

————. 2013i. Meadow Lake, CY, Saskatchewan (Code 4717052) (table). *Aboriginal Population Profile, 2011*. National Household Survey. Statistics Canada Catalogue no. 99-011-X2011007. 13 Nov. Accessed 3 Sept. 2014. http://www12.statcan.gc.ca/nhs-enm/2011/dp-pd/aprof/index.cfm?Lang=E.

————. 2013j. Saskatoon, CY, Saskatchewan (Code 4711066) (table). *Aboriginal Population Profile. 2011*. National Household Survey. Statistics Canada Catalogue no. 99-011-X2011007. 13 Nov. Accessed 3 Sept. 2014. http://www12.statcan.gc.ca/nhs-enm/2011/dp-pd/aprof/index.cfm?Lang=E.

————. 2014a. *Aboriginal Peoples in Canada: First Nations People, Métis and Inuit*. National Household Survey. Statistics Canada Catalogue no. 99-011-4. 28 Mar. http://www12.statcan.gc.ca/ nhs-enm/2011/as-sa/99-011-x/99-011-x2011001-eng.cfm#a5.

————. 2014b. *The Canadian Population in 2011: Age & Sex*. Statistics Canada Catalogue no. 98-311-x-2011001. 14 Jan. http://www12.statcan.gc.ca/census-recensement/2011/as-sa/98-311-x/ 98-311-x2011001-eng.cfm#a3.

————. 2014c. "Census Profile." 9 May. http://www12.statcan.gc.ca/census-recensement/2011/ dp-pd/prof/index.cfm?Lang=E.

Storey, Keith. 2009. "Help Wanted: Demographics, Labor Supply and Economic Change in Newfoundland and Labrador." Presentation at "Challenged by Demography: A NORA Conference on the Demographic Challenges of the North Atlantic Region," Alta, Norway, 20 Oct.

Weinstein, Jay A. 1976. *Demographic Transition and Social Change*. Morristown, NJ: General Learning Press.

Wenzel, George. 2009. "Subsistence and Conservation Hunting: A Nunavut Case Study", In M. Freeman and L. Foote, eds, *Inuit, Polar Bears and Sustainable Use: Local, National and International Perspectives*. Edmonton: CCI Press: 51–64.

Yukon Bureau of Statistics. 2010. *Yukon Migration Patterns, 1999–2008*. www.eco.gov.yk.ca/stats/pdf/ migration2008.pdf.

- 5 -

Resource Development, Megaprojects, and Northern Benefits

The Canadian North is an essential part of the global economy. The lifeblood of the North's economy is based on the export of its energy and mineral production. Yet, because of its remote location, permafrost, and limited transportation infrastructure, the cost of resource development in the North is much higher than in southern Canada. The result is that some projects, such as mineral exploitation in northern Ontario's Ring of Fire, are stymied because of the absence of a transportation infrastructure while existing megaprojects generate few spinoffs for the North. Overcoming these obstacles to northern development and capturing more spinoff benefits are critical to the North's long-term economic growth and diversification. Accommodating Aboriginal peoples in the resource development scenario and the concept of northern benefits is a related issue.

But what about the future for non-renewable resource development? The Conference Board of Canada predicts a rosy future for resource development in Canada's North over the next decade: "The economic forecast predicts the doubling of northern metal and non-metallic mineral output, from $4.4 billion in 2011–12 to $8.5 billion in 2020" (Rhéaume and Caron-Vuotari, 2013: 4). This, of course, assumes an ever-increasing global demand for energy and commodities, plus favourable prices for these commodities. While this prediction may well come true, past history of commodity prices and the global business cycle suggest that it could be a bumpy ride indeed.

Megaprojects undertaken by large international corporations dominate the economic landscape. The oil sands are but one example where the vast amounts of capital required to extract the resource limit the active players to the very largest companies in the world. Further North, especially in the Arctic, costs of energy and mineral developments are much higher than in other regions of Canada, partly because of high labour and transportation costs and partly because the demanding physical geography increases the cost of construction and the maintenance of buildings and equipment. For those reasons, projects are large-scale undertakings based on world-class deposits where companies can afford to have a long-term horizon. Not surprisingly, then, multinational companies and Crown corporations are the leading firms involved in the northern economy. Megaprojects, some think, are the engine of northern regional growth, but these huge industrial efforts create a very narrow economic base subject to a construction version of the boom-and-bust cycle. As well, the resource sector accounts for only a small portion of those employed in the North.

Without a doubt, megaprojects do accelerate northern industrial growth for a short period of time, but such projects do little to broaden the frontier economy; and, with a few exceptions, they fail to strengthen the local Aboriginal economies over the long haul. Since the nature of regional development has a much broader agenda than simply increasing economic output, it follows that megaprojects are best described as "engines of economic growth" rather than "engines of regional development."

Unlike southern Canada, the North has several structural weaknesses that severely limit a widening of the northern economy as a result of spinoff effects from megaprojects. Three key weaknesses are:

1. Large economic leakage to southern Canada, thus limiting the positive economic benefits of megaprojects to the North.
2. Restricted northern transportation network with its few highways and heavy reliance on air and water transportation, thus curtailing the spatial spinoffs of economic benefits to northern businesses and its scattered workforce.
3. Inexperienced workers and few trained trades people force companies to import both semi-skilled and skilled workers from the south, thus hampering the participation of the largely Aboriginal **potential labour force.**

Can these weaknesses be overcome or at least modified? To a large degree, the solution lies in a trained workforce plus a massive shift in both the private and public sectors of the economy. For example, public investment in northern infrastructure and education needs to go well beyond that associated with the "Northern Vision" of the Diefenbaker government. Second, the corporate sector must adapt its behaviour and thus become a more socially responsible partner in its northern operations with regard to the First Nations, Métis, and Inuit workforce. Such a shift to on-the-job training has been begun by a few companies but much more needs to happen. A brief synopsis of the past century provides some positive findings and, if the trend continues, hope for a brighter future. To simplify matters, this synopsis is divided into two time periods: then and now.

Then

Private corporations shaped the direction of resource development for much of the twentieth century. At the same time, the federal government was content to leave development of the North to these corporations. This laissez-faire approach minimized Ottawa's investment and left two issues under the carpet. One was the North's transportation system and the other was the settling of land claims.

At this time, provincial governments paid no attention to Aboriginal title. Crown land was used by Aboriginal peoples to hunt and fish at the pleasure of the Crown. The concept of Aboriginal title had not yet taken on legal status so mining and timber licences were easily granted to corporations. During this period, Aboriginal peoples were still living on the land and environmental organizations were just forming. Consequently, corporations paid little attention to the environmental and social consequences of their projects and freely discharged their wastes into streams and the atmosphere. An excellent example began in 1947 when US iron and steel interests funded the Iron Ore Company of Canada to develop the vast iron deposits in northern Quebec and Labrador by building a rail–sea transportation system capable of supplying iron ore to American steel plants.[1] By the early 1950s, a massive iron mining and transportation system was completed, but neither the Innu and Naskapi tribes nor their lands were respected. Twenty years later, the federal government passed a few basic rules and regulations designed to protect the environment from excessive damage (see Chapter 6).

Now

By the mid-twentieth century, Canadian society and its governments took a different view of resource development. First, Ottawa abandoned its laissez-faire policy towards the North and began an active role in stimulating resource development by building highways to resources. But beyond this economic thrust, a change in attitude towards

Figure 5.1   Then and Now: Who Is at the Decision-making Table?

While political cartoonist Bruce MacKinnon has exaggerated the shift in those at the decision-making table for new oil exploration projects, Aboriginal leaders have more to say about proposed resource development on their ancestral lands than ever before. This shift could result in a more equitable sharing of resource wealth between Aboriginal northerners and resource companies.

Source: Bruce Mackinnon/Artizans.com

environmental and social consequences of resource development emerged. No longer did society consider only the economic benefits of northern projects; their costs—to the environment and to Aboriginal peoples—became part of the conversation. The **James Bay and Northern Quebec Agreement** in 1975 was a landmark event. By the twenty-first century, proposed resource projects had to pass rigorous environmental assessment and proponents were required to consult with and in some cases gain approval of Aboriginal groups directly affected by the project. The Mary River Project in northern Baffin Island, while similar in many ways to the construction effort of the Iron Ore Company of Canada in northern Quebec in the late 1940s, differed strikingly in its legal obligations for the company to prepare an environment assessment statement and to strike an impact and benefits agreement with the Inuit of Baffin Island.

Vignette 5.1   Transportation: The North's Gordian Knot

Economic development of Canada's North has long been hampered by the absence of a modern transportation system. The "Northern Vision," but particularly the Roads to Resources program of the Diefenbaker government, was one attempt at severing this knot. In 1958, his government made the historic decision to build a 671-kilometre highway across the permanently frozen terrain from Dawson City to Inuvik. This costly undertaking faced many physical challenges, including the need to place the highway on a gravel berm so as to insulate the permafrost and thus prevent the ice in the ground from melting and causing subsidence of the highway.

But in such a huge area of permafrost, the cost of providing a modern transportation system is beyond the financial capacity of Ottawa and its provincial counterparts. In a few cases, world-class mineral deposits have caused private companies to tackle this challenge. The most recent one, Baffinland, is investing some $4 billion to develop the Mary River iron mine and its transportation system. This system involves an enormous investment in a railroad (now postponed until the 2020s because of plunging iron ore prices), port facilities, and vessels able to plough through the Arctic ice to carry high-grade iron ore from the mine site in northern Baffin Island to various European Union destinations. In most cases, however, private companies are attracted to precious minerals, such as diamonds, gold, silver, and platinum, which cost much less to transport to global markets. For this reason, most mining operations beyond Canada's transportation network extract highly valued minerals where air links and **ice roads** keep transportation investments and operating costs relatively low.

Energy has its own transportation challenges. For hydroelectric generating plants, high-voltage transmission lines from northern BC, Manitoba, and Quebec take surplus power to the United States. Oil takes a different transportation approach—pipelines and rail. Pipelines offer the lowest cost per barrel of oil but the destination is fixed. Until now, oil and gas pipelines have taken these products either to central Canada or to the United States. The Alliance Pipeline, completed in 2000, carries natural gas from Fort St John in northeast BC to Chicago. The one exception is the Kinder Morgan pipeline from the oil sands to Vancouver, where the bitumen is loaded onto supertankers for foreign markets. As more bitumen is produced in Alberta's oil sands, oil companies are looking at Asian markets where prices are higher than in the US. Plans for oil and gas pipelines to both the west and east coasts have been announced, but construction, if it occurs, is still in the future.

Supreme Court of Canada

While the Supreme Court of Canada recognized Aboriginal title in 1973 in the *Calder* case, the Supreme Court justices realized that each project proposed on lands with Aboriginal title could have different levels of impact. Therefore, the Court called on the interested parties to negotiate solutions. These solutions have been slow to come. For example, Aboriginal leaders are pressing for a share of royalties from megaprojects that occur on their traditional land base. A recent example is the Meadowbank gold mine in Nunavut, which lies within the traditional Inuit lands but outside the territory claimed under the Nunavut Land Claims Agreement. These lands are defined as Crown lands, where mineral resources fall under the jurisdiction of the federal government. On 20 June 2010, officials of the company, Agnico-Eagle, pledged to respect the land and its people at a ceremony at Baker Lake. Leading Inuit dignitaries participated in this ceremony. Even though the life of this mine is estimated at 19 years, Jose Kusugak, president of the Kivalliq Inuit Association, declared that Baker Lake is now "the happiest community in Nunavut because of the hope generated by Agnico-Eagle's Meadowbank gold mine," while the mayor of Baker Lake, David Aksawnee, said, "I see a lot of people working on this site. They're well-fed now and I'm happy to see that" (Bell, 2010). This "happiness" was made possible by three factors:

1. The Nunavut Land Claims Agreement (NLCA) includes a provision for a 5 per cent share of subsurface royalties to flow to the government of Nunavut.
2. An impact and benefits agreement between the company and Kivalliq Inuit Association spells out the company's contractual obligation to provide employment, training, business opportunities, and funding arrangements.
3. The company recognizes that mining on Inuit traditional land entails economic and social obligations so that it must demonstrate respect for the land and its people.

Happiness for Baker Lake soon turned to some "sadness" for the company. While this Quebec-based firm accepted the challenge of turning a chronically underemployed Inuit workforce with no mining experience into productive workers, it was no easy task. If successful, the seven Kivalliq communities would have an economic base and the company would lower its need to fly in southern workers. The company was banking on an on-the-job training program and a steady paycheque to ensure a stable Inuit workforce. By 2011, Inuit workers were bringing $16 million home from their two-weeks-on and two-weeks-off rotations (Braden, 2013). But the local workers had a high rate of "no-show" when the time came to board the aircraft for another two weeks at the camp. While the company recognized that regular employment was not well embedded in Inuit culture, they did not give up on local hires. Instead, the company made its training program more demanding; and, most importantly, it helped the new workers adjust to the time away from home, personal finances, the pace of camp life, and safety issues. It reduced turnover and increased reliability. The mine manager, Dominique Girard, reported that "Inuit turnover of 25 per cent in 2011 dropped to 16 per cent in 2012 and so far this year, a very impressive eight per cent" (Braden, 2013). Perhaps Agnico-Eagle is on the right path, where the company makes a "social commitment" to ensure participation by an ill-prepared Aboriginal labour force. This

---

Vignette 5.2 Corporate Responsibility

In 1939, Peter Drucker wondered how the vast power held by multinational corporations could be justified within the tradition of Western political philosophy. As a leading social philosopher, Drucker recognized that the corporation was not inherently conservative, as were the state, the church, and the army. On the contrary, the corporation, especially the multinational corporation, must constantly transform itself to meet new circumstances. Drucker believed that the corporation was the ideal candidate to lead a rapidly changing modern industrial society. Indeed, if corporations refuse to take responsibilities for solving social and environmental problems besetting the global community, who else will have the power to do so?

---

corporate adjustment to a difficult social situation may have been just what Peter Drucker had in mind for corporate responsibility (Vignette 5.2). Yet, over 70 years earlier, **North Rankin nickel mine** employed Inuit who had just moved from the land to settlements. So what happened to **corporate memory**?

The Role of Megaprojects in Northern Development

Megaprojects are huge industrial undertakings that transform small areas of the northern hinterland into industrial nodes with the sole purpose of exporting a commodity to other parts of the world. In one sense, megaprojects hold the key for resource development in the Canadian North. Much time, however, is needed for the seeds of diversification to take root in the North's seemingly unyielding geography. Megaprojects must overcome construction in remote sites where permafrost and long cold winters add to their challenges. Yet, these two physical barriers can be turned into a transportation advantage. Winter (or ice) roads, for example, provide a low-cost means of trucking bulky supplies and diesel fuel to remote sites.

Given all these challenges, only governments and large corporations, especially multinationals, have the capacity to finance, design, construct, and operate megaprojects. Multinational corporations also have excellent political connections that often lead to various forms of government support, including price supports, tax concessions, and outright grants. Provincial Crown corporations are another important business entity operating in the North. Crown corporations do have a substantial advantage in that they have a monopoly. In the case of hydroelectricity, they have public funding to develop electrical power, sell it at set prices above costs, and export surplus power to other provinces and US states.

The full beneficial impact of megaprojects on the North (see Table 5.1) is lost through a high level of economic leakage, thus limiting the effect on local and regional growth. During the construction phase, business opportunities abound, but the small northern business sector cannot satisfy the needs of the construction firms, and economic spinoffs generated by the construction are lost to firms in southern Canada. Equipment for the oil sands projects is built in southern Canada or in foreign centres. The explanation is simple: the North, and in some cases Canada, does not have the manufacturing capacity. Similarly, the North does not have a skilled labour force and companies employ many southern workers. Another form of leakage stems from air commuting, which results in southern workers and their families spending their wages in their home communities in southern Canada. The net result is that virtually all of the multiplier effects derived from wages accrue to southern Canada. The same pattern holds true for business contracts awarded to firms for specific equipment and services. In the case of the Norman Wells Project, the major contract to build the pipeline was awarded to an established firm in Edmonton, Interprovincial Pipelines Ltd.

Megaprojects, then, are not an economic panacea for the North. Economic leakage is one reason. Negative side effects provide other reasons. In the past, some have had disruptive impacts on Aboriginal societies, and most such projects pose a serious threat to the environment. The link between industrial waste products and Aboriginal peoples occurs in the consumption of country food. Unfortunately, the toxic wastes enter the rivers and streams, causing fish and wildlife to ingest them, and then when Aboriginal peoples consume the fish and game their health is jeopardized. Past examples in the twentieth century, such as the mercury contamination of the English–Wabigoon River system by a pulp plant at Dryden, Ontario, and the resulting devastating impact on the residents of Grassy Narrows and Whitedog reserves, should alert corporations in the twenty-first century to the dangers of toxic pollution (see Chapter 7 for a more detailed account).

But is the "corporate memory" working? A recent example concerns waste effluence entering the Athabasca River from the oil sands operations. The "treated" effluence from **tailings ponds** is released into the Athabasca River. Largely anecdotal evidence suggested for some time that this effluence had negatively affected aquatic life and the health of Aboriginal people at Fort Chipewyan, where the river flows into Lake Athabasca, but this evidence had not been substantiated through scientific study (Kelly et al., 2009; University of Alberta, 2010; CBC News, 2010b). Now, however, the impact on the environment and wildlife is documented in a recent study, *Environmental and Human Health Implications of the Athabasca Oil Sands*. This report is the first comprehensive study of the possible link between the oil sands and the health of Fort Chipewyan residents. Much of the illness experienced by Fort Chipewyan residents— including high rates of cancer—is related to their consumption of country food contaminated by industrial wastes. In a press release, the head of this scientific study, Dr Stéphane McLachlan (2014a), stated:

> As heavy metals and carcinogenic hydrocarbons flow downstream from bitumen extraction sites, traditional, wild-caught foods like moose and fish becoming increasingly risky for consumption, thus jeopardizing a tradition of living off the land that has been preserved by Cree and Dené people for thousands of years. But if Fort Chipewyan residents choose to avoid toxins by purchasing store-bought foods, they must pay exceptionally high prices for low-quality ingredients or highly processed items.

Megaprojects have two phases: construction and operation. Of the two, the construction phase places the greatest pressure on the affected community and surrounding area.

The construction phrase creates an intense work atmosphere where thousands of workers live in trailers, work long hours, and take home big paycheques. The affected community is under immense pressure from the constant noise from the equipment and trucks, from the construction dust and air pollution, from the number of new families and their demand for community services, and from the shortage of housing. During construction, the common characteristics of megaprojects are:

- Construction costs alone exceed $1 billion.
- The construction period exceeds two years.
- Substantial short-term employment and business opportunities are generated.
- Outside labour and businesses play a central role because local workers and firms are unable to satisfy demands from the northern megaproject.
- The sheer size of the construction workforce could overwhelm existing community services and infrastructure, hence the emphasis on flying southern workers to the site.
- Construction work changes and sometimes damages the local environment.
- The local transportation system is expanded to meet the needs of the megaproject.

A basic question facing the North is that the very nature of the global economy emphasizes corporate profits, not environmental and social responsibilities. Given the fragile nature of the northern environment and the presence of large numbers of Aboriginal people who are just finding their footing in the market economy, do these firms have an obligation to encourage social development and to protect the environment? Peter Drucker thought so (Vignette 5.2).

Vignette 5.3 Booming Fort McMurray

Megaprojects do result in economic growth. The Alberta oil sands have created the modern oil town of Fort McMurray. Like all resource towns, the boom conditions create a dark side. Along with the arrival of large numbers of construction workers, social problems arise. Fort McMurray continues to experience "the price of prosperity." Back in 2002, Tom Barrett of the *Edmonton Journal* (2002) wrote:

The latest economic boom to sweep through Fort McMurray has brought with it a predictable increase in crime. Crime, crack, and cocaine are problems for a community that plays as hard as its works. Take a lot of disposable income, add a huge population of single men without roots in the community and you can expect some problems, says RCMP Staff Sgt. Scott Stauffer. Money buys cocaine, ecstasy and a host of softer drugs, plus plenty of liquor. Heavy drinking often leads to fights and traffic accidents.

More than a decade later, Barrett's comments still hold true. While Fort McMurray has gained much social infrastructure, such as schools and recreation centres, it still is growing rapidly. Newcomers, by their sheer numbers, place enormous pressure on the community's housing stock, infrastructure, and recreational facilities.

Vignette 5.4 Regional Multiplier

In neo-classical economic theory, the most common form of economic impact analysis is based on the Keynesian concept of the multiplier effect. The multiplier effect is a measure of the economic impact of a new development, such as a factory or mine, on the local or regional economy. A high regional multiplier translates into a region heading towards economic diversification.

There are three types of impacts: the direct impact of the wages, salaries, and profits of the new development; the indirect impact from payments to regional industries supplying goods and services to the new firm; and the induced impact, which is the increase in payments to retail stores and their regional suppliers brought about by the spending of the new income.

The regional multiplier is expressed mathematically as $1/(1-s)$ where s is the marginal propensity to consume goods and services within the region. Goods and services purchased outside the region represent economic leakage. For example, let us assume that the induced impact is determined by a regional multiplier of 1.5. This multiplier indicates that one-third of every dollar spent on supplies and wages by the owners of the new enterprise occurs within the region. It is calculated from the expression $(1/1-s)$ where $1/(1-0.33) = 1.5$. Arriving at the total annual income impact involves applying the multiplier to the total expenditures from direct and indirect impacts within the region. If this annual amount was $4 million, then by applying the multiplier of 1.5, the indirect impact is $2 million and the total impact is $6 million ($4 million direct impacts and $2 million indirect impacts).

In the case of the Alberta-Pacific pulp project, which began production in 1993, the multiplier was used to calculate the number of anticipated indirect jobs (jobs not directly associated with the project). In this example, the multiplier was assumed to be low—1.2—and the number of jobs in the northern region was calculated as follows: 600 direct jobs x 1.2 = 720 total jobs (120 indirect jobs).

Megaprojects in the Twenty-First Century

By the twenty-first century, Canada's North had attracted companies from around the world. First came American companies, then European firms arrived, and finally an Asian presence put its stamp on northern resource development. The Alberta oil sands were the main attraction. These huge reserves of oil have attracted the major oil companies, including ExxonMobil (United States); Royal Dutch Shell (Netherlands); BP (United Kingdom); Total SA (France); Chevron Corporation (United States); ConocoPhillips (United States); Statoil (Norway); and PetroChina (China). These global companies often form Canadian subsidiaries or purchase a share in existing companies, thus spreading risk and sharing profits. Following a similar strategy, China, through its China National Offshore Oil Company (CNOOC), purchased **Nexen** in February 2013. CNOOC thus gained new energy holdings in the **Long Lake Project** in the Alberta oil sands.

The trend is towards fewer but very large corporations. In the global mining community, mergers in 2006 and 2007 left only four major players (Xstrata, BHP Billiton, Rio Tinto, and Vale) dominating global coal, iron, and nickel production. To counter this move, China, Japan, and South Korea have encouraged their state-owned companies to

Table 5.1 Megaprojects in the North

Project	Resource	Production Date	Major Market	Ownership	Transportation
Quebec/Labrador iron ore project	iron ore	1954	United States	The Iron Ore Company of Canada	rail and lake carriers
Great Canadian oil sands plant	oil sands	1967	United States	Sun Oil Company	pipeline
Syncrude Canada plant	oil sands	1978	United States	Syncrude Canada Ltd	pipeline
Peace River hydroelectric project	electricity	1980	United States	BC Hydro	transmission lines
Northeast Coal Project	coal	1984	Japan	Denison Mines	rail and ocean ship
Norman Wells Oil Project	oil	1985	United States	Esso Resources Canada	pipeline
James Bay Hydroelectric Project	electricity	1986	United States	Hydro-Québec	transmission lines
Peace River pulp plant	pulp	1990	Japan	Daishowa	rail/ship
Alberta-Pacific forest plant	pulp	1993	Japan	Mitsubishi and Oji Paper Company	rail/ship
Ekati Diamond Project	diamonds	1998	United States	BHP (now Dominion Diamond Corp.)	aircraft
McArthur River Project	uranium	2000	United States	Cameco Corporation	truck
Alliance Gas Pipeline	natural gas	2000	United States	Alliance Pipeline Inc.	pipeline
Suncor Millennium Project	oil sands	2001	United States	Exxon Mobil Corp.; Royal Dutch/Shell Group; Petro-Canada	pipeline
Diavik Diamond Project	diamonds	2003	United States	Rio Tinto	aircraft
Voisey's Bay mine	nickel	2005	United States	Inco (now Vale)	ship
Horizon Oil Sands Project	oil sands	2008	United States	Canadian Natural Resources Ltd	pipeline
Snap Lake mine Victor mine	diamonds iron	2008	United States	De Beers	aircraft
Victor mine	diamonds	2008	United States	De Beers	aircraft
Bloom Lake mine	iron	2009	China	Wisco and Consolidated Thompson	rail/ship
MacKay and Dover projects	oil sands	2010	United Sstates	PetroChina	pipeline
Long Lake project	oil sands	2013	United States	CNOOC	pipeline
Cigar Lake mine	uranium	2014	United States	Cameco	truck
Mary River Project	iron	2015	EU	ArcelorMittal and Nunavut Iron	ore ship

---

Vignette 5.5 China's Strategy of Diversifying Imports

|||||||||||||||||||||||||||||||||||||||||||||||||||||||||||||||||||||||||||||||||||||||||||||||||||||||||||||||||||||||||||||||||||||||||||||||||||||||||||||||||||||||

Wuhan Iron and Steel Corp. (Wisco), China's third-largest steelmaker, seeks to diversify its sources of iron ore away from giants such as Vale and Rio Tinto. In 2009, Wisco grabbed the opportunity to invest in the Bloom Lake project, a nearly completed iron ore mine in the Quebec–Labrador Iron Ore Trough. The developer, a relatively small Canadian mining firm called Consolidated Thompson, needed capital to complete the mine. As well, Wisco purchased half of Bloom Lake's annual output of eight million tonnes at market rates. The first shipment of iron ore concentrate left Sept-Îles on 27 July 2010 for China. With over a 100-year ore supply at Bloom Lake, Wisco has successfully diversified for the long haul.

Sources: Gibben (2009); Consolidated Thompson (2010).

---

secure a stake in foreign resources. For instance, China's bursting economy needs to import increasing amounts of energy and raw materials to sustain its growth and to satisfy its consumer demand. Accordingly, Chinese long-term strategy called for investments in the Alberta oil sands. China also needs raw materials. China imports huge quantities of iron ore through BHP Billiton and Vale. One Chinese steel company invested in a new iron mine in northern Quebec to diversify its sources of iron ore (Vignette 5.5).

Resource Development: Theory and Reality

The economic history of Canada is largely based on the exploitation of resources in opening Canadian frontiers to a market economy. In southern regions of Canada, resource development initiated a process of economic growth by exploiting major resources (furs, timber, codfish, wheat, minerals) that, in some cases, led to further economic activities and eventually to the diversification of regional economies. Canadian scholars have recognized and analyzed this historic pattern of regional development in Canada. In 1930, Harold Innis first observed that Canada's regional development was triggered by resource development that, over time, matured into a more diversified economy. Known as the staple thesis, Innis's approach presented a fresh interpretation of regional economic history and geography—from a Canadian perspective.

Mel Watkins (1963) transformed Innis's thesis into a more conventional theory of regional development. Regional development is not an automatic process of the market economy. In fact, the most positive outcome of the staple thesis is a broadening and maturing of the resource economy, but in the worst-case scenario the resource economy flounders and eventually slides into a staple trap where, instead of development, underdevelopment takes hold (Watkins, 1977). Hayter and Barnes (2001: 37) put it more gently: "as Innisians emphasized, there is nothing automatic about such diversification."

While the staple theory provides a broad historical framework to explain regional development, what about current reality? For example, the price of oil—around US$60 a barrel in June 2015 after having been much higher in 2014—is unlikely to return to levels experienced in the last decade of the twentieth century when a barrel of oil was worth less than US$20. As pointed out in Vignette 5.7, energy prices have risen to new levels but other commodity prices may have hit their peak in 2011 or 2012. Since price bumps and dips are inevitable, the question is: Will the current range of prices remain above the previous business cycle? For the resource-oriented North, higher commodity prices are good news, making existing operations more profitable and encouraging more investment. On the other hand, lower prices spell bad news for resource development.

Each region of Canada has distinct geographic factors that make it unique. The North has five factors that both distinguish it from other regions of Canada and hamper its regional growth. All affect the capacity of megaprojects to lead to a mature, diversified northern economy.[2]

1. Natural factors provide special challenges to developers. These costly physical challenges that must be overcome include permafrost, a fragile environment, and sea ice.
2. High transportation costs are the Achilles heel for northern projects. One reason is a very limited existing transportation infrastructure; another is the great distance required to haul supplies and equipment to construction sites and then to ship resources to world markets.
3. Northern benefits are relatively small because powerful economic leakages drain megaproject construction spinoffs to southern business and its labour force.
4. Transformation of the North into a diversified economy takes more than the construction of large-scale industrial projects. While an important start, the critical economic mass necessary to build on the short-term economic boost from megaprojects does not exist in the North, thus the path to a mature, diversified economy is far in the future. A more dismal interpretation is that its geography may never allow this to happen and the North may fall into Watkins's staple trap.
5. Transformation demands the participation of the local population.[3] The inability of the Aboriginal workforce to participate more fully represents a major stumbling block to diversification and the economic inequality that exists in the North (Vignette 5.6).

Vignette 5.6   The State of the Northern Aboriginal Workforce

The Aboriginal component of the northern labour force and business community is just beginning to gain the experience, skills, and capital necessary to compete in this northern version of the market economy. As Poelzer (2009: 448) observed:

> The supply-versus-demand problem has two dimensions: First, the inadequate supply of northern residents qualified to meet the labour demand in the public and private sectors; and second, the insufficient supply of diverse educational opportunities, particularly at the bachelor's-degree level.

This supply-versus-demand problem represents a gap remaining between Aboriginal and non-Aboriginal Canadians. Governments and companies are aware of this gap and the need to close it. How big is the gap? Taking Nunavut as an example, Berger (2006: 18) pointed out that "there are in the vicinity of 1,500 jobs that could be claimed by Inuit had they the necessary skills." Berger might have added: "and still hold on to their ever-evolving Canadian/Inuit identity."

Vignette 5.7   A New Dance Floor for Commodity Prices?

||||||||||||||||||||||||||||||||||||||||||||||||||||||||||||||||||||||||||||||||||||||||||||||||||||||||||||||||||||||||||||||||||||||||||||||||||||||||||||||||||||||

In the first years of the twenty-first century, global prices for energy and resources jumped significantly. Is this part of the normal boom-and-bust commodity price system whereby prices rise when the world economy is expanding but fall when the world economy contracts? Or is something else happening? In short, has the world's economy reached a higher plateau where the demand for energy and resources will fluctuate, but within a much higher range of prices? The argument goes like this: global demand for resources has increased due to the rapidly expanding economies in Asia, especially in China and India. The increased demand has led to record price jumps. Suddenly, resource-producing regions—at the same level of production—are benefiting from higher prices. While the world business cycle will no doubt continue to expand and contract, some believe that the price levels for resources will be maintained at a higher range of commodity prices. As Brian Oleson, a University of Manitoba agribusiness professor, has said: "We are entering a new era. It's almost as if the platform [for energy, grain, and mineral prices] has been raised and we're all dancing on a new dance floor" (Greenwood, 2007). But that new dance floor still has its troublesome commodity cycle of lows and highs. The spot price in US dollars for uranium demonstrates this cycle, with a high of $95/lb in 2007 and 2008 to a low of under $10/lb in 2000. More recently, its price reached a high of $72/lb in early 2011, but since then the price has declined. As of July 2014, the spot price for uranium had fallen to just over $28/lb. Other commodities, such as gold and iron, reached their peak price in 2011. The price of iron ore in January 2011 was US$181 per metric ton but its price slid downward to US$47 per ton by April 2015. Gold was valued at US$1,837/oz. in July 2011 but its price had dropped to US$1,188/oz. by April 2015. Oil, too, took a sudden and sharp price hit. As of mid-2014, oil prices ranged from $101/barrel for **West Texas Intermediate** to $106 for **Brent**. But a year later, oil prices had fallen significantly: as of the end of June 2015 the WTI price was $58.73/barrel while the Brent price was set at $62.97.

Resource Base

Canada's northern resource base—except for forestry and wildlife—is best understood through its geology, while its exploitation responds to global prices. As discussed in Chapter 2, the geomorphic regions of Canada are associated with particular energy and mineral wealth. The Canadian Shield, for instance, is associated with hard-rock minerals such as diamonds, gold, iron, nickel, and uranium while the Interior Plains and Arctic Lands are associated with vast sedimentary basins that contain huge quantities of oil and natural gas.

The resource base divides into non-renewable and renewable resources. Non-renewable resources are mineral and petroleum deposits. These resources have a fixed life—once a mine opens, for example, the first shovelful presages its death. Two examples illustrate the variation in longevity of gold mines. On the one hand, the Giant gold mine near Yellowknife operated for 56 years (1948 to 2004). On the other hand, the gold deposit at the Meadowbank mine near Baker Lake, which opened in 2010, is estimated to last less than 20 years. While the extraction of these resources stimulates the northern economy, the long-term impact is limited because of their finite nature. Then, too, global prices for commodities can fall, resulting in a depressed industry. In 2014, the

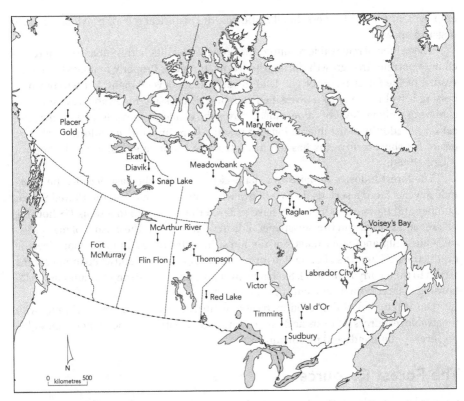

Figure 5.2 Major Mines, 2014

downward trend in uranium prices cast a dark shadow over this industry. Work on two proposed mines in northern Saskatchewan has been suspended because of low global prices (Table 5.2). These three issues raise questions regarding the sustainability of non-renewable resources; the extent to which the exploitation of these resources offers the potential for economic diversification; and the prospects for providing long-term

Table 5.2 Mineral and Oil Production in the Territorial North, 2014

Mineral Type	Yukon	Northwest Territories	Nunavut	Territorial North
Gold and silver	190	0	643	833
Diamonds	0	1,794		1,794
Oil	0	406	0	406
Copper	164	2	0	166
Tungsten	0	85	0	85
Lead/zinc	66	0	0	66
Other	9	5	0	14
Total	429	2,292	643	3,364

Sources: Natural Resources Canada (2015); NWT Bureau of Statistics (2015).

employment for northerners. In all instances, the historical record provides a largely negative answer.

The principal renewable resource, the boreal forest, lies in the Subarctic where the climate permits tree growth during the warm summers. Properly managed, the harvesting of the forest resource can lead to sustainable development, but without processing it will not lead to a diversified economy. Processing renewable resources, by achieving a higher level of value added, moves the economy towards greater diversification. In addition, sustainable logging practices are imperative. Unfortunately, the desire for short-term gains has led to destructive resource exploitation in the past (Clapp, 1999).

The forest industry remains in a depressed state. For the past decade, the forest industry has faced a serious decline in exports to its major market, the United States. Lumber and newsprint shipments have suffered the most, due to a weak US housing market and a decline in the newspaper industry. The result is the closure of many sawmills and pulp and paper mills. For the forest industry to regain its previous level of production, US imports of Canadian forest products, especially softwood lumber, must return to their early 1990s level. Such a recovery is linked to a strong rebound in the US housing market from its current record low.

Our discussion of the resource base now turns to specific cases, beginning with renewable resources—forests and water—and ending with two types of non-renewable resources—diamonds and petroleum.

The Forest Resource

Canada has a portion of the circumpolar forest that forms a "green halo" near the top of the globe. As the dominant natural vegetation found in the Subarctic, the boreal forest extends from Newfoundland and Labrador to Yukon. The forest represents both a key timber resource and an environment full of wildlife. Here, biodiversity reaches its peak within the North (Chapter 2).

The commercial part of the boreal forest lies in the southern section where mature stands exist. Unfortunately, foreign markets for forest products have contracted over the past two decades. Employment has declined and many mills have closed. The economic future of the timber industry looks bleak; forestry activities in much of the boreal forest have waned, and this has led to population losses.

Most logging takes place in British Columbia, Alberta, Saskatchewan, Manitoba, Ontario, Quebec, and Newfoundland and Labrador. In the territories, Yukon and the Northwest Territories have small logging operations in their portion of the boreal forest. With few exceptions, access to forest lands requires a company to obtain a timber lease from a provincial or territorial government. First Nations have obtained timber leases on lands where they claim Aboriginal title and, by forming joint ventures with established forestry firms, have become involved in logging and lumber production. Caught in the North American downturn, few private and First Nations forest enterprises have had success. Located in the heart of the boreal forest, the Lubicon Lake Nation (a Cree band)[4] and the Meadow Lake Tribal Council (a combination of Cree and Dene bands) have followed different paths, but neither has met with much success.

Vignette 5.8 The Resource Economy and Boom-and-Bust Cycles

With its narrow resource base, the northern economy is vulnerable to wide fluctua-tions, caused partly because of the nature of non-renewable resources (mines die) but also because of shifts in global demand. These world shifts are part of the global business cycle. This cycle—a normal part of the market economy—affects all parts of the world but the downturn is especially troubling to resource hinterlands that have few other economic activities to fall back on. If the downturn lasts for an extended period, workers and their families will depart, thus aggravating the eco-nomic situation.

The impact of the **boom-and-bust cycle** occurs at three levels. At the primary level, a small mine begins with boom-like conditions during the construction phase, followed by an operations phase when a more modest level of economic activity within the immediate area takes place. Eventually, the resource is exhausted and the operation closes, marking the bust phase of non-renewable development. A second type of cycle is associated with megaprojects. Their impacts follow a similar pattern with two exceptions. First, the economic and social impacts spread over a larger area—a regional impact rather than a local one. Second, the impacts last for a much longer period of time. The third type of boom–bust cycle that affects the North is due to global economic downturns that cause resource com-panies to halt production or to close their operations, at least until commodity prices increase again.

For the past 10 years, the forest industry has been contracting, causing high levels of unemployment in the small forest-oriented communities. This has led to out-migration, which compounds the economic difficulties facing forest communities. A measure of the decline is provided by the downward trend in the annual timber har-vest. In 2004, Canada's forests reached a peak annual harvest of about 200 million cubic metres of timber, but since then the depressed state of the industry has seen the annual figure fall well below this figure and shows no signs of recovery. In 2012, for example, the annual harvest had fallen to 152 million cubic metres (Natural Resources Canada, 2014). Approximately 70 per cent of this harvest is in softwood lumber, with the remainder going to the pulp and paper industry. This decline was due to a drop in exports to the United States, where traditionally nearly three-quarters of Canada's forestry output was sold.

Like many small towns along the southern edge of the northern coniferous forest, Meadow Lake, Saskatchewan, provides a microcosm of the impact of low prices and falling demand on boreal forest communities. Meadow Lake has struggled to maintain its forest industry, including seeking public funds for its forest companies to survive, but the Saskatchewan government was unwilling to provide more subsidies, causing Millar Western Forest Products to close its pulp mill in 2007. In addition, the town's sawmill suffered from the high US duties on softwood lumber from 2002 to 2006, fol-lowed by the catastrophic drop in US demand for softwood lumber. One unique feature of this forest complex is the involvement of the Meadow Lake Tribal Council, which holds timber rights to a 3.3 million-hectare forest management licence area. The tim-ber from this forest supplies the NorSask sawmill, owned and operated by the Meadow

Lake Tribal Council. In 1992, the Tribal Council formed a partnership with Millar Western to supply wood pulp to the Millar Western pulp mill. With the disappearance of the US market, the pulp and sawmills fell on hard times. Both ceased production in 2007, marking the end to this promising partnership. Since 2012, hopes have been pinned on the Meadow Lake Bioenergy Centre, a 36 MW biomass power plant. This sustainable project would use the waste wood products from the NorSask sawmill to generate electrical energy that would be sold to SaskPower. Preliminary work on the site is completed but construction, as of June 2015, had not begun.

Vignette 5.9 Resource Differences in the Arctic and Subarctic

The resource geographies of the Arctic and Subarctic differ sharply. The Subarctic has several advantages over the Arctic:

- Its resource base is much broader and includes a wider range of resources— major energy, mineral, timber, and water resources.
- Its geographic location makes it more accessible to the American market.
- Its natural environment presents less of a barrier to industrial projects because of a milder climate, a longer river/lake navigation season, and a land less affected by permafrost.

All of these advantages translate into lower costs of resource development in the Subarctic than in the Arctic. For instance, Fort McMurray has excellent highway access to the rest of Alberta. In addition, sustainable development prospects are more promising in the Subarctic than in the Arctic because of the greater array of renewable resources in the Subarctic, especially the boreal forest. Climate change could tip the scales in favour of the Arctic if the polar ice cap melts and open water exists during the summer.

Mineral Resources

Canada is one of the world's leading mineral producers and exporters. Nearly half of this production comes from northern mines. The annual value of Canadian mineral production is over $50 billion (Rhéaume and Caron-Vuotari, 2013: Table 1).

Within the world of mineral exploration, the discovery of commercial mineral deposits has an element of suspense and surprise—a prospector never knows when or if he/she will hit pay dirt. Diamonds are a case in point. Until diamonds were discovered in the Canadian Shield of the Northwest Territories, most geologists did not believe that such deposits existed in Canada (see Vignette 5.10). The timing was perfect for the Northwest Territories because, in the 1990s, the central pillar of its resource economy—gold—was coming to an end.

Gold Mining

The value of gold mining in Canada in 2014 was $6.8 billion (Natural Resources Canada, 2015). Almost all comes from the Canadian North. According to Natural

Resources Canada, just over 75 per cent of the gold produced in 2014 came from northern mines in Ontario ($3.4 billion) and Quebec ($1.7 billion) with the other provinces making up the Provincial North contributing another 12 per cent for a total of 87 per cent. While the Territorial North accounts for only 11 per cent of the value of gold production in Canada, or $769 million, gold mining dominates the resource industry of Yukon and Nunavut.

In the late nineteenth century, the discovery of placer gold in the famous Klondike goldfields[5] kicked off mining in the Territorial North. After the discovery of gold near Yellowknife, gold mining was the basis of the economy of the Northwest Territories for over five decades, but those days ended with the closure of the Giant mine near Yellowknife in 2004. The Lupin gold mine in Nunavut, like many other gold mines, had a short lifespan, from 1982 to 2004. The recently opened Meadowbank mine in Nunavut and the promise of one or more in Yukon signal that gold may again play an important role in the Territorial North. For example, the discovery of a large gold deposit near the Klondike River has amazed geologists, and if a mine is developed it will certainly kick-start Yukon's sluggish resource-driven economy (Figure 5.3).

Gold, like other precious metals, figures prominently in the northern mining world because this highly valued mineral can more easily overcome the high cost of transportation to world markets. Gold production in the Provincial North has long been an economic factor. Val d'Or in Quebec and Timmins in Ontario are among the oldest gold mines in the North. Red Lake lies in northwest Ontario, and since 1926 this mine has produced much of Ontario's gold.

Gold mining is strongly affected by price swings. Ten years ago, the gold industry was depressed because of low prices and many mines closed. From 2005 to 2011,

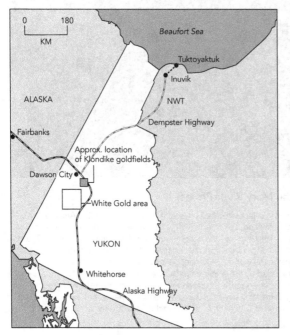

Figure 5.3 Yukon's White Gold Discovery: A New Klondike?

After more than a century, gold fever came back to the Klondike following a discovery in 2004 by Shawn Ryan, a local prospector, who found what might turn out to be the source of the Klondike placer gold near the confluence of the White and Klondike rivers just south of Dawson City. Ryan's discovery, coupled with record high prices for gold, means that an underground mine is almost assured in the White Gold area, while further north Kinross Gold Corporation plans to open its Eagle gold mine soon, subject to approval of its environmental statement and the signing of an impact and benefits agreement by the Na-Cho Nyak Dun First Nation. "Almost" is a common word in Yukon. By May 2015, these two mines remained only dreams.

Sources: Bouw (2010); Thompson (2011); Kinross (2013).

however, the price of gold increased from$US400/oz. to a record high of nearly $US2,000/oz. The number of mines increased. By July 2014, the price of gold had slipped to around $US1,300/oz. (Nasdaq, 2014).

What is the future for gold? The last decade saw a "mini-gold boom" due to prices that increased five times from 2005 to 2011. Looking to the future, unsettling political events, such as in the Middle East and Ukraine, could get out of hand and trigger a sharp rise in the price of gold. On the other hand, prices could follow the decline of other commodity prices as the global economy cools.

Iron Mining

Major iron mining operations have long existed in northern Quebec and Labrador. In this part of the Provincial North, iron ore mining takes place near Labrador City and Wabush in Labrador and at Fermont, with open-pit mines at Bloom Lake, Mount Wright, and Fire Lake in Quebec. Buoyed by high prices, which reached $US180/tonne in January 2011, Baffinland proceeded with its Mary River Project in the Arctic and

Figure 5.4   The Campsite for the Mary River Project

The huge iron deposit known as the Mary River Project is located on northern Baffin Island in the Arctic. ArcelorMittal is the principal share owner of Baffinland, the company developing the open-pit iron ore mine. The iron deposit contains over 60 per cent iron, making it one of the richest in the world. In comparison, the rich iron ore deposits in the Labrador Trough are around 30 per cent iron. Added to its advantage of high-grade ore, Mary River's proven reserves are among the largest in the world. The challenge is to get the ore to market from this isolated site located in permafrost and surrounded by the Arctic ice pack. The eventual plan calls for a railway to a port at Steensby Inlet and ice-breaking bulk carriers to transport the ore to European markets. A new short-term plan has involved mining and trucking a smaller quantity of ore to Milne Inlet, where it has been stockpiled beginning in the fall of 2014, with an expectation of shipment beginning in the 2015 open-water season.

Source: Baffinland Exploration Camp, Mary River, July 2012. Photographer: Marc Pike. © Baffinland Iron Mines

planned to begin shipping ore to markets in the European Union in 2015 during the season of open water. This ambitious megaproject involves the creation of a road/rail/port transportation system on Baffin Island, Nunavut (Figures 5.2 and 5.4). Slumping iron prices that fell to $US80/tonne by mid-2014 caused rethinking of the Mary River Project while existing iron companies slashed costs by reducing the size of their workforce or by halting production (Marotte, 2014; Parsons and Keeling, 2015). Nevertheless, this ambitious project remains on course.

Uranium Mining

Northern Saskatchewan's Canadian Shield contains some of the world's richest uranium ore deposits and its mines account for 20 per cent of world production (see Figure 5.5 and Table 5.3). The large size and high grade of these ore bodies make them more economical to mine than deposits in other countries. Some of these deposits are mined by open-pit techniques while others are mined using hard-rock mining techniques. For example, the McArthur River and Cigar Lake mines represent the world's largest high-grade uranium mines. Ore grades within the deposit are 100 times the world average. At the Cigar Lake mine, production began in March 2014. While this deposit is one of the richest uranium ore bodies in the world, many challenges had to

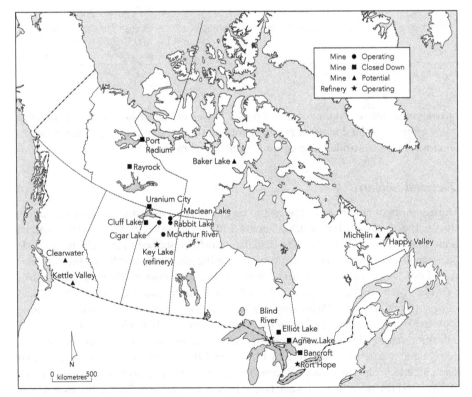

Figure 5.5 Uranium Mines: Past, Present, and Potential

Source: From "Uranium: A Discussion Guide," Canadian Coalition for Nuclear Responsibility, http://www.ccnr.org/nfb_uranium_0.html.

Table 5.3 Uranium Mines and Mills in Northern Saskatchewan, 2014

Facility	Licensee	Licence	Type
Cigar Lake Mine	Cameco Corporation	Operation	Licensed to 2021
Cluff Lake Mine Site	AREVA Resources Canada Inc.	Decommission	Decommissioned
Key Lake Mill	Cameco Corporation	Operation	Licensed to produce up to 7,200,000 kg of uranium per year; licensed to receive ore slurry from McArthur mine
McArthur River Mine	Cameco Corporation	Operation	Licensed to mine up to 7,200,000 kg of uranium per year
McClean Lake Mine/Mill	AREVA Resources Canada Inc.	Operation	Licensed to produce up to 3,629,300 kg of uranium per year
MidWest Joint Venture Mine	AREVA Resources Canada Inc.	Proposed	Suspended due to low uranium prices
Millenium Mine	Cameco Corporation	Proposed	Suspended due to low uranium prices
Rabbit Lake Mine	Cameco Corporation	Operation	Licensed to produce up to 6,500,000 kg of uranium per year

Source: Canadian Nuclear Safety Commission (2014).

be overcome before production could begin. The challenge of such a remarkably high grade of ore was the danger of exposing miners to its radiation. The solution was to employ remote-controlled underground mining techniques.

Diamond Mining

The discovery of diamonds in the 1990s was the most exciting and unexpected mining story in the Northwest Territories (Vignette 5.10). Until the late twentieth century, most geologists did not believe that the Canadian Shield contained diamonds. In 1987, De Beers geologists found the Victor Mine kimberlite cluster in northern Ontario. Four years later, two prospectors, Charles Fipke and Stewart Blusson, discovered diamonds near Lac de Gras in the mineral-rich **Slave Geological Province** of the Canadian Shield some 300 km northeast of Yellowknife. Other diamond deposits were found in the same geological zone, consisting of ancient kimberlite pipes within volcanic formation, but only three deposits (Ekati, Diavik, and Snap Lake) turned into operating mines. A fourth mine, Gahcho Kué, was approved by Ottawa in 2014 and its production is scheduled begin in 2016 (Vignette 5.10). Ekati was the first mine to open, in 1998; Diavik followed in 2003; and the mine at Snap Lake began production in 2008. Each mine took about 10 years from discovery to production, reflecting another feature of megaprojects, that only very large companies have the financial strength to tackle such projects. Ekati and

Diavik are open-pit mines and Snap Lake is an underground mine. Each deposit has an estimated 20- to 30-year supply of diamonds, with peak production estimated at 2013 and shut-down of the last of the three mines around 2030 (Hoefer, 2009: 408).

Within a year of Ekati opening, diamonds joined oil as the backbone of the resource economy of the Northwest Territories. By value of production, number of employees, and spinoff effects, these three mines constitute the heart of the northern economy. Ekati and Diavik are located about 300 km northeast of Yellowknife while Snap Lake is some 200 km from Yellowknife. All mines are linked to Yellowknife by a winter ice road. In 1998, the Ekati mine accounted for $500 million worth of diamonds. By 2001, its production was valued at $847 million. Just across the border in Nunavut, another diamond mine—Jericho—began production in 2006, but it abruptly suspended operations on 6 February 2008. In 2009, the annual value of diamond production and exports from the Northwest Territories was $1.5 billion (Table 5.4). Snap Lake came into production in 2008. Winspear Resources discovered this kimberlite deposit in 1997, and De Beers Canada bought the deposit in the fall of 2000.

Outside of the Northwest Territories, several promising finds exist, including the Shore Gold deposit just north of Prince Albert, Saskatchewan. However, only one diamond mine, the De Beers-operated Victor mine in Ontario, is functioning, and that mine may exhaust its ore body by 2019. By 2019, another diamond mine, the Renard Project located in Quebec, may begin operations.

The quality and production of Canadian diamonds are very high. By 2009, Canada's diamond production ranked third in the world by value and third by volume of production (Canada, 2014). However, the diamond mines in the Northwest Territories have removed the most accessible ore by open-pit mining. By 2010, tunnel mining had

Table 5.4   Value of Diamond Shipments from the Northwest Territories

Year	$ millions
2002	793
2003	1,588
2004	2,097
2005	1,763
2006	1,567
2007	1,765
2008	2,084
2009	1,448
2010	2,030
2011	2,053
2012	1,615
2013	1,561
2014	1,794

Sources: Bureau of Statistics, NWT (2010: 37); Natural Resources Canada (2015).

Figure 5.6 Snap Lake Diamond Mine

The Snap Lake mine site, located in the lake-strewn landscape of the Canadian Shield, began production in 2008. The mine is owned and operated by De Beers. The site is situated in the natural vegetation transition zone between the boreal forest and the tundra.

Source: De Beers Canada (2007). Reprinted by permission of De Beers Canada Inc.

begun, suggesting that the diamond boom was over (Vanderklippe, 2010). The importance of the Gahcho Kué mine is also tied to the fact that two of Canada's major diamond mines—Diavik and Ekati—are approaching the end of their productive lives, with Diavik's closing scheduled for 2020. Ekati should remain in production unit 2031 but its annual output is expected to decrease. The Gahcho Kué mine site, some 80 km southeast of Snap Lake, could offset the anticipated drop in production from the other two diamond mines.

Diamond Processing

Processing resources remains an elusive goal. Yet, economics normally calls for processing to take place near major markets. Diamond processing is one exception and represents a small step in the direction of regional development.

While most raw diamonds are exported to the United Kingdom, Botswana, and Belgium for sorting and gem quality diamonds are sent to one of the main diamond cutting and trading centres in Antwerp, Mumbai, Tel Aviv, New York, China, Thailand, or

Vignette 5.10   Timetable for Diamond Mines in the Northwest Territories and Nunavut

The Northwest Territories is the diamond centre for North America. By 2006, two major mines, Ekati and Diavik, accounted for 12 per cent of the world's diamond production. During the 2000s some of these diamonds were sorted, cut, and polished in Yellowknife, but these operations have since closed. Diamonds have been mined in the Northwest Territories only recently, as the following timeline shows. In 2006, Jericho diamond mine, located near the now closed Lupin gold mine in Nunavut, began production. The high cost of operating in the Arctic caused this diamond mine to close after two years.

Date	Event
1990	Geologists did not believe that the geological structure necessary for kimberlite and hence diamond deposits existed in the Canadian Shield.
1991	Charles Fipke and Stewart Blusson discover diamonds near Lac de Gras, Northwest Territories, sparking the largest staking rush in Canadian history.
1993	BHP opens an exploration camp near the original discovery just north of Lac de Gras.
1998	BHP's Ekati open-pit diamond mine (Panda) opens; BHP and the NWT government sign an agreement whereby diamond sorting and valuation of some Ekati diamonds will take place in Yellowknife.
1999	Sirius Diamonds Ltd opens a cutting and polishing facility in Yellowknife.
2000	Ottawa approves the Diavik proposal and construction of the Diavik mine begins; Deton' Cho Diamonds Inc. and Arslanian Cutting Works NWT Ltd open cutting and polishing facilities in Yellowknife.
2001	BHP opens a second open-pit mine called Misery.
2003	Diavik mine starts production.
2006	Jericho mine, the first diamond mine in Nunavut, begins operations but closes in 2008.
2008	The De Beers-owned Snap Lake underground mine begins production.
2012	Diamond cutting and polishing firms in Yellowknife have ceased operations.
2013	Open-pit mining is replaced by tunnel mining at Diavik and Ekati, indicating higher costs and less production. Dominion Diamond Corporation, a Toronto firm, acquires controlling interest in the Ekati mine from BHP Billiton.
2014	Gahcho Kué mine operated by De Beers plans to begin production.
2030	Likely end of diamond production in the four mines.

Sources: Adapted from Hoefer (2009) and NWT Diamonds Timeline, www.gov.nt.ca/RWED/diamond/timeline.htm.

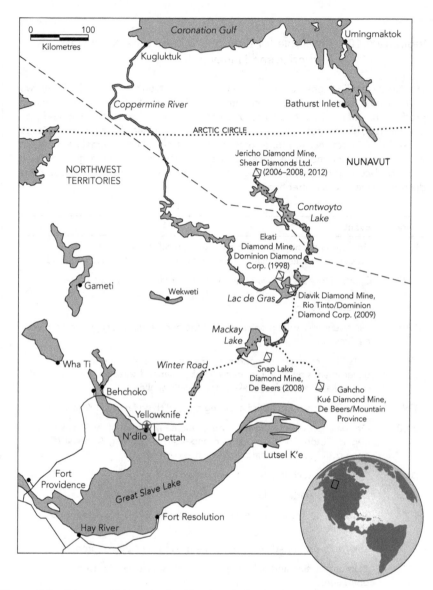

Figure 5.7 Diamond Mines in the Northwest Territories and Nunavut

Source: Diavik Diamond Mines Inc.

Johannesberg, a small portion was allocated to the new diamond sorting and cutting industry in Yellowknife and Sudbury. The NWT government insisted that some processing take place in the Northwest Territories. Accordingly, BHP Billiton Diamonds, Diavik Diamond Mine, and De Beers's Snap Lake operation sold rough diamonds to manufacturers in the Northwest Territories as part of an understanding between producers and the territorial government to promote and support value-added diamond activity in the Northwest Territories (NWT, 2007). At the peak, from 2003 to 2006, four factories had set up shop in Yellowknife's Diamond Row district—Sirius, Arslanian Cutting Works, Laurelton, and Canada Dene Diamonds. These firms imported their skilled diamond cutters from around the world, with many in Yellowknife from Armenia. These firms employed about 200 people. Unfortunately, this value-added activity did not take hold, and by 2012 no diamond cutting and polishing company remained open in Yellowknife. In Ontario, diamonds from the Victor mine supply Sudbury's diamond processing operation, but the Ontario diamond mine is scheduled to close by 2019.

This quick turnabout raises the question: Are efforts to capture value-added activities viable? Sudbury's diamond-processing plant apparently has flourished because the Ontario government insisted that 10 per cent of better quality diamonds produced at De Beers's Victor mine in Ontario's Subarctic go to the Sudbury plant to improve their capacity to compete against lower cost centres such as India. John Kaiser, a market analyst, was quoted by Gagnon (2010) as saying:

> Diamond processing plants in the Northwest Territories and Ontario is one of these make-work businesses that make the local politicians proud of themselves that they have some sort of industry . . . in their local area, but it's really outside the economic box. On the other hand, without government intervention in the marketplace, the North will receive few benefits from mega resource projects.

The North, then, is confronted with a trade-off—companies pay more for diamond processing in the North but northern communities obtain employment and training benefits, plus a few highly skilled diamond cutters and polishers who have immigrated from global diamond-processing centres.

Nickel Mining

The Canadian Shield remains a storehouse for base and precious metals. Nickel is found in the Thompson Nickel Belt in Manitoba; the Sudbury Basin of Ontario; and the Cape Smith Fold Belt in the Ungava Peninsula of Quebec. While the Thompson Nickel Belt lies in the Subarctic, two other geological structures associated with nickel deposits are found in the Arctic. One is situated in Nunavik, the Inuit homeland in northern Quebec. The Raglan mine owned by Xstrata Nickel began production in 1997. Recently, another nickel mine—just 20 kilometres south of the Xstrata mine—began production. The Nunavik Nickel Mine is owned by Canadian Royalties, now a wholly owned subsidiary of the Chinese nickel company Jilin Jien following a successful takeover bid initiated in 2009. This mine started production in 2012. The other Arctic nickel area lies along the Labrador coast at latitude 56°N, where the Voisey's Bay mine is situated. The Raglan mine and its Jilin Jien neighbour are linked by all-weather roads to an airstrip and to the concentrate, storage, and ship-loading facilities at Deception Bay on Hudson Strait. The current mine life for the Raglan mine is estimated at more than 30 years (Welch, 2010). The mine is a fly-in/fly-out operation with employees working for 28 days, followed by 14 days off. Its workforce totals approximately 500, with only

Figure 5.8   Raglan Nickel Mine

The all-weather gravel road that extends 100 km to Deception Bay is shown in the foreground. Since the mine is in Nunavik, the company has made an impact and benefits agreement with Makivik, the Inuit company that manages the financial aspect of the James Bay and Northern Quebec Agreement.

Source: Courtesy Glencore

1 per cent, or five workers, being Inuit. The ore is crushed, ground, and treated on site at the mill to produce a nickel-copper concentrate trucked 100 kilometres to Deception Bay, shipped to Quebec City, and then moved by train to Sudbury for smelting. The final refining process takes place at Xstrata's Norwegian refinery where pure nickel is produced. In contrast, the newer Chinese-owned mine in 2014 began shipping nickel concentrate directly to China via the Northwest Passage (see Chapter 8).

At Voisey's Bay, one of the world's richest nickel deposits was discovered by accident—the prospectors were searching for diamonds. The size of this deposit is staggering, with 150 million tons of proven nickel ore plus smaller deposits of cobalt and copper. This deposit has an estimated lifespan of 30 years. The ore is located in three sites: the Ovid, the Eastern Deeps, and the Western Extension. Apart from the claims of the Innu and Inuit of Labrador, the ownership of this deposit has changed hands several times. Vale, a giant Brazilian mining company, purchased Inco in 2006 and therefore the Voisey's Bay nickel deposit. In 2009, the value of nickel produced from Voisey's Bay mine reached nearly $500 million (Natural Resources Canada, 2010). From open-pit mines, the ore is shipped by freighter and then by rail to its refinery in Sudbury, Ontario. The government of Newfoundland and Labrador obtained a commitment for local processing in its 2002 agreement with Inco (now Vale). In 2005, Inco opened its demonstration hydromet processing facility at Long Harbour, which

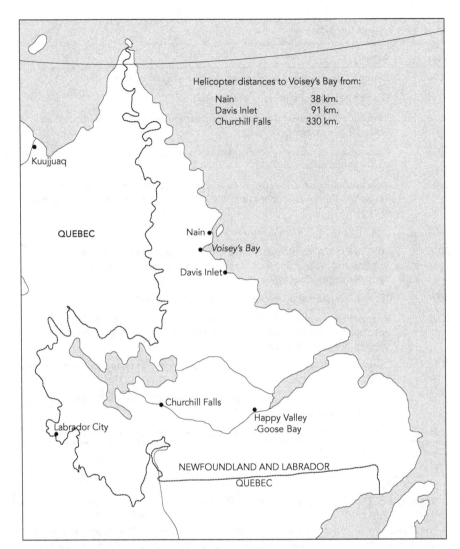

Helicopter distances to Voisey's Bay from:

Nain	38 km.
Davis Inlet	91 km.
Churchill Falls	330 km.

Figure 5.9   Map of Labrador and Voisey's Bay Nickel Mine

By 2018–19, the multi-billion dollar mining operation at Voisey's Bay will ship its ore to its processing plant at Long Harbour. While the boundary between Quebec and Labrador represents the longest provincial boundary in Canada, it remains the most contentious. In 1927, the Judicial Committee of the Privy Council in Great Britain ruled that the boundary in Labrador that separated its two Dominions in North America, Canada and Newfoundland, followed the watershed divide. When Newfoundland joined Confederation in 1949, its boundary in Labrador was confirmed in the Terms of Union. Still, this boundary has not been accepted by Quebec and it remains unsurveyed and unmarked on the ground.

Source: "Voisey's Bay: An Introduction," at: www.arcticcircle.uconn.edu/SEEJ/Voisey/intro.html.

proved to be commercially successful. As a result, Vale began construction of a nickel concentrate mill at a cost of nearly $4.25 billion. The plan was to ship the ore for processing to Long Harbour in Placentia Bay by 2013, but construction delays push this date to 2018–19 and Vale agreed to pay the government $400,000 for this concession (Roberts, 2015).

Table 5.5 Voisey's Bay Timeline

Year	Activity
1994	Diamond Fields discovers potential "elephant" nickel deposit.
1996	Diamond Fields sells its nickel site to Inco.
June 2002	Inco signs a $2.9 billion development agreement with the Newfoundland and Labrador government.
2002–3	Preparation work clearing land and building construction facilities at Voisey's Bay and Argentia begins.
2003	Construction commences on the open-pit mine at Voisey's Bay, the Inco Innovation Centre in St John's, and the hydro-metallurgical demonstration plant in Argentia.
2004	Inco's Innovation Centre completed.
2006	Voisey's Bay begins production and Argentia pilot plant receives first concentrates. Vale, a Brazilian mining company, by purchasing Inco, owns and operates Voisey's Bay mines.
2007	Testing of hydromet process at Argentia pilot plant starts.
2008	Vale decides whether to build a commercial hydromet processing facility or to build a matte processing facility that will produce finished nickel from concentrate smelted elsewhere.
2014	Long Harbour Nickel Processing Plant construction completed at a cost of nearly $5 billion.
2018	Vale decides whether to expand the open-pit mine at Voisey's Bay into an underground mine.
2036	With ore exhausted, Vale closes its Voisey's Bay mining operation.

Source: Adapted from Hasselback (2002).

Vignette 5.11   Northern Ontario's Ring of Fire

The Canadian Shield contains vast, undiscovered minerals. One recent discovery took place in the remote James Bay Lowland of northern Ontario. Known as the Ring of Fire, this huge deposit consists of chromite, copper, nickel, platinum, and other minerals. The size of the deposit is similar to the Sudbury ore body, suggesting that mining could last for over 100 years. However, Ontario is still years away from established mines in the region because of the lack of land-based transportation. Given the bulky nature of the ore bodies, a railway is ideal but a highway would suffice to reach southern refineries and markets. Not only is the estimated cost of such a railway high—around $2 billion (Sudol, 2011)—but provincial and federal governments have shown no interest in contributing to a mining railway. Other barriers to the Ring of Fire project exist, including the concerns of the Webequie and Marten Falls First Nations about the lack of consultations to determine their role in the development; and, given the potential risks of such mining to the environment, the proponents face a major hurdle in gaining approval for their yet-to-be-prepared environmental impact statement.

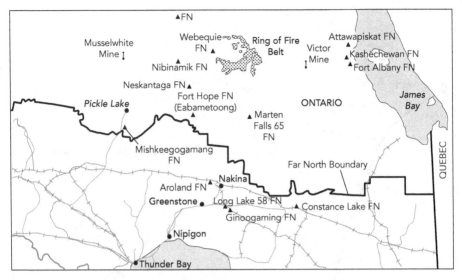

Figure 5.10   Map of the Ring of Fire Belt

Like many rich mineral deposits in the Canadian North, access to the Ring of Fire remains a problem. With a railway out of the question, an all-season road across the marsh-like terrain of the Hudson Bay Lowlands to this remote ore body is the next best option and its estimated cost of $1 billion is half that of a railway (CBC News, 2014). This cost figure may be low because global warming is causing the ice in the permafrost to melt and the ground to subside. Nevertheless, so critical is access that, in 2013, Cliffs Natural Resources, which was planning on developing the huge chromite deposit at the Ring of Fire, suspended its site development. Cliffs Natural Resources will not commence its site operations until an all-season public highway is completed.

Source: CBC News (2014).

Petroleum Resources and Production

The North contains most of Canada's petroleum reserves in a variety of sedimentary basins. The largest basin consists of the Alberta oil sands. Further north, known petroleum reserves—with both proven and unproven reserves—are well beyond the edge of commercial exploitation because of their remote location. The exception is the Norman Wells oil field.[6] With oil prices soaring in the 1970s, Esso Resources Canada saw an opportunity to expand production at its Norman Wells oil field by constructing a pipeline to the national pipeline system in northern Alberta and then, starting in 1985, shipping much of the new production to American markets (Figure 5.11). Norman Wells production hit a high in 1995, but since then the annual output has declined (Table 5.6).

The most accessible Arctic reserves are found in the Mackenzie Delta and the Beaufort Sea. In the late twentieth century, exploratory drilling operations had already found vast quantities of oil and natural gas. Petroleum exploration began in the mid-1960s. Nearly 200 exploration wells have been drilled in the region, resulting in the discovery of nearly 60 oil and gas fields. The largest discoveries in the Mackenzie Delta include the Taglu, Parsons Lake, and Niglintgak natural gas fields. The Amauligak oil field represents the largest offshore discovery. More deposits remain to be found. In 2007, for example, Devon Energy Corporation discovered a huge oil pool in the Beaufort Sea. The petroleum reserves of the Beaufort Sea and Mackenzie Delta amount to over 10 per cent of Canada's

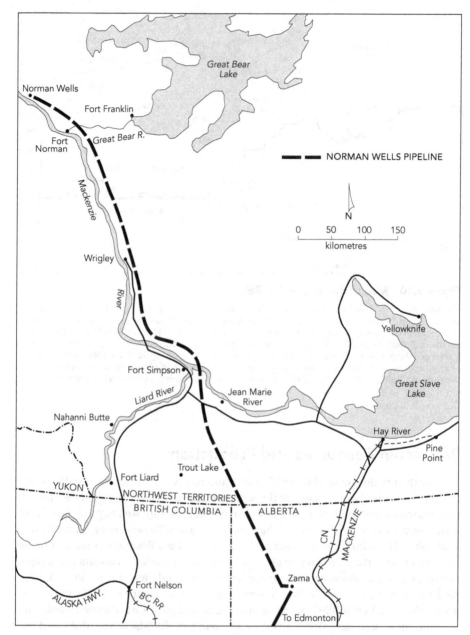

Figure 5.11 Norman Wells Pipeline Route

The Norman Wells oil field continues to produce oil, which is carried south by the first completely buried oil pipeline in permafrost terrain. This huge construction undertaking, known as the Norman Wells Oil Expansion and Pipeline Project, faced two challenges. First, the project had to stickhandle around irate Dene whose land claims were still outstanding and who had just seen their dream of a separate territory called **Denendeh** opposed by Ottawa. Second, the buried pipeline route was in the zone of discontinuous permafrost and thus would encounter both frozen and unfrozen ground. The Great Slave Lake Railway shown on this map extends from Grimshaw, Alberta, to Hay River, NWT. While the branch line to Pine Point no longer exists, this railway represents the northernmost rail line in Canada. Construction of this railway began in 1962 under the Roads to Resources program of the Diefenbaker government.

Source: Bone (1984: 63). Courtesy of Aboriginal Affairs and Northern Development Canada

Table 5.6 Production and Value of Norman Wells Petroleum, 1981–2014

Year	Oil Production (000 m³)	$ Value (millions)	$ Value/m³
1984	175	20	114.3
1985	1,148	195	169.9
1990	1,864	248	133.1
1995	1,590	206	129.6
2000	1,536	382	208.3
2005	1,090	425	389.9
2010	841	457	523.4
2011	680	396	582.4
2012	756	n.a.	571.4
2013	654	406	620.7
2014	642	n.a.	n.a.

Source: NWT Bureau of Statistics, *Statistics Quarterly*.

petroleum reserves. With the exception of local gas production for the town of Inuvik, these petroleum deposits still await commercial development. The Mackenzie Gas Project (MGP) failed to get off the ground. Central to this proposed project was a 1,220-km pipeline to markets in the United States, but with gas prices so low and the US awash in natural gas, the MGP has been shelved indefinitely. The supply of natural gas in North America has increased due to a new technology—the so-called **horizontal drilling** and **hydraulic fracturing** (fracking) system. The discoveries of "elephant-size" shale deposits of natural gas have turned the supply/demand equation on its head with natural gas prices declining and Arctic gas deposits unable to compete with these southern discoveries.

The least accessible reserves lie deep beneath the Canadian sector of the seabed of the Arctic Ocean, while even larger possible reserves are suspected to exist in the international area or unclaimed zone of the Arctic Ocean. These latter reserves may not reach the exploitation stage this century. As discussed in more detail in Chapter 8, the scramble by nations bordering the Arctic Ocean for ownership of the international section of the Arctic Ocean is due, in large part, to these sedimentary structures that may contain much more oil and natural gas than exists in the rest of Canada (Gautier et al., 2009).

Alberta Oil Sands Production

The Alberta oil sands represent one of the largest petroleum deposits in the world (Figures 5.12 and 5.13). In fact, the oil reserves of this deposit rival the size of conventional oil reserves found in the Middle East. According to Alberta Energy (2014), Alberta ranks third, after Saudi Arabia and Venezuela, in terms of proven global crude oil reserves, with 13 per cent of total global reserves (1,342 billion barrels). Until 2015, Alberta's economic growth led all provinces because of the development of the oil sands. In particular, Alberta has benefited greatly from royalties, capital investments, and employment. Within Alberta, this growth has focused on Edmonton and Fort McMurray (the Regional Municipality of Wood Buffalo). Fort McMurray's population jumped from 43,000 in 2001 to over 65,000 in 2011, an increase of 57 per cent (Bone, 2014: Table 8.12). By 2014, "Fort Mac" itself may have reached 75,000 (Fort McMurray Chamber of Commerce, 2015). How the city will fare with the construction slowdown in the oil patch and much lower prices/royalties is a critical question. A larger question is how the slowdown will affect the rest of Canada.

Over the past 20 years, the oil sands have had an enormous impact on Canada's economy and its exports. For instance, the International Monetary Fund calculated that energy exports jumped from 11 per cent of Canada's merchandise trade in 1991 to 26 per cent in 2013—and most of that increase is attributed to the oil sands (Lusinyan et al., 2014: 4). Investments in the oil sands total billions of dollars and Alberta and, to a lesser degree, Ontario and Quebec are the major beneficiaries because they have the access and capacity to supply many of the products needed for oils sands projects (Richardson, 2007; Lusinyan et al., 2014). Demand for labour has cast a wider net, with many coming to work at Fort McMurray from Atlantic Canada (Jensen, 2013).

The economic effects of oil sands development fall into three categories: employment, capital investment, and indirect benefits. In a sense, the oil sands developments match Innis's argument about resource development leading to economic diversification. As well, there is a spatial dimension to the oil sands in that these projects draw Canadian workers from across the country. In Atlantic Canada, working in Fort McMurray and leaving the family behind is common practice and has been called "the Big Commute" (CBC News, 2007). As well, each project requires processing facilities, i.e., upgraders and heavy oil refineries, as well as vast amounts of building supplies, equipment, and specialized products. Companies throughout Canada, especially in western Canada, Ontario, and Quebec, benefit from these spinoff effects.

Development of the oil sands is expensive and involves huge capital investments. Three processes are involved:

- extracting the bitumen by either open-pit mining or a steam injection system;
- upgrading the bitumen into a heavy oil;
- refining the heavy oil into a variety of synthetic products.

Production, whether by open-pit mining or **steam-assisted gravity drainage** (SAGD), takes place in three areas of Alberta: the Athabasca area stretches over 40,000 km² near Fort McMurray; the Cold Lake area extends over 22,000 km²; and the Peace River area represents the smallest deposit of bitumen at 8,000 km² (Figure 5.12).

Do these oil sands projects represent a class beyond megaprojects? Joseph (2010) makes a case for calling them "gigaprojects" because of the huge investment and the scale of the projects. With six operating open-pit mines, nine mines under construction, and 17 steam-driven wells, plus more on the drawing board, extracting bitumen provides an economic burst with benefits that spread well beyond Alberta (Alberta, 2011). The economic impact is enormous. With 20 oil sands companies located in Alberta, capital—especially foreign capital—has rushed into the oil sands in record amounts to grab a piece of one of the world's largest oil reserves (Manta, 2011). Canadian labour also benefits. Indeed, as many as 10,000 trades workers from Newfoundland and Labrador regularly commute to Fort McMurray and Cold Lake for high-paying jobs. But open-pit mining, the creation of

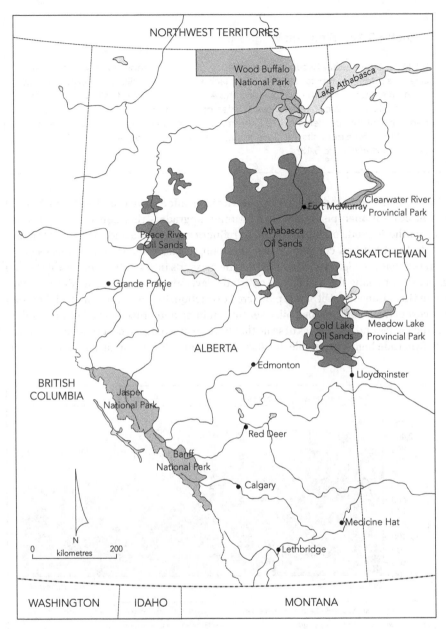

Figure 5.12 Alberta's Oil Sands Deposits

Source: Einstein (2006).

huge tailings ponds, atmospheric pollution, and the release of effluent into the Athabasca
River represent an incalculable cost to the environment (see Chapter 6).

In 1967, Great Canadian Oil Sands (a predecessor company of Suncor) began the
world's first oil sands mine. The Syncrude mine followed in 1978 and is now the largest mine by geographic area. In 2003, Shell Canada opened its Albian Sands mine at

Vignette 5.12   Bitumen: What Is It?

Bitumen is a naturally occurring combination of hydrocarbons mixed with sand and clay. This pitch-like substance is viscous, black, sticky, and tar-like. Early explorers saw bitumen seeping from the banks along the Athabasca River and called it oil tar. The Alberta oil sands lie beneath an overburden of 5–50 metres of clay and barren sand. Beneath that overburden are 40–60 metres of bitumen atop relatively flat limestone rock. While some deposits are close to the surface, most lie beneath some 50 metres of overburden (Figure 5.13).

Muskeg River, followed in 2008 by the Horizon mine owned by Canadian Natural Resources. All open-pit mines have bitumen upgraders that convert the oil sands into synthetic crude oil for shipment to refineries in Canada and the United States.

In the 1960s, oil companies were hesitant to tackle the oil sands because they believed that the high cost of converting the oil sands into synthetic crude oil was too great, thus making the viability of oil sands development improbable. Even in 1967, when Great Canadian Oil Sands pioneered extraction, its costs were much higher than the cost of most conventional oil. Over time, mining and processing oil sands resulted in efficiencies and economies of scale that have greatly reduced costs and made Suncor and Syncrude highly profitable. In 2008 Nexen opened its in-situ facility at Long Lake,

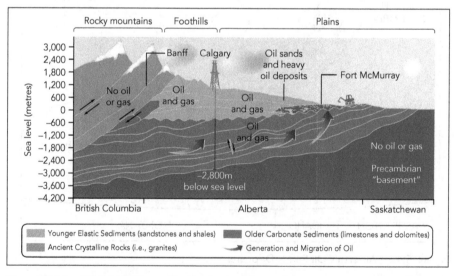

Figure 5.13   Oil Sands Viewed in Geological Cross-Section

Oil is a liquid hydrocarbon derived primarily from simple marine plants and animals. Trapped in porous sedimentary rocks, oil moved more than 100 km eastward and upward from its origin. Eventually, the oil reached and saturated large areas of sandstone at and just below the surface of what is now northern Alberta and Saskatchewan. Plans to exploit the deeper Saskatchewan deposits call for in-situ extraction techniques known as steam-assisted gravity drainage.

Sources: Oil Sands InfoMine (2008); Centre for Energy (2007).

40 km southeast of Fort McMurray. This plant contains the most advanced technology currently available. As a result, Nexen (2010), now owned by CNOOC, the Chinese state firm, combines steam-assisted gravity drainage with cogeneration and upgrading, thus reducing its operating costs to average about $25/barrel, a figure well below what other companies are able to achieve. The key is cogeneration, which reduces natural gas inputs. Even so, producing oil from these projects remains more expensive than conventional oil production in Canada and certainly much more expensive than the cost of producing Middle East oil. As well, the separation process requires large quantities of energy and water, which raises both cost and environmental concerns.

Alberta oil sands production continues to grow. In 2008, Alberta's production of bitumen reached an all-time high of 1.3 million barrels per day (bbl/d), but a new record was set in 2012 with an output of 1.9 million bbl/d. In fact, Alberta Energy forecast that production will rise to 3.8 million bbl/d in 2022 (Alberta Energy, 2010, 2014). However, the key to this expanded production is greater access to world markets, and that access will involve new pipelines to ocean ports.

Six open-pit mines (Suncor's Millennium and Steepbank mines; Syncrude's Mildred Lake and Aurora mines; Shell's Muskeg River; and Canadian Natural Resources' Horizon mine) account for nearly 60 per cent of current production. The remainder of oil sands production comes from SAGD operations—10 in the Athabasca oil sands; six in the Cold Lake reserves; and one in the Peace River deposit (Strategy West, 2010). Current upgrading capacity in Alberta is approximately 1,209,000 bbl/d of bitumen with synthetic crude oil output at approximately 1,037,500 bbl/d (Alberta Energy, 2010). Six **upgrader plants** are operating and one is under construction (Alberta Energy, 2014). Two operating upgraders are situated in Alberta's **Upgrader Alley**—one in Edmonton (Suncor's Strathcona upgrader) and the other at Fort Saskatchewan (Shell's Scotford upgrader). The remaining four upgraders are near their respective open-pit mines (Syncrude at Mildred Lake; Suncor at Base and Millennium; Nexen at Long Lake; and Canadian Natural Resources at Horizon). The North West Redwater Partnership upgrader, located just north of Edmonton, broke ground in 2013 with a completion date of 2017 (North West Redwater Partnership, 2014).

Best to Process or Export?

The old argument about Canadians being hewers of wood and drawers of water has lost much of its validity. Still, the export of Canadian bitumen for processing elsewhere does rub a nerve. While the existing upgraders refine about 60 per cent of Alberta's oil sands production by turning bitumen into lighter crude, that percentage is expected to decline as new oil sands production outweighs the capacity to process it (Tait, McCarthy, and Vanderklippe, 2011). Early in the twenty-first century, oil companies proposing additional oil sands production decided not to build upgraders but to export diluted bitumen to US heavy oil refineries because of cost advantages in the short run (Vignette 5.13). The oil industry uses the same argument for shipping bitumen to Asian countries such as China. Consequently, two proposed west coast pipelines (**Northern Gateway** and **Trans Mountain Expansion**) have been proposed to ship raw bitumen offshore for refining. At the other end of the country, the Canada East pipeline is intended to export bitumen to foreign buyers, probably in the European

---

Vignette 5.13   Keystone Pipeline and Its Proposed Expansion

A $12 billion pipeline completed in June 2010 transports Athabasca diluted bitumen to heavy oil refineries in Illinois. When oil sands companies looked at the $8 billion to $15 billion cost of building a small to large upgrader, they preferred to pay a "rent" for space on a pipeline. The proposed Keystone XL extension to the Gulf coast of Texas would result in more bitumen processed in the United States than in Canada. The source of bitumen for the Keystone XL pipeline would come from oil sands projects planned or now under construction. However, in February 2015 President Barack Obama vetoed a bill that would have given the go-ahead to the pipeline, stating that he could not allow Congress to impose its will on the extended assessment process being led by the US State Department. The US Environmental Protection Agency has opposed Keystone from the outset, because Alberta's "dirty oil" has adverse impacts on the US environment and because such a project, and its source, flies in the face of the US **clean energy policy**.

Global oil pricing geography means that Alberta oil receives the lowest price, known as the West Texas Intermediate price (the benchmark for North American crude oil pricing), while the price along the Gulf coast is closer to the higher North Sea Brent price (the benchmark for European crude oil pricing). The highest oil prices occur in the Asian market. Much of the difference can be attributed to transportation costs.

Source: TransCanada (2013).

Union; but the plan is also to ship some to refineries in Montreal, Quebec City, and Saint John, New Brunswick. The Alberta government has strongly supported the building of pipelines and has actively promoted the **Keystone XL**, Northern Gateway, Trans Mountain Expansion, and Canada East projects. Still, Alberta sees the advantage of processing bitumen within the province. In 2009, the province announced its **Bitumen-Royalty-in-Kind program** (BRIK). In doing so, the province chose to collect its royalties in kind, i.e., in bitumen, rather than in cash to ensure a sufficient supply of bitumen for proposed upgraders in Alberta (Alberta Treasury Board, 2010). The response to this government policy has been tepid with only one small upgrader, the North West Redwater Partnership upgrader, under construction at a cost of $8.5 billion with a completion date of 2017 (Alberta's Industrial Heartland, 2014b).

Foreign Investment in Megaprojects

Foreign investment is one characteristic of megaprojects. This is particularly true in the Alberta oil sands, where the investment is unusually high for each project and the construction period long. For companies, this means no return on capital investment until the bitumen moves to market. In Canada's North, foreign companies play a dominant role because they have the financial resources and, particularly in the case of state-owned companies, because a considerable amount of the energy or raw material will eventually flow to manufacturing industries in their home countries.

Until the twenty-first century, the majority of investors were either large Canadian companies (Cenovus, Husky, Suncor, and Syncrude) or multinational American-controlled

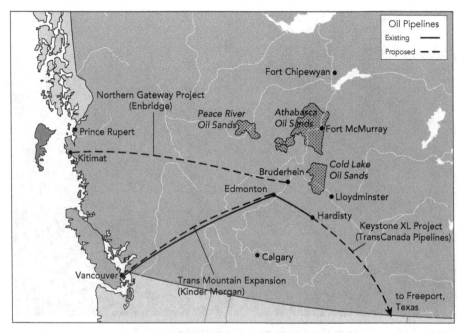

Figure 5.14   Map of Alberta Oil Sands and Pipelines

The controversy over the construction of three new pipelines from the oil sands to ocean ports has complicated matters. Opposition from environment groups and First Nations has powerful political ramifications.
While Ottawa approved the Northern Gateway project, its construction remains in doubt. The Trans
Mountain Expansion, still not submitted to the National Energy Board, faces similar opposition. A bill that
passed in Congress to approve the Keystone XL project was vetoed by President Obama over concerns
about "dirty oil" and the need to allow the State Department to complete its vetting process. If built, the
Keystone XL pipeline will allow Canadian oil sands producers to expand their production by providing more
access to the large refining markets found in the American Midwest and along the US Gulf coast and to
world markets. The pipeline, if ever approved, will have capacity to transport 830,000 barrels of oil per day to
Texas and Midwest refineries

companies (Chevron and Imperial Oil). Husky Oil, while based in Calgary, has Hong Kong
billionaire Li Ka-Shing as its controlling shareholder. China and other Asian state-owned
companies are challenging the North American stranglehold on investment and ownership of the production from the Alberta oil sands. While the privately owned companies
focus on profit, the state-owned companies have other goals. For example, Asian state-
owned companies are pursuing a strategy of buying resources around the world as a
hedge against inflation and devaluation of US currency. For example, from 2007 to 2011,
Chinese state-owned companies (PetroChina, Sinopec, and China Investment Corp.)
invested heavily in oil sands projects to a total of $11.7 billion, making up about 16 per cent
of the total investments of $73.6 billion in that time period (De Souza, 2012). Nikki Skuce
(2012), a member of the BC environmental group Forest Ethics Advocacy, conducted an
analysis on oil sands ownership based on data acquired from Bloomberg Professional
(29 Jan. 2012) and production data acquired from *Oilsands Review* (Jan. 2012) and found
extensive foreign ownership (see Table 5.7). Added to that list is the $15.1 billion purchase
of Nexen by the China National Offshore Oil Company [CNOOC] in December 2012 (Puzic,
2012). At the same time, Ottawa also approved the Malaysian state-owned company
Petronas's $6 billion purchase of Alberta's Progress Energy.

Figure 5.15   Oil Sands Development in the Fort McMurray Area

The Syncrude oil sands complex north of Fort McMurray turns oil sands into oil. Huge quantities of water and energy are required; toxic waters from the upgrader are placed in tailings ponds and later released into the Athabasca River, which may be affecting the aquatic life and the health of residents of Fort Chipewyan, many of whom eat fish on a daily basis. In 2014, a study indicated health problems faced by First Nations peoples at Fort Chipewyan were related to the game and fish they consume (McLachlan, 2014b).

Source: Ron Garnett/AirScapes.ca.

Risks are high for new projects in the oil sands where profit margins are thin. By 2014, the Chinese state companies were soon confronted with the deep drop in oil prices, rising costs of construction in Alberta, and growing environmental requirements imposed by the federal and Alberta governments. The Nexen purchase by CNOOC in 2012 looked very promising, but the returns for Nexen were reportedly only 3.5 per cent in 2014 while the overall return on investment for CNOOC was 11 per cent (Jones, 2014). One drag on its oil sands operations relates to the unresolved technical challenges facing its Long Lake project, making full production and strong profits far in the future.

On the subject of foreign takeovers in the oil patch, then Prime Minister Stephen Harper declared that:

Those acquisitions are "the end of a trend," rather than a beginning. When we say that Canada is open for business, we do not mean that Canada is for sale to foreign governments. [He went on to state] that his government has been concerned for some time that "a series of large-scale controlling transactions by foreign state-owned companies could rapidly transform (the oil sands) industry from one that is essentially a free market to one that is effectively under control of a foreign government." (Puzic, 2012)

Table 5.7 Ownership of Oil Sands Operations, 2012

Companies with foreign headquarters

Statoil: 99.83 per cent foreign ownership

Mocal Energy: 99.33 per cent foreign ownership

Murphy Oil: 99.23 per cent foreign ownership

Royal Dutch Shell: 98.49 per cent foreign ownership

Devon Energy: 98.44 per cent foreign ownership

ConocoPhillips: 97.83 per cent foreign ownership

Companies with Canadian headquarters

Petrobank Energy Resources: 94.8 per cent foreign ownership

Husky Energy: 90.9 per cent foreign ownership

MEG Energy: 89.1 per cent foreign ownership

Imperial Oil: 88.9 per cent foreign ownership

Nexen: 69.9 per cent foreign ownership

Canadian Natural Resources Limited: 58.8 per cent foreign ownership

Suncor Energy: 56.8 per cent foreign ownership

Canadian Oil Sands: 56.8 per cent foreign ownership

Cenovus: 54.7 per cent foreign ownership

Source: Skuce (2012).

Natural Gas: The Energy Source of the Future?

In a world concerned about greenhouse gas emissions, the clean-burning nature of natural gas, compared to coal and oil, greatly reduces the emission of greenhouse gases to the atmosphere. Conversion of coal-burning thermal electric plants continues, and even the North American trucking industry is switching from diesel fuel to natural gas. The current low price for natural gas makes conversion more attractive. Ontario, for example, has closed its coal-burning thermal plants and replaced them with natural gas plants to reduce air pollution.

Natural gas from northern British Columbia and Alberta is already flowing into the energy-starved Chicago market. Huge deposits of natural gas in northern Alberta and BC were discovered but lacked the means to reach US markets. In 2000, the completion of the Alliance Pipeline solved that transportation problem, thus allowing natural gas from this region to reach the Chicago market. Natural gas deposits in the Mackenzie Delta appeared to have the inside edge and plans were well advanced through the Mackenzie Gas Project, but the new technology of fracking allowed vast reservoirs of natural gas trapped in shale formations to enter the commercial market. The result is that the commercial viability of Arctic gas has been pushed well into the middle of the twenty-first century. Meanwhile, Asian markets may receive natural gas from the huge deposits in northeast British Columbia.

With Arctic gas on hold, the largest deposits of natural gas in Canada at the Liard and Horn river basins shale gas fields in northeast BC are on the edge of development. Still, a pipeline and ocean transportation challenge confronted investors because of the high cost of the investment. While no fewer than 19 proposals have surfaced, three are serious contenders. The most advanced project is the Woodfibre LNG project that expects to start shipping liquefied natural gas from its Squamish site by 2017 (Woodfibre LNG, 2015). The other two companies that have signed project development agreements with the BC government are Pacific NorthWest LNG with its export facility near the port of Prince Rupert, which expects exports will commence in 2018; and the AltaGas Douglas Channel LNG Consortium, which is working towards a final investment decision in late 2015.

Given all this enthusiasm, the BC government is pushing the LNG button hard for two reasons. First, other countries with vast gas reserves are also looking at the Asian liquefied natural gas market. The fear is that if BC does not move quickly, the Asian market will be lost. The second reason is that Premier Christy Clark is banking on LNG exports for a major economic and political boost. The BC government (British Columbia, 2014), perhaps optimistically, calls for five LNG plants to be constructed in the province between 2015 and 2024, which would create a total industry investment of $175 billion and create up to 100,000 jobs: 58,700 direct and indirect construction jobs, 23,800 permanent direct and indirect jobs for operations, and thousands more of induced jobs as a result of households having more income. This economic activity would contribute up to a trillion dollars to the province's GDP.

Water Resources

One of Canada's most valuable resources is water. Much of that resource lies in the Canadian North. The Mackenzie River, for instance, is one of the world's longest rivers. Consequently, the rivers and lakes of the Canadian North contain enormous potential for hydroelectric power development. However, market conditions have dictated that only those resources in the Subarctic warrant development. Crown corporations have taken the lead and undertaken massive hydroelectric projects in three provinces—Quebec, Manitoba, and British Columbia. The famous and controversial **Churchill Falls hydro project** in Labrador was a private project built by British Newfoundland Development Corporation and now managed by Nalcor Energy, the provincial energy corporation. In the case of Churchill Falls, the nearby iron ore mines provided the market for much of its energy, which is committed to Hydro-Québec until 2041. In other instances, the power is shipped south to Canadian consumers with the surplus sold to American utilities. At first, these huge projects were seen as a solution to both energy shortages and air pollution problems. As it turned out, hydroelectric development had an impact on the natural environment by transforming it into an industrial landscape, not only at the immediate dam sites but by flooding large areas that had been Native hunting and fishing grounds. In turn, since the natural environment was the basis of the traditional economy of Aboriginal peoples living in these areas, their lives and traditional diets have been irrevocably altered by spending more and more time as village dwellers rather than as hunters on the land (see Chapter 6 for more details on environmental impacts and Chapter 7 for the effect on the Quebec Cree).

Vignette 5.14   Long-Distance Transmission of Electric Power

Development of remote hydroelectric resources depends on the transmission of electrical energy via high-voltage power lines to southern markets. Without access to large markets, large-scale hydroelectric projects would not be viable. For many years, this potential power sat idle due to the high cost of transporting electrical energy to major industrial areas. Technological advances in electric power transmission have made remote hydroelectric projects commercially attractive. In 1903, for example, the transmission of electrical power some 140 km from Shawinigan Falls to Montreal was considered an engineering feat. Today, Hydro-Québec's high-voltage transmission lines from its power stations in the James Bay region to the cities and industries in southern Quebec are almost 10 times longer than the line between Shawinigan Falls and Montreal.

Most hydroelectric power is generated in two geomorphic areas, the Cordillera and the Canadian Shield, which have ideal natural conditions for generating electricity: a large volume of water from heavy annual precipitation and terrain with steep drops in elevation. Potential hydroelectric sites still exist in northern British Columbia, Quebec, Manitoba, Newfoundland and Labrador, and Saskatchewan. Ontario, which has the greatest need for more energy, lacks similar sites in its Canadian Shield because of relatively low elevations.

Major hydroelectric projects require large water reservoirs. To smooth out fluctuations in river flows, storage dams and water diversions supplement the water supply to the reservoirs. Ensuring a reliable flow of water throughout the year maximizes the installed generating capacity of the hydroelectric power plant. Even with a complex system of reservoirs, storage dams, and diversions, a long period of below-average precipitation can reduce river flow and hence electrical power generation. In the 1980s, the Subarctic experienced below-average precipitation so that hydroelectric power production at installations in northern Manitoba and Quebec was adversely affected. The following decades saw a return to average precipitation and the restoration of water levels. However, these variations in annual precipitation, while unpredictable, are part of a natural hydrological cycle. Such long-term variations should send a warning to those dependent on hydroelectric energy and to provinces enjoying revenues from the sale of such power.

Of all the regions of Canada, Quebec has the greatest potential for the production of hydroelectricity. The reason lies in the physical geography of Quebec's portion of the Canadian Shield. The James Bay Project occupies a huge area of the Canadian Shield in northern Quebec. In a sense, the Canadian Shield in Quebec acts as a huge reservoir with its many lakes. The river system consists of three major rivers and several smaller ones—all flow into James Bay. The main attraction of these rivers for hydroelectric development is threefold:

1. They originate in the uplands of the Canadian Shield in the interior of northern Quebec, where elevation ranges between 500 and 1,000 m, thus providing a natural drop for generating electrical power.
2. The Canadian Shield contains large lakes that are easily transformed into huge reservoirs.
3. The climate of northern Quebec has a high annual amount of precipitation.

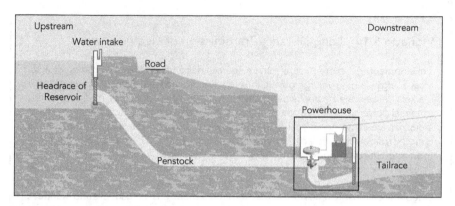

Figure 5.16 Diagram of Hydroelectric Generation

Source: Adapted from Hydro-Québec (2007b).

The negative aspects of such huge projects are related to three factors. First, much land is submerged to create reservoirs and river flows are altered (and in some cases diverted) from their seasonal rhythm to control the release of water from the reservoirs to meet the high demand for electricity in winter. Second, the hydroelectric landscape harms wildlife, which was the primary source of food for Aboriginal hunting societies and remains an important source. Third, ownership of Crown land is not as clear-cut as once thought. Aboriginal peoples who occupy these lands have claims that must be addressed before construction proceeds. However, until the James Bay and Northern Quebec Agreement in 1975, Aboriginal claims to the land were ignored.

The nature of hydroelectric projects is shown in Figure 5.16. Water is stored upstream in a reservoir, and the water (headrace) flows into the penstock (a tunnel leading to the powerhouse or generating station). Electrical power is created as the water drives the turbines in the powerhouse. After passing through the powerhouse, the water (called tailrace) flows into the river. From the reservoir to the river, the drop in elevation and the amount of water released are critical to generating power.

Crown Corporations and Hydroelectric Power

With the exception of Nova Scotia, each provincial government has created a Crown corporation to develop and market hydroelectric and thermal electrical sites. In this way, provincial Crown corporations play a major role in the development of water resources in northern areas of provinces. Provincial hydroelectric Crown corporations in British Columbia, Manitoba, and Quebec have built huge hydroelectric projects to produce low-cost energy for their consumers and to sell surplus energy to the United States. While the James Bay Project is by far the largest hydroelectric facility in North America and the second largest in the world (after the Three Gorges project in China), other large hydroelectric sites are found in Canada's Subarctic, including Churchill Falls in Labrador, the Manicouagan–Outardes complex on the Quebec North Shore, the Sir Adam Beck Hydroelectric Generating Stations on the Niagara River in Ontario, the Gordon Shrum Dam on the Peace River in northern British Columbia, and the Columbia River facility in the southern part of British Columbia.

Two regions in the Provincial North where provincial Crown corporations have developed northern power sites for electric power are the Nelson River Basin and the James Bay Basin. Newfoundland and Labrador, through its provincial corporation, Nalcor Energy, is also in the process of becoming a significant player with the development at Muskrat Falls on the Lower Churchill River. Yet high construction costs and concerted objections from local citizens, Aboriginal groups, and environmental groups cast some doubt on its prospects. This project calls for the transmission of power by subsea cables from Labrador to the island of Newfoundland and from Newfoundland to Nova Scotia.

Nelson River Basin

The Nelson River drainage basin extends over nearly 100,000 km², bringing waters from the Rocky Mountains to Hudson Bay as well as from the Red River and Winnipeg River basins. While the Red River is mainly in the United States, the Winnipeg River flows from the Lake of the Woods in Ontario (Figure 5.17). These two rivers and the Saskatchewan River empty their waters into Lake Winnipeg. As a consequence, Lake

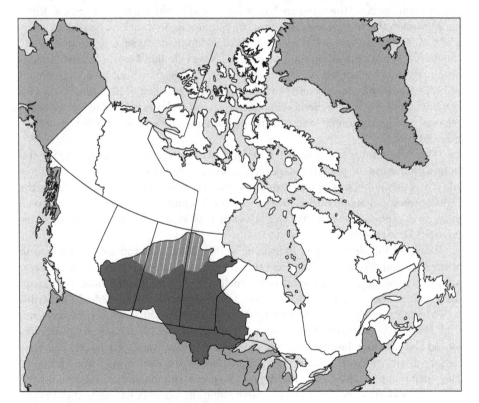

Figure 5.17 Nelson River Drainage Basin

The Nelson River watershed and the watershed of the upper Churchill River, which was diverted to the Nelson River. The purpose of the diversion was to increase the flow of water through the generators of the hydroelectric plants along the Nelson River.

Source: "Canada Drainage Basins," *The National Atlas of Canada*, 5th edn (Ottawa: Natural Resources Canada, 1985, at: atlas. nrcan.gc.ca/site/english/maps/archives/5thedition/environment/water/mcr4055.

Winnipeg serves as a giant natural reservoir for the hydro stations on the Nelson River. Manitoba Hydro regulates the flows into and out of Lake Winnipeg to maximize hydro-electric production throughout the year. The Grand Rapids hydro station on the Saskatchewan River regulates the flow of those waters into Lake Winnipeg. Grand Rapids dam created a large reservoir known as Cedar Lake that stores the water of the Saskatchewan River before it is released into Lake Winnipeg.

The Nelson River itself is 644 km in length, stretching from Lake Winnipeg to Hudson Bay. Much of its journey to the sea is through the rugged Canadian Shield, thus providing many ideal sites for hydroelectric dams. In addition, waters from the Churchill River (not the same river as that in Labrador!) have been diverted into the Nelson River to increase the volume of water. A series of dams and power stations dot the Nelson River and are known as the Nelson River Project. While much smaller than the James Bay Project, the Manitoba hydroelectric effort was but one hydroelectric development begun by provincial Crown corporations in the early 1960s.

In 1961, Manitoba Hydro completed the first stage of the Nelson River Project with the completion of the Kelsey dam and power station. A decade later, the Quebec government declared the waters of James Bay would be harnessed. One critical differ-ence between the two projects was that the Manitoba First Nations had taken treaty and their ability to share in the hydroelectric development was severely limited. While the disruption of both Aboriginal communities was similar, Manitoba First Nations obtained a much narrower arrangement—known as the Northern Flood Agreement of 1977—than the James Bay and Northern Quebec Agreement of 1975, which went far beyond compensation for direct damage to the land and way of life and into issues of regional self-government and future agreements based on "nation-to-nation" negotiations.

The Nelson River hydroelectric development, which now includes five power sta-tions with more in the planning stage (Figure 5.18), represents the major source of elec-tric power in Manitoba. Surplus production, like in northern Quebec, is exported to the United States and neighbouring provinces. Ontario's energy shortage may result in an increase in exports. Long-term power sale contracts were signed with Northern States Power of Minneapolis.

Demand for electricity is increasing within Manitoba and its US markets in Minnesota and Wisconsin; Manitoba Hydro continues to expand its northern water power. Accordingly, the Wuskwatim hydroelectric project located near Thompson was completed in 2012. Remarkable about this hydro project is that Manitoba Hydro broke from its past practices in 2008 by forming a partnership with Nisichawayasihk Cree Nation (Wuskwatim Power Limited Partnership, 2010). This partnership has moved the province beyond the sour relations between Manitoba Hydro and First Nations associated with flooding of traditional lands and destroying fishing grounds. The partnership sees Manitoba Hydro providing construction and management servi-ces to Wuskwatim Power Limited Partnership in accordance with the Project Development Agreement (PDA) signed in June 2006. This PDA ensures economic bene-fits to Nisichawayasihk members in jobs and training, with the prospect of becoming Manitoba Hydro employees.

Manitoba Hydro's most ambitious undertaking is at the Conawapa site on the lower Nelson River. The Conawapa Generating Station would be the largest hydroelectric

Figure 5.18   Jenpeg Generating Station

Located on the upper reaches of the Nelson River, Jenpeg produces hydroelectric power and is used to control the outflow of water from Lake Winnipeg into the Nelson River.

Source: Manitoba Hydro (2007).

project in Manitoba. The nearest large markets are southern Ontario and Minnesota. Market conditions are not yet right for such a huge investment and Manitoba Hydro has delayed the start of construction for the foreseeable future (Manitoba Hydro, 2010). Adding to its construction costs, a long and expensive transmission line would be required. However, the markets for such power are looking elsewhere for their future needs, making additional exports of Manitoba electricity somewhat problematic. For instance, Ontario is turning to gas-generated power and possibly nuclear power while US customers have the option of gas-generated electric energy. Gas, thanks to successful fracking in the United States, is in great surplus and this surplus has driven gas prices down, making thermal-generated electrical power very attractive.

The James Bay Project

In 1971, Quebec Premier Robert Bourassa announced plans to turn the "wilderness" of northern Quebec into an energy landscape consisting of a series of hydroelectric dams, reservoirs, and power stations. Hydro-Québec was assigned the task of implementing this project. The provincial strategy went beyond producing power. Hydro-Québec had two other objectives—to accelerate the growth of Quebec's industrial economy; and, obviously, to make this huge investment financially viable. The first objective called for Hydro-Québec to offer low-cost power to firms locating in Quebec. Second, the financial viability was ensured by the sale of surplus electricity at market prices to utilities in New England, and these profits in turn helped pay for the project. In fact, without long-term agreements with American utilities in place, the first phase of the James Bay Project might not have happened. The first site was focused on La Grande River and its lakes. Later, the remaining two large basins in

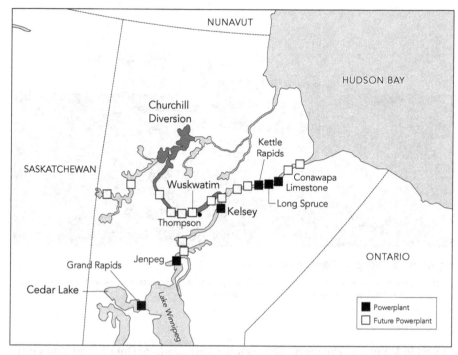

Figure 5.19   Power Plants on the Nelson River in Northern Manitoba

The Nelson River is ideally suited for producing hydroelectric power. This river forms the heart of the hydroelectric system in Manitoba. The most recent power plant, Wuskwatim, was completed in 2012 and much of its power goes to Thompson and its nickel mine and smelter. Prospects for another power plant are now slim because of the availability of natural gas in the United States. The first power plant in northern Manitoba, Grand Rapids (1960), was built on the Saskatchewan River as its waters enter Lake Winnipeg. The other power plants are Kelsey (1961), Kettle Rapids (1974), Long Spruce (1979), Jenpeg (1979), just north of Lake Winnipeg on the upper arm of the Nelson River, and Limestone (1990). Conawapa, if built, would be the largest of these hydroelectric projects.

northern Quebec—the Great Whale River Basin and the Nottaway–Broadback–Rupert River Basin—would be developed.

The first phase, known as La Grande Project, began in 1972, but was halted by legal action launched by the Quebec Cree and Inuit. At that time, Quebec, like other provinces, did not recognize Aboriginal rights to their traditional lands. The Quebec court ruled that the two parties must reach an agreement. In 1975, the James Bay and Northern Quebec Agreement became the law of the land. Construction then moved ahead, and La Grande Project Phase I was completed in 1985 at a cost $13.7 billion.

The Great Whale River Project was to be the second phase. Announced in 1985, the strategy of Hydro-Québec was again to seek long-term contracts for its electricity and to use the initial payments to cover the construction costs. Maine, New York, and Vermont all negotiated contracts with Hydro-Québec. Although hydroelectric power is touted as environmentally friendly energy, environmental organizations, including the Sierra Club and the New York Energy Efficiency Coalition, and the Quebec Cree under the leadership of Chief Matthew Coon Come strongly opposed the development of the Great Whale River Project and launched a public relations campaign in New York claiming that the project would cause great harm to the land and people. In 1992, the state of New York cancelled two contracts with Hydro-Québec totalling $17.6 billion.

Figure 5.20   Eastmain River, Its Reservoir, and the Eastmain-1 Power Station

With the signing of the Paix des Braves in 2002, construction of the Eastmain-1 Project began almost immediately. In 2006 the Eastmain-1 reservoir was filled, and during the same year the Eastmain-1 power station, shown above, was producing power.

Source: Photo Hydro-Québec

These contracts were critical for the project. While the relatively low price of natural gas was the principal factor affecting this decision, environmental concerns related to the transformation of 1,667 km² of aquatic habitat in northern Quebec also played an important role. Ironically, much of the natural gas purchased by thermal electricity plants in New England came from western Canada via the extension of the TransCanada Pipeline network in Ontario to the Duke Pipeline network in New York. In 1994, the Quebec government announced that the Great Whale River Project would not proceed until the demand and price for its electricity improved.

While the Great Whale River Project remains on hold, the Nottaway–Broadback–Rupert River project to the south came into the spotlight at the beginning of the twenty-first century (Figure 5.20). A major turning point in relations between the Quebec Cree and the Quebec government sparked interest in developing all resources in the traditional Cree territory. In 2002, the two parties signed the Paix de Braves Agreement (see Vignette 7.9), which called for the development of the natural resources of northern Quebec, a transfer of $3.6 billion in annual payments spread over 50 years to the Cree Regional Authority, and the sharing of the profits from these undertakings. This agreement with the Cree, who consented to the immediate construction of the Eastmain-1 Project on the Eastmain River and the partial diversion of the Rupert River (subject to the applicable environmental legislation and protection stipulated in the *James Bay and Northern Quebec Agreement*), cleared the way for harnessing the Eastmain River and partially diverting the Rupert River north to La Grande Project, where the water now passes through the La Grande power stations.

Conclusion

Two questions were posed at the beginning of this chapter: Can resource development in the North lead to economic stability and diversification? Is sustainable development possible in an economy based on non-renewable resources?

On the surface, given the nature of non-renewable resources and the depressed situation for the forest industry, the answer to both questions is no. Yet, the continued evolution of the resource economy towards greater involvement of Aboriginal peoples and their institutions might change the outcome. Resource-sharing might be such a catalyst? Then the bigger question is: Could the mineral industry somehow take on some or all of the attributes of the fur trade, when a partnership existed?

With the North as a resource hinterland, northern economic stability and diversity are unlikely to follow the pattern that emerged in southern Canada. Yet, a northern version is possible. Back in 1939, Peter Drucker wondered how the vast power held by multinational corporations could be justified within the tradition of Western political philosophy (Vignette 5.2). Part of the answer may lie in joint ventures such as those between Aboriginal organizations and multinational corporations. Within this business structure, economic and social development could then coalesce, leaving governments to play their traditional supporting role. As the James Bay and Northern Quebec Agreement illustrates, times change, paradigms shift, and old ways are left behind. The Paix des Braves, a bold step beyond the JBNQA, is but one example where a crisis atmosphere provided an arena for new thinking and an agreed balancing of interests. Yet, the apparent failure of the Mackenzie Gas Project, which did involve Aboriginal partners, and slumping commodity prices indicate the danger of relying on megaprojects to move the North towards the goals of economic stability and diversity and sustainable development.

Taking a different look at sustainability, let us ask the question: How critical is the human resource issue to achieving sustainability? As Greg Poelzer (2009: 427) has noted, education that bridges both Aboriginal and non-Aboriginal worlds forms a critical foundation for a sustainable North: "Education is key to facing the complex dualities, enormous challenges and tremendous opportunities in the contemporary Canadian North." For the long term, then, focusing on the basic social building blocks, such as education, may well be a more reliable route to the goal of sustainable development than the uncertain, piecemeal approach of mega resource projects.

Challenge Questions

1. Resource projects involve energy and minerals. Does the extraction of these resources suggest a relationship to the North's geomorphic regions shown in Figure 2.12?
2. Megaprojects should have a greater impact on the northern economy. What structural weaknesses prevent this hinterland from fully benefiting from megaprojects and thus beginning to diversify its economy?
3. Does the failure of a northern diamond processing industry in Yellowknife forebode the future of northern diversification?
4. Whether you consider the oil sands with reserves lasting at least 100 years or the Meadowbank gold mine with reserves lasting less than 20 years, do you think the North can avoid the staple trap?

5. In the early 1990s, the economy of the Northwest Territories depended heavily on gold mining. With the closure of its gold mines, diamond mining replaced gold. Does this "replacement" of one non-renewable resource for another represent a northern version of "sustainable development"?

6. In the late twentieth century, the James Bay Project was bitterly opposed by Quebec Cree. How did the twenty-first-century version of the James Bay Project take on a new life and win their active support?

7. If you were the Prime Minister of Canada, what argument would you make to approve or reject the two proposed west coast bitumen pipelines and the Canada East one?

8. While the federal government approved of the Mackenzie Gas Project, it is no longer a viable project. Why not?

9. Peter Drucker believed that the corporation, rather than the state, was the ideal candidate for the leadership of a rapidly changing modern industrial society. What resource companies best demonstrated such leadership?

10. Will the newly elected NDP Alberta government continue to support the Keystone XL, the Northern Gateway, and Trans Mountain Expansion pipeline projects? After all, these pipelines would transport more and more bitumen to heavy oil refineries in the United States and China, thus diminishing the prospects for processing bitumen locally, which would increase the economic diversification of Alberta.

Notes

1. Following World War II, the rich Mesabi Iron Range in Minnesota was near exhaustion, and US iron and steel plants began to look abroad for alternative sources. Iron ore deposits in northern Quebec and Labrador were close at hand but required enormous investment in infrastructure. The previously worthless ore bodies of the Labrador Trough, known for over a half-century, suddenly became a valued resource. By 1947, plans were laid by six American iron and steel companies for large-scale open-pit mining around Knob Lake in northern Quebec and at other locations in the region. Several mining towns were built, including Gagnon and Schefferville in Quebec and Labrador City and Wabush in Labrador. In 1954, Schefferville was connected by a 571-km railway to the port of Sept-Îles on the north shore of the St Lawrence River. Iron concentrate was shipped by rail to Sept-Îles and then by ship from Sept-Îles to the iron and steel plants in Ohio and Pennsylvania. Boom conditions took hold until the 1980s, when a downturn in the demand for US steel resulted in a sharp drop in iron ore shipments and lower-priced foreign steel from Asian firms began to displace American steel. Several mines closed, and the result was a depressed northern mining landscape of abandoned operations.

2. *Natural factor:* Sea ice provides one example of the challenge of physical geography of the Arctic Ocean. Two former Arctic mines provide an insight into the transportation challenges posed and the need for co-operation between mining firms. Polaris (on Little Cornwallis Island) and Nanisivik (on the northern tip of Baffin Island) were lead/zinc mines that shipped their ore through the eastern end of the Northwest Passage (Lancaster Sound) to Germany. But because the Arctic ice pack blocked Lancaster Sound for most of the year prior to 2007, both mines had to store their ore for 10 months and then ship it by a specially designed ship, the MV Arctic. Even so, floating ice was often present and the double-steel reinforced hull of the MV Arctic proved most effective to deal with floating ice but not with the pack ice. In 2002, both mines closed when their commercial ore bodies were exhausted.

Transportation factor: Transportation infrastructure is sorely lacking in the North. In Nunavut, for example, few people—less than 40,000—occupy this largest territory in Canada, yet no part of Nunavut is connected to the rest of Canada by road or rail, forcing companies like Agnico-Eagle to rely on air and sea transportation. In the northern reaches of provinces, provincial governments have tried to overcome this distance barrier by building modern transportation routes into their hinterlands. The cost of creating a northern transportation network similar to that found in southern Canada is well beyond the capacity of Ottawa and the provinces. Richard Rohmer's Mid-Canada Development Corridor across the Subarctic communities hinged on a northern railway, which Rohmer believed would act to open the Subarctic to a wide range of economic developments similar to the CPR in western Canada. Pearson's Liberal government refused to provide any funding and Rohmer's ambitious plan ended.

The age-old argument of the federal government's role in providing transportation infrastructures goes to the heart of Canada's history. Under the terms of Confederation, the federal government had no responsibility for highway construction in the provinces and zero interest in such investments in the territories. While politically popular, the high construction costs soon dampened even the most ardent politician. The first federal investment in northern transportation was a result of political pressure from western farmers that caused Ottawa to build the Hudson Bay Railway to Churchill, Manitoba. Completed in 1929, it was designed to provide an alternative route to European markets for western grain. Prairie farmers believed that the shorter Hudson Bay rail route to major European customers would reduce the total shipping costs. The new railway never became an important export route for grain because of the high cost of marine insurance in the ice-ridden waters of Hudson Strait and Hudson Bay. Instead, the Hudson Bay Railway was instrumental in unlocking the mineral, forest, hydro, and fishing resources of northern Manitoba.

By the late 1950s Ottawa began to play a role in northern transportation, thus abandoning its laissez-faire policy. The reason was straightforward: the federal government saw northern development as a way to strengthen the nation and to ensure Canadian sovereignty. With Prime Minister Diefenbaker's Roads to Resources program (see Chapter 3) for northern areas of the provinces and a similar program in the territories, roads were built to help private companies reach world markets. One example was the road built to connect the asbestos mine at Cassiar (now closed) in northern British Columbia with the port of Stewart. Other highways were constructed to provide major northern centres with a land connection to southern Canada. For example, the Mackenzie Highway from Edmonton to Hay River was extended to Yellowknife. In all, from 1959 to 1970, over 6,000 km of new roads were built under this program at a cost of $145 million (Gilchrist, 1988: 1877).

In the 1920s, the British Columbia government sought to stimulate resource development by building the Pacific Great Eastern Railway (PGE), but the high cost of construction in the Cordillera and limited funds in the provincial treasury prevented its completion until 1952. Hence, the PGE was known as "the railway that begins nowhere and ends nowhere." By connecting Prince George to the port of North Vancouver, the railway opened the vast forests of British Columbia's interior to global markets, and in time the PGE reached further north to Fort St John and further east to Dawson Creek and the Peace River country. This railway was renamed British Columbia Railroad in 1972 and BC Rail in 1984. Canadian National Railway purchased BC Rail in 2004.

Northern benefits factor: The concern that northerners were not benefiting from resource projects was a key focus of the Berger Inquiry. With the public awareness of this shortcoming in northern development, resource companies were required to make special allowances for attracting northern workers and for ensuring contracts go to northern businesses. This process began with the Norman Wells Oil Development and Pipeline Project of the early 1980s. Unlike southern Canada, northern businesses and workers are incapable of responding to the needs of large-scale industrial projects. Consequently, large southern firms supply the demand for equipment and supplies and companies turn to southern labour markets where skilled workers are readily available. The Norman Wells Expansion Project exemplified this Aboriginal/non-Aboriginal workforce difference in experience and training. During the construction phase from 1982 to 1985, the largely non-Aboriginal residents of the small community of Norman Wells (about 650 inhabitants) enjoyed the economic boom. The workforce in the three largely Dene communities along the pipeline route had low participation rates.

3. By the 1970s, the North was transformed into a resource hinterland for the world economy. At the same time, great social changes were taking place in the Aboriginal community. One change was their relocation to small settlements, which exposed Aboriginal people, but especially their children, more intensely than ever before to Canadian culture and the Canadian economy through the institutions established by the state in these settlements. Other changes were a weakening of their land-based economy, a growing participation in the wage economy, and a greater dependency on government.

    Looking back at the early pattern of resource development, few northern Aboriginal people were prepared for the market economy, and consequently they could gain little from development and not much more from job and business opportunities. This issue was certainly a major theme in the 1977 Berger Report. In some developments, such as the James Bay Project in northern Quebec, the Cree saw their traditional land-based economy negatively affected. In the twenty-first century, however, resource projects are taking place after most land claims have been settled, and companies are negotiating "private" impact and benefits agreements with locally affected Aboriginal groups (Bradshaw and Prno, 2007; Slocombe, 2000). With these agreements in place, coupled with recognition by companies of a need to provide northern benefits, resource projects are now more likely to have a positive impact and such impacts may lead to economic diversification within the local area and even beyond.

4. The Lubicon Lake Nation has one chief while the Lubicon Lake Band has another. This political divide has haunted the Lubicon in their current search for a land claim agreement with Ottawa. How did the Lubicon stumble into these troubled waters? First, the tribe was missed in the 1899 Treaty 8 negotiations. Second, efforts to obtain a land claim agreement with Ottawa have failed, largely over who represents the Lubicon. Without a treaty, the Alberta government felt free to award timber leases for the boreal forest claimed by the Lubicon to five forest companies. At the same time, the government permitted oil companies to conduct both drilling and fracking on these lands. Not unexpectedly, the awarding of these leases before the Lubicon Lake Nation settled their land claim has created a bitter and smouldering dispute. One such dispute is with the Japanese firm, Daishowa, whose pulp plant is located in Peace River, Alberta. This firm continues to harvest timber from Lubicon traditional land. This dispute continues, with the Lubicon Lake Nation arguing in a lawsuit that the Alberta government cannot grant forest and mineral rights on land the Lubicon never surrendered (Weber, 2014). Added to this violation of their traditional lands, the Lubicon Lake Nation went to court in March 2015 over the oil spill on its territory. However, without a legal claim to the land, the Alberta courts have been reluctant to register a decision. Perhaps only the Supreme Court of Canada has the power to unravel this tangled jurisdictional dilemma.

5. The Klondike gold rush of 1897–8 was the first major mining operation in the North but was the opposite of megaprojects, which are controlled by corporations. The Klondike gold rush, on the other hand, drew thousands of individual prospectors into the North

seeking the gold nuggets found in the streambeds of the Klondike River and its tributaries. Placer mining involved the recovery of auriferous deposits found in streambeds by simple and inexpensive technology called panning or surface sluicing. This technology enabled seasoned prospectors and greenhorn amateurs to recover the most accessible nuggets and fine sand-like gold particles from tributaries of the Klondike River. Within a few years, placer gold became harder and harder to find. Gold was buried in frozen ground well below the surface but individuals had neither the capital nor the organization necessary to mine this gold. This type of mining was much more capital-intensive, requiring expensive dredging and hydraulic equipment along with separate, highly organized water-supply systems and electric power supplies. Large companies gradually took over the gold-mining industry in Yukon. The Yukon Gold Corporation was the largest of these operations. To supply its mining sites on Bonanza and Eldorado creeks with water and electrical power, it built a hydroelectric dam and a 100-km-long water distribution system (Rae, 1968: 99–100). This mining was limited to the summer months when the frozen ground (permafrost) could be thawed and the gravel sorted. During the long summer days, the dredges operated around the clock, although the existence of permafrost made hydraulic operations more difficult and time-consuming.

6. The Norman Wells oil field, discovered in 1920, has proven oil reserves of around 100 million cubic metres of light grade oil (Esso, 1980). A small refinery was erected at the site and oil was produced to meet the needs of the residents of the Mackenzie Valley. In 1925, operations ceased because the demand for oil was too small. In 1932, production recommenced because the mining operation at Port Radium on Great Bear Lake needed fuel oil. In 1936, the gold mine at Yellowknife also required fuel oil. Given the slow growth of a market for oil in the Mackenzie Basin, production increased very slowly until, for a short time during World War II, the Canol Project brought oil from Norman Wells by pipeline to Whitehorse and then on to Fairbanks, Alaska. After the war, the refinery at Whitehorse was closed, the new wells at Norman Wells were capped, and the pipeline was abandoned. By 1947, the pipeline, pumping equipment, and support vehicles were sold as surplus war assets. Imperial Oil purchased the refinery and moved it to Edmonton, where it processed oil from the newly discovered Leduc field in central Alberta. Norman Wells returned to supplying customers in the Mackenzie Valley. From 1946 on, the local market grew as Indians and Métis moved into settlements where their houses were heated by fuel oil. The demand for fuel oil increased as community infrastructures grew and more mines appeared.

References and Selected Reading

Abele, Frances, Thomas J. Courchene, F. Leslie Seidle, and France St-Hilaire. 2009. *Northern Exposure: Peoples, Powers and Prospects in Canada's North.* Institute for Research on Public Policy, vol. 4. Quebec City: Marquis Book Printing.

Alberta. 2011. "Oil Sands Projects in the Regional Municipality of Wood Buffalo." *Labour Market Information* (Feb.). www.woodbuffalo.net/linksFACTSOG.html.

Alberta Energy. 2010. "Fact and Statistics." 16 Aug. www.energy.alberta.ca/OilSands/791.asp.

————. 2014. "Fact and Statistics." www.energy.alberta.ca/OilSands/791.asp.

Alberta's Industrial Heartland. 2014a. "Project Status." 27 June. http://www.industrialheartland.com/index.php?option=com_content&view=article&id=130:project-status&catid=53&Itemid=160.

————. 2014b. *Alberta's Oil Sands Projects and Upgraders.* Jan. http://www.energy.alberta.ca/LandAccess/pdfs/OilSands_projects.pdf.

Alberta Treasury Board. 2010. *Responsible Actions: A Plan for Alberta's Oil Sands.* www.treasuryboard.gov.ab.ca/ResponsibleActions.cfm.

Antwerp Facets. 2010. "Financial Crisis Shuffles Rankings of Diamond Producers." 18 Aug. www.antwerpfacetsonline.be/financial-crisis-shuffles-rankings-diamond-producers.

Barrett, Tom. 2002. "Big Salaries, Boredom Lead to Trouble." *Edmonton Journal* (Special Fort McMurray Economic Report), 10 May, EJ15.
Bell, Jim. 2010. "Mine's Alchemy Turns Nunavut Poverty into Hope." *Nunatsiaqonline*, 20 June. www.nunatsiaqonline.ca/stories/article/98789_agnico-eagles_alchemy_turns_nunavut_poverty_into_hope/.
Berger, Thomas R. 2006. *The Nunavut Project: Nunavut Land Claims Agreement. Implementation Contract Negotiations for Second Planning Period 2003–2013*. Conciliator's Final Report. Ottawa: Indian and Northern Affairs Canada. www.ainc-inac.gc.ca/al/ldc/ccl/fagr/nuna/lca/nlc-eng.asp.
Bone, Robert M. 1984. *The DIAND Norman Wells Socio-economic Monitoring Program*. Report 9-84. Ottawa: Department of Indian Affairs and Northern Development.
————. 1998. "Resource Towns." *Cahiers de Géographie du Québec* 42, 116: 249–59.
————. 2014. *The Regional Geography of Canada*, 6th edn. Toronto: Oxford University Press.
Bouw, Brenda. 2010. "A Eureka Moment for a Stubborn Prospector." *Globe and Mail*, 16 Aug. www.theglobeandmail.com/report-on-business/industry-news/energy-and-resources/eureka-moment-for-stubborn-prospector/article1675008/.
Braden, Bill. 2013. "Agnico Eagle Leads Nunavut into Modern Mining Era." *Canadian Mining Journal*, 1 Oct. http://www.canadianminingjournal.com/news/agnico-eagle-leads-nunavut-into-modern-mining-era/1002568842/?&er=NA.
Bradshaw, Ben, and Jason Prno. 2007. "Program Evaluation in a Northern, Aboriginal Setting: Assessing Impact and Benefit Agreements." Presentation at the annual meeting of the Canadian Association of Geographers, 30 May.
British Columbia, Minister of Natural Gas Development. 2014. *LNG 101: A Guide to British Columbia's Liquefied Natural Gas Sector*. http://www.gov.bc.ca/mngd/doc/LNG101.pdf.
Byrd, Craig. 2006. *Diamonds: Still Shining Brightly for Canada's North*. Ottawa: Statistics Canada, Catalogue no. 65-507-MIE, No. 007. www.statcan.ca/english/research/65-507-MIE/65-507-MIE2006007.pdf.
Canada. 2014. "Explore for More: Mining Facts and Figures." http://www.acareerinmining.ca/en/industry/factsfigures.asp.
Canadian Coalition for Nuclear Responsibility. 2005. Map of Uranium Mining Activities in Canada, from "Uranium Discussion Guide." www.ccnr.org/uranium_map.html.
Canadian Nuclear Safety Commission. 2014. "Uranium Mines and Mills in Canada." 11 June. http://www.nuclearsafety.gc.ca/eng/uranium/mines-and-mills/index.cfm#OperatingUraniumMinesandMills.
CBC News. 2007. "The Big Commute." 29 Oct. www.cbc.ca/nl/features/bigcommute/.
————. 2010a. "Vale Ramping Up Long Harbour Work." 28 June. www.cbc.ca/canada/newfoundland-labrador/story/2010/06/28/vale-long-harbour.html.
————. 2010b. "Oil Sands Poisoning Fish, Say Scientists, Fishermen." 16 Sept. www.cbc.ca/canada/calgary/story/2010/09/16/edmonton-oilsands-deformed-fish.html.
————. 2014. "Ontario Leaders Tout Ring of Fire's Potential and Promise Action." 29 May. http://www.cbc.ca/news/canada/toronto/ontario-votes-2014/ontario-leaders-tout-ring-of-fire-s-potential-and-promise-action-1.2657380.
Centre for Energy. 2007. "How Are Oilsands and Heavy Oil Formed?" www.centreforenergy.com/generator.asp?xml=/silos/ong/oilsands/oilsandsAndHeavyOilOverview03XML.asp&template=1,1,1.
Clapp, R.A. 1999. "The Resource Cycle in Forest and Fishing." *Canadian Geographer* 42, 2: 129–44.
Consolidated Thompson. 2010. "News and Media." 9 Sept. www.consolidatedthompson.com/s/NewsReleases.asp.
De Beers Canada. 2007. "Snap Lake: Project Development—Phase One." www.debeerscanada.com/files_2/snap_lake/snaplake-development_photo-enlarge_a.html.
De Souza, Mike. 2012. "Majority of Oil Sands Ownership and Profits Are Foreign, Says Analysis." National Post, 10 May. http://business.financialpost.com/news/majority-of-oil-sands-ownership-and-profits-are-foreign-says-analysis.
Diavik Diamond Mines Inc. 2007. "Diavik Diamond Mine." www.diavik.ca/loc.htm.
Drucker, Peter F. 1939. *The End of Economic Man: A Study of the New Totalitarianism*. New York: John Day.
————. 1993. *Post-Capitalist Society*. New York: Harper Business.
Duerden, Frank. 1992. "A Critical Look at Sustainable Development in the Canadian North." *Arctic* 45, 3: 219–25.
Duffy, Patrick. 1981. *Norman Wells Oilfield Development and Pipeline Project: Report of the Environmental Assessment Panel*. Ottawa: Federal Environmental Assessment Review Office.

Editorial Board. 2014. "Delay the Approval of Conawapa." *Winnipeg Free Press*, 24 Mar. http://www. winnipegfreepress.com/opinion/editorials/delay-the-approval-of-conawapa-251846721.html.

Einstein, Norman. 2006. "Image: Athabasca Oil Sands." *Wikipedia.* en.wikipedia.org/wiki/Image:Athabasca_ Oil_Sands_map.png.

Esso. 1980. *Norman Wells Oilfield Expansion: Development Plan.* Calgary: Esso Resources Canada Ltd.

Fekete, Jason. 2010. "China Pumps More Investment into Alberta Oilsands." *Vancouver Sun*, 13 May. www. vancouversun.com/business/China+pumps+more+investment+into+Alberta+oilsands/3025041/ story.html.

Fenge, Terry. 2009. "Economic Development in Northern Canada: Challenges and Opportunities." In Abele et al. (2009: 375–93).

Fort McMurray Chamber of Commerce. 2015. *Demographics and Census.* http://www.fortmcmurray-chamber.ca/demographics.html.

Gagnon, Jeanne. 2010. "More Quality Rough Diamonds Needed." *Northern News Service Online*, 16 June. http://www.nnsl.com/frames/newspapers/2010-06/jun16_10d.html.

Gautier, Donald L., et al. 2009. "Assessment of Undiscovered Oil and Gas in the Arctic." *Science* 324, 5931: 1175–9

Geological Survey of Canada. 2001. *Pipeline–Permafrost Interaction: Norman Wells Pipeline Research.* sts. gsc.nrcan.gc.ca/permafrost/pipeline.htm.

———. 2007. "Norman Wells Pipeline Research." gsc.nrcan.gc.ca/permafrost/pipeline_e.php.

Gibben, Robert. 2009. "China Makes Strategic Investment in Quebec Iron Ore." *Financial Post*, 7 Sept. www.financialpost.com/story.html?id=1969515.

Gilchrist, C.W. 1988. "Roads and Highways." In James H. Marsh, ed., *The Canadian Encyclopedia*, 2nd edn. Edmonton: Hurtig, 1876–7.

Greenwood, John, 2007. "Surging Prices Separates Wheat from the Chaff", *National Post*. 15 June. FP1.

Hasselback, Drew. 2002. "Voisey's Bay Deal Is Done, Next Comes Big Writedown." *National Post*, 12 June, FP7.

Hayter, Roger, and Trevor J. Barnes. 2001. "Canada's Resource Economy." *Canadian Geographer* 45, 1: 36–41.

Hoefer, Tom. 2009. "Diamond Mining in the Northwest Territories: An Industry Perspective on Making the Most of Northern Resource Development." In Abele et al. (2009: 395–414).

Hydro-Québec. 2005. "La Grande Rivière." www.hydroquebec.com/generation/hydroelectric/la_grande/ index.html.

———. 2006. "Eastmain-1: Stage 3: Cofferdams and Dam." www.hydroquebec.com/eastmain1/en/ batir/etapes_photos.html?list=4.

———. 2007a. "James Bay Project: Geographic Location." www.hydroquebec.com/visit/ virtual_visit/>.

———. 2007b. "Construction Projects in Quebec." www.hydroquebec.com/eastmain1/en/batir/fiche_6. html.

Innis, Harold A. 1930. *The Fur Trade in Canada.* Toronto: University of Toronto Press.

Jang, Brent. 2014. "B.C. on Track to Have Woodfibre LNG Project Running by 2017: Clark." *Globe and Mail*, 6 May. http://www.theglobeandmail.com/news/british-columbia/woodfibre-lng-project-to-be-running-by-2017-premier-clark/article18503077/#dashboard/follows/.

Jensen, Laurel. 2013. "How the Oil Sands Industry is Benefitting Companies in Other Parts of Canada." *Alberta Oil*, Sept. http://www.albertaoilmagazine.com/2013/09/our-home-and-native-sands/.

Joint Review Panel. 2009. *Foundation for a Sustainable Northern Future: Report of the Joint Review Panel for the Mackenzie Gas Project*, Dec. www.ngps.nt.ca/report.html.

Joseph, Chris. 2010. "The Tar Sands of Alberta: Exploring the Gigaproject Concept." Presentation at Prairie Summit Regina for the Canadian Association of Geographers, 4 June.

Kelly, E.N., J.W. Short, D.W. Schindler, P.V. Hodson, M. Ma, A.K. Kwan, and B.L. Fortin. 2009. "Oil Sands Development Contributes Polycyclic Aromatic Compounds to the Athabasca River and Its Tributaries." *Proceedings, National Academy of Sciences* 106: 22346–51.

Kinross. 2013. "White Gold Exploration Project." http://www.kinross.com/operations/dp-white-gold,-yukon.aspx.

Légaré, André. 2000. "La Nunavut Tunngavik Inc.: Un examen de ses activités et de sa structure adminis-trative." *Études/Inuit/Studies* 24, 1: 197–224.

Lusinyan, Lusine, Julien Reynaud, Tim Mahedy, Dirk Muir, Ivo Krznar, and Soma Patra. 2014. *The Unconventional Energy Boom in North America: Macroeconomic Implications and Challenges for Canada.* International Monetary Fund, 15 Jan. http://www.imf.org/external/pubs/ft/scr/2014/cr1428.pdf.

Mackenzie Gas Project. 2005. *Environmental Impact Statement (EIS) in Brief.* www.mackenziegasproject.
com/theProject/regulatoryProcess/EISInBrief/EISInBrief.html.
————. 2007. "Mackenzie Gas Project Overview." www.mackenziegasproject.com/theProject/over-
view/index.html.
McLachlan, Stéphane M. 2014a. "Press Conference Held to Discuss Alarming Results of Fort Chipewyan
Health Study." 10 July. http://acfnchallenge.wordpress.com/2014/07/10/press-conference-held-to-
discuss-alarming-results-of-fort-chipewyan-health-study/.
————. 2014b. *Environmental and Human Health Implications of the Athabasca Oil Sands for the
Mikisew Cree First Nation and Athabasca Chipewyan First Nation in Northern Alberta.* 7 July. Winnipeg:
Environmental Conservation Laboratory, University of Manitoba.
Manitoba Hydro. 2007. "Jenpeg Generating Station." www.hydro.mb.ca/corporate/facilities/gs_jenpeg.
shtml.
————. 2010. "Conawapa Generating Station." www.hydro.mb.ca/projects/conawapa.shtml?WT.
mc_id=2608.
Manta. 2011. "Oil Sands Companies in Alberta." Mar. www.manta.com/world/North+America/Canada/
Alberta/oil_sand_mining--E313708D/.
Marotte, Bertrand. 2014. "Labrador Iron Halts Mines amid Steel-Industry Slump." *Globe and Mail,* 2 July.
http://www.theglobeandmail.com/report-on-business/industry-news/energy-and-resources/labra-
dor-iron-mines-suspends-operations-amid-falling-ore-prices-high-costs/article19409363/.
Nasdaq. 2014. "Latest Price & Chart for CBOT Gold 100 oz." http://www.nasdaq.com/markets/gold.
aspx?timeframe=10y.
Nassichuk, W.W. 1987. "Forty Years of Northern Non-Renewable Natural Resource Development." *Arctic*
40, 4: 274–84.
Natural Resources Canada. 2010. "Mineral Production of Canada, by Provinces and Territories." 1 Nov.
mmsd.mms.rncan-nrcan.gc.ca/stat-stat/prod-prod/ann-ann-eng.aspx.
————. 2014. "Forest: Statistical Data." 25 Oct. http://cfs.nrcan.gc.ca/statsprofile.
————. 2015. "Minerals and Metals: Annual Statistics." 24 July. http://www.nrcan.gc.ca/
mining-materials/statistics/8848.
Nexen. 2010. "Long Lake Phase One." www.nexeninc.com/en/Operations/OilSands/LongLake/
PhaseOne.aspxSAGD.
Northern Gas Pipelines. 2005. Maps Archive. www.arcticgaspipeline.com/Reference/Maps/APGmap7-02.jpg.
Northern Miner. 2013. "Vale Commissions Hydromet Nickel Plant at Long Harbour." 16 Dec. http://www.
northernminer.com/news/vale-starts-commissioning-novel-hydrometallurgical-nickel-plant-in-newfo
undland/1002757473/?&e=rq0wMrp3vyWrlxu0q82vM20&ref=enews_NM&utm_source=NM&utm_
medium=email&utm_campaign=NM-EN12172013.
North West Redwater Partnership. 2014. *The Project.* https://www.nwrpartnership.com/project.
Northwest Territories (NWT). 2007. *Diamond Facts: 2006 Diamond Industry Report.* www.iti.gov.nt.ca/
diamond/pdf/diamondfacts_2006.pdf.
Northwest Territories (NWT) Bureau of Statistics. 2015. *Statistics Quarterly* 37, 2 (Mar.). http://www.
statsnwt.ca/publications/statistics-quarterly/sqmar2015.pdf.
Parsons, Jon, and Arn Keeling. 2015. "Crisis in Labrador West: Boom, Bust, Repeat." The Independent.ca.
1 June. http://theindependent.ca/2015/06/01/crisis-in-labrador-west-boom-bust-repeat/.
Poelzer, Greg. 2009. "Education: A Critical Foundation for a Sustainable North." In Abele et al. (2009: 427–65).
Procter, R.M., G.C. Taylor, and J.A. Wade. 1984. *Oil and Natural Gas Resources of Canada, 1983.* Geological
Survey of Canada Papers 83–31. Ottawa: Minister of Supply and Services.
Quebec. 2011. "*Plan Nord,*" 6 July. www.plannord.gouv.qc.ca/english/messages/index.asp.
Rae, K.J. 1968. *The Political Economy of the Canadian North: An Interpretation of the Course of Development
in the Northern Territories of Canada to the Early 1960s.* Toronto: University of Toronto Press.
Rhéaume, Gilles, and Margaret Caron-Vuotari. 2013. *The Future of Mining in Canada's North.* Conference
Board of Canada, Jan. http://www.miningnorth.com/_rsc/site-content/library/Final-13-201_
FutureofMining_CFN.pdf.
Richardson, Lee. 2007. "The Oil Sands: Toward Sustainable Development: Report of the Standing
Committee on Natural Resources." *Parliament of Canada.* Mar. http://www.parl.gc.ca/
HousePublications/Publication.aspx?DocId=2614277&Language=E&Mode=1&Parl=39&Ses=1.
Roberts, Terry. 2015. "Vale Gets Extension for Exporting Voisey's Bay Ore." CBC News. 24 Feb. http://www.cbc.
ca/news/canada/newfoundland-labrador/vale-gets-extension-for-exporting-voisey-s-bay-ore-1.2969111.

Skuce, Nikki. 2012. "Who Benefits? An Investigation of Foreign Investment in the Tar Sands." Forest
  Ethics Advocacy. http://forestethics.org//sites/forestethics.huang.radicaldesigns.org/files/FEA_
  TarSands_funding_briefing.pdf.
Slocombe, D. Scott. 2000. "Resources, People and Places: Resource and Environmental Geography in
  Canada." *Canadian Geographer* 44, 1: 56–66.
————. 2008. *Canadian International Merchandise Trade*. Catalogue no. 65-001-xɪв. http://www5.
  statcan.gc.ca/olc-cel/olc.action?ObjId=65-001-X&ObjType=2&lang=en&limit=0&fpv=1130.
Statistics Canada. 2012. "Land and Freshwater Area, by Province and Territory." Table 15.6. 20 Dec.
  http://www.statcan.gc.ca/pub/11-402-x/2012000/chap/geo/tbl/tbl06-eng.htm.
Sudol, Stan. 2011. "Rails to the Ring of Fire." *Toronto Star*, 30 May. www.thestar.com/printarticle/998685.
Strategy West. 2010. *Existing and Proposed Commercial Oil Sands Projects*, Sept. www.strategywest.com/
  downloads/StratWest_OSProjects_201009.pdf.
Tait, Carrie, Shawn McCarthy, and Nathan Vanderklippe. 2011. "From Oil Sands to Gas Pumps: Alberta
  Looks to Its Energy Future." *Globe and Mail*, 18 Nov. http://www.theglobeandmail.com/report-on-
  business/industry-news/energy-and-resources/from-oil-sands-to-gas-pump-alberta-looks-to-its-
  energy-future/article4200973/#dashboard/follows.
Thompson, John. 2011. "Victoria Gold Proposes New Mine." *Yukon News*, 4 Jan. yukon-news.com/
  business/21162/.
TransCanada. 2013. "Keystone XL Pipeline Project." 27 Nov. http://www.transcanada.com/keystone.html.
————. 2014. "A Proposed Oil Pipeline from Alberta to Nebraska." http://keystone-xl.com/about/
  the-keystone-xl-oil-pipeline-project/.
University of Alberta. 2010. "Oilsands Mining and Processing Are Polluting the Athabasca River, Research
  Finds." *Science Daily*, 30 Aug. www.sciencedaily.com/releases/2010/08/100830152536.htm#.
US Energy Information Administration. 2010. "Canada: Oil." www.eia.doe.gov/cabs/canada/Oil.html.
Vanderklippe, Nathan. 2010. "The North Scrapes the Bottom." *Globe and Mail*, 9 Apr., в9.
Watkins, Melville H. 1963. "A Staple Theory of Economic Growth." *Canadian Journal of Economics and
  Political Science* 29: 160–9.
————. 1977. "The Staple Theory Revisited." *Journal of Canadian Studies* 12, 5: 83–95.
Weber, Bob. 2014. "Lubicon vs. PennWest: Band Files Lawsuit against Alberta Energy Firm over Fracking."
  Canadian Press, 2 Jan. http://www.huffingtonpost.ca/2013/12/02/lubicon-pennwest-lawsuit_n_
  4372856.html.
Welch, Michael J. 2010. "Raglan, 2010." 18 Mar. www.infomine.com/minesite/minesite.asp?site=raglan.
Wellstead, Adam. 2007. "The (Post) Staples Economy and the (Post) Staples State." *Canadian Political
  Science Review* 1, 1: 8–25.
Woodfibre LNG. 2015. "The Project." http://www.woodfibrelng.ca/the-project/about-the-project/.
Wuskwatim Power Limited Partnership. 2010. "About the Partnership." www.wuskwatim.ca/partnership.
  html.
Yukon Bureau of Statistics. 2013. *The Yukon Statistical Review, 2012*. Whitehorse: Yukon Government
  Executive Council Office. http://www.eco.gov.yk.ca/stats/pdf/Annual_Review_2012.pdf.

- 6 -

Environmental Impact
of Resource Projects

Setting the Stage

Resource projects have an impact on the environment. The North, with its fragile environment, is a particularly vulnerable region. In its role as a provider of energy and raw materials to global markets, the Canadian North has seen harsh impacts on its landscape from industrial projects (Figure 6.1). The challenge facing governments is how to encourage resource development without fouling the environment. The record shows that, while progress has occurred, companies still need to be held accountable in a more forceful way.

With the passing of environmental legislation starting in the 1970s,[1] industrial impacts on the North have abated but they are not eliminated. Even more important, in 1992 the federal government, following the lead of the 1987 Brundtland Report (World Commission on Environment and Development, 1987: 43), passed the **Canadian Environmental Assessment Act**, which calls for **sustainable development**. In 2012, Ottawa "streamlined" the Canadian Environmental Assessment Act to accommodate major resource projects by reducing the scope of this Act and by setting time limits on the review process (Canadian Environmental Law Association, 2014).

Provinces and territories accept the concept of sustainable development but are reluctant to pass environmental legislation that is "too stringent" because additional cost to companies might halt their resource development plans. The economies of provinces and territories depend heavily on resource development. These governments, with the exception of Nunavut, also directly benefit from **resource revenues**. Alberta obtains the most resource revenue of any province/territory thanks to its oil sands developments. In 2012–13, **royalties** from companies operating in its oil sands reached a staggering $3.56 billion (Alberta, 2014a). The Territorial North relies heavily on energy and mining projects to support the economy and, in the case of Yukon and the Northwest Territories, to generate revenues. The question is: Where is the point of an acceptable **trade-off** between resource projects and a sustainable environment?

Why Is the North So Vulnerable to Industrial Impacts?

The North has been hit hard by past industrial and global pollution (Figures 6.1 and 6.2 and Vignette 6.1). Unlike warmer lands, the North's cold environment slows biological processes that reduce some types of pollution. Two physical factors delaying biological processes are:

- The North's cold climate means that its resulting biological regime requires a much longer time to repair itself from various forms of industrial pollution.
- Permafrost, a unique natural feature of the North, is easily disturbed by industrial construction projects. Removal of the vegetation cover triggers gelifluction and subsidence. Gelifluction occurs when the upper layer of soil thaws during the warmer months so that water-saturated soil and rock debris move downslope, resulting in slumping. Melting of ice-rich terrain results in the settling of the ground, resulting in an irregular, hummocky landscape termed thermokarst topography (Vignette 2.7). Oddly enough, the retreat of permafrost due to global warming has an unexpected and costly consequence for the reclamation work at the Giant mine near Yellowknife (see Figure 6.2 for more information).

Vignette 6.1 Pollution from Abroad

As part of the global circulation system, air currents pass over the North and a portion of the airborne pollution in these currents falls on the North. Similarly, ocean water currents bring pollution to northern shores. Gradually but inevitably, these chemical compounds enter the aquatic and terrestrial food chains. Combined with local pollution, the process of biomagnification creates serious health hazards for those who consume large quantities of wild game and fish.

The Ugly Past

In 1962, Rachel Carson's epoch-making book, *Silent Spring*, revealed the extent of harm caused to the environment and people by the chemical stew of pollutants from industrial and agricultural enterprise. Soon, the media and environmental organizations exposed more examples of industry disposing of toxic wastes in a haphazard manner. The ugly truth was no longer hidden and society demanded that governments enact environmental legislation. Out of the uproar came an explosion in public interest in the environment. Local environmental groups sprang up as soon as a threat to the environment emerged in the media, but more importantly, membership in conservation organizations like the Sierra Club soared because of public concern about the environment. These **non-governmental organizations** (NGOs) led the fight against companies and governments that proposed projects harmful to the environment. They often focused on a single issue, such as fossil fuels and global warming. The **Sierra Club** (2014), for example, is "leading the charge to move away from the dirty fossil fuels that cause climate disruption and toward a clean energy economy."

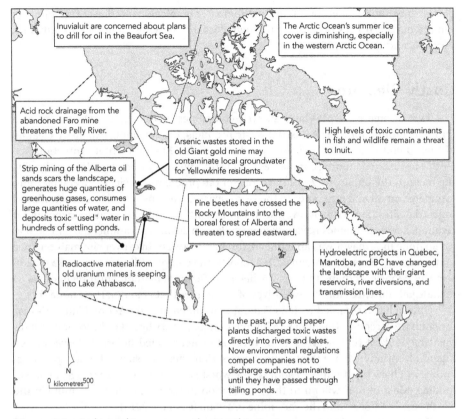

Figure 6.1   Industrial Impacts on the North's Environment

The scars of past industrial damage to the northern environment remain a problem. Cleanup costs have fallen to the federal and provincial governments. The most expensive cleanup operations in the Territorial North are at two abandoned mining sites, the former Giant gold mine near Yellowknife and the open-pit lead/zinc mine at Faro, Yukon. In the case of the Giant mine, the fear is that residues of the arsenic-tainted wastes could seep into the water supply for the city of Yellowknife, while **acid rock drainage** from the Faro mine could allow sulphuric acid to escape into the nearby Pelly River and poison its water and destroy the aquatic life, including spawning grounds for Alaskan salmon stocks.

NGOs are a powerful social force that, while sometimes advocating extreme positions, marshal public attention and, in turn, place pressure on governments, which ultimately have the power to reject resource proposals. In part, pressure from the Canadian public resulted in Canada enacting in 1999 the Canadian Environmental Protection Act (CEPA). This Act is the cornerstone of Canada's environmental legislation. Prior to the passing of CEPA, hundreds of thousands of tons of highly toxic chemicals, including arsenic and cyanide, were left at abandoned mine sites in northern Canada. While other chemicals, such as mercury, are dangerous to human health and threaten wildlife, the most dangerous site is the former Giant gold mine near Yellowknife, Northwest Territories. While a costly cleanup program is underway, huge quantities of arsenic trioxide, a white, odourless, tasteless powder, are still exposed to the elements. Added to the mess left by mining operations, exploration parties have abandoned buildings, equipment, chemicals, and diesel fuel tanks at remote northern sites because of the high cost of removing them by air (Duhaime, Bernard, and Comtois, 2005).

Costly Cleanup

The lack of environmental laws allowed companies to treat the environment as a dumping ground for their industrial waste products, thus avoiding expensive cleanup of their polluted work sites. Even after the Canadian Environmental Protection Act, 1999, was enacted, mining companies could still walk away from site cleanup, leaving the mess to the public. By 2005, the scope of the problem was clear—over 3,000 contaminated small and large sites in the North existed on Crown lands from abandoned industrial sites. Surprisingly, even with the recently introduced security deposits for cleanup, a few companies have pulled up stakes. In 2013, for example, the Jericho diamond mine in Nunavut was abandoned, forcing Ottawa to take over site maintenance of the waste tailings. As Morgan (2015) reports, the company did not pay its $2 million security deposit. Perhaps the answer lies in stiffening environmental regulations to include jail sentences for the executives of companies that abandon their mine sites.[2]

Ottawa has responded with its Federal Contaminated Sites Action Plan, a 15-year program with a commitment of $3.5 billion (Canada, 2013). By 2011, the federal government recognized that nearly half of these sites abandoned before 1999 warranted urgent attention because they could cause serious health issues. The new program, known as Phase II, attempts to address the worst sites with an additional $1 billion of funding. Most of that will go to the remediation programs at the Giant mine in the Northwest Territories and the Faro mine in Yukon (Canada, 2013).

Remediation of highly contaminated sites is a costly process. In 2013, the total number of highly contaminated sites on federal lands totalled 1,601 (Canada, 2013). These sites vary from mineral exploration companies leaving small amounts of toxic waste at the drill site, including oil drums and drilling equipment, to massive contamination such as at the former DEW Line radar sites. In the latter case, the removal of high levels of PCBs and contaminated soil posed an expensive problem. In 1998, the cleanup cost of the DEW Line stations was estimated at $230 million, but this figure was far too low, mainly because the scope of the cleanup was much greater than originally estimated. In 2014, the new cost figure is more than double, at $525 million (National Defence and the Canadian Armed Forces, 2014). An even more expensive cleanup remains in progress at the former Giant mine site. The scope of that remediation is presented in Vignette 6.2.

Impact on Aboriginal Canadians

In northern Canada, Aboriginal peoples still consume large quantities of **country food**. In part, this pattern of harvesting and consuming food from the land goes back in time to their traditional lifestyle and remains an important element in their culture. For this reason, degradation of the environment affects the mental and physical health of Aboriginal peoples much more than other Canadians, who consume relatively little if any country food.

Vignette 6.2   Canada's Worst Mining Mess

The worst example of a contaminated site is the former Giant gold mine at Yellowknife. Royal Oak Mines closed its doors in 1999. The company then went into receivership, leaving the Canadian taxpayer stuck with the bill for the cleanup of the 237,000 tonnes of lethal arsenic trioxide **tailings** (Figure 6.2). Most tailings are stored underground in the former mine shafts. Royal Oak Mines assumed that permafrost would return to the underground mining shafts and thus secure the tailings, but a warmer climate has prevented this natural freezing process. Now, the concern is that unfrozen arsenic stored underground could seep into the Yellowknife water supply. Already the arsenic-laced dust lying on the surface has blown over the surrounding land and waters and, in this way, has affected the country food eaten by the Yellowknives Dene First Nation residents. The **Giant Mine Remediation Project**, part of the federal government's effort to clean up abandoned mine sites, calls for artificially freezing the arsenic materials stored underground and disposing of the contaminated mine buildings and ground surface (Aboriginal Affairs and Northern Development, 2010, 2013). Since surface workers are directly exposed to arsenic dust, they must wear full **hazmat suits** and breathe supplied air. When both surface and underground work is completed, this remediation project will likely cost more than $1 billion. As well, additional funds, possibly $2 million/year, are necessary to undertake 25 years of post-remediation monitoring to ensure that the poisonous toxic remnants remain secure (Keeling and Sandlos, 2012; Weber, 2013; Pearson, 2014).

Aboriginal Canadians have suffered from the effects of industrial pollution in various ways, including polluted drinking water sources, the contamination of wild animals and fish that had been primary food sources, and loss of valuable hunting grounds through flooding and logging. One well-documented instance of industrial pollution that had devastating effects on the health of Aboriginal peoples stands out. The Ojibway living along the English–Wabigoon River system near Kenora, Ontario, by consuming fish that contained high levels of mercury, showed signs of Minamata disease, a form of mercury poisoning.[3] This sad story, as recounted by Shkilnyk (1985), involved the discharge of toxic chemicals with a high content of mercury into the Wabigoon River by a paper mill at Dryden, Ontario. Some 130 km downstream on the English River, many Ojibway living in the Whitedog and Grassy Narrows communities were affected by methyl mercury poisoning (Roebuck, 1999).

The food provided by Mother Earth is understood as a sacred trust by Aboriginal peoples (Vignette 6.3), and consequently, scientific reports about the contamination of wildlife are not readily accepted (Myers and Furgal, 2006; Tyrrell, 2006). Inuit ontology supports the concept of a trust relationship between the Inuit and the animal world. As an Inuk woman explained, "We do not inflict injustices on the animals and trust that they in turn are good for us, we who do not wrong them in some way" (Poirier and Brooke, 2000: 87).

The involvement of Aboriginal organizations in the process of environmental assessment and management is now a fact of life. Peter Usher argues that **traditional**

Figure 6.2 Abandoned Giant Mine near Yellowknife

Between 1948 and 2004, this mine provided the principal economic raison d'être for the existence of Yellowknife. But was it worth it? When the good times ended, a dangerous legacy of 237,000 tonnes of arsenic trioxide remained stored in the abandoned mine shafts. On the surface, arsenic dust remains in the deteriorating buildings and old flues that carried the waste product to the underground shaft. Such dust, picked up by winds, exposes residents of Yellowknife and the Yellowknives Dene First Nation to arsenic poisoning. With the mining company unable to pay for the cleanup, the federal government was left with the most expensive cleanup job of an abandoned mine in Canadian history.

Source: Photo by Tannis Toohey/Toronto Star via Getty Images

ecological knowledge (TEK) must be an essential element in environmental assessment and management (Usher, 2000; see also White, 2006). Traditional ecological knowledge or TEK is oral history passed down from one generation to the next about a people's relationship to the environment and wildlife. Such knowledge is an essential source of information about the natural environment and the relationship of Indigenous people to the land and to each other. Stated differently, TEK provides an Aboriginal perspective about how resource development should be assessed and managed (Ellis, 2005; Woo et al., 2007; Christensen and Grant, 2007; Natcher, 2007). TEK, like all forms of knowledge, is modified over time to meet new circumstances.

How did Aboriginal peoples become partners in assessing resource projects? The primary change agent was comprehensive land claim agreements. Comprehensive land claim agreements have been instrumental in making Aboriginal peoples partners in monitoring resource projects. Co-management agreements, as part of modern land claim settlements, provide Aboriginal organizations a measure of control over proposed industrial projects on their traditional lands, as in the case of the Inuvialuit Settlement Area—a vast area of Crown and Inuvialuit land in the Western Arctic of just over 90,000 km². With the first comprehensive land claim agreement in 1984 the

Vignette 6.3 Conflicting World Views

World views are derived from cultural beliefs and values. Western and Aboriginal cultures represent two distinct ways of viewing and understanding the world, yet the Western paradigm of viewing the natural environment as there for the taking has dominated the relationship between cultures since Europeans first arrived in North America. Aboriginal peoples have a different paradigm, steeped in their long-standing relationship with the land that evokes a different spiritual view of the place of humans in the natural world. While Western culture sees humans at the top of a hierarchical arrangement of living creatures, Aboriginal culture views the world from a holistic perspective where all life is seen as a series of relationships among equals. Can this non-Western approach to the land and wildlife help in addressing the issues noted above and in formulating a gentler and more respectful approach to the environment when the prospect of resource development enters the equation? Culture, whether Western or Aboriginal, does not stand still. Since contact, Aboriginals have used industrial products. Trucks and snowmobiles are commonly used today to hunt wildlife.[4] At the Ninth Circumpolar General Assembly in Kuujjuaq, Nunavik, Sheila Watt-Cloutier (2002: 2), then president of Inuit Circumpolar Conference Canada, succinctly described the Inuit view of the relationship between the natural world and health:

> As Inuit, we think in holistic ways. We know that everything is interrelated—the threads of our lives are woven into a garment that is inherently sustainable. Our culture reflects our values, spirit, economy, and health. Our land and natural resources sustain us, and the health of these resources affects our health. If we use and develop these resources with respect, our environment will remain healthy and so will we. The process of the hunt is invaluable, [for] through it we learn what is required to survive and how to gain wisdom—the key to living and acting sustainably.

Inuvialuit created two environmental co-management units—the Environmental Impact Screening Committee and the Environmental Impact Review Board. In both cases, the Inuvialuit name half of the members of these two environmental units, thus ensuring local input and control. Such input and control were demonstrated when an offshore oil drilling proposal in the Beaufort Sea was rejected (Bone, 2003). Exposure of Aboriginal workers to toxic industrial products occurred prior to industry being more closely regulated since the 1970s, but such blatant industrial problems are much less likely to occur today. For example, Aboriginal and other workers were involved in dangerous work in the uranium industry. The worst example occurred in the 1950s when Sahtu Dene men were hired to help in transporting sacks of radioactive ore. The uranium ore was first shipped by barge from the mine at Port Radium on the eastern shore of Great Bear Lake to the Sahtu Dene community of Deline, NWT, on the lake's western shore, and from there it was transported by various means to the Mackenzie River.[5] Years later, the extremely high incidence of cancer among these workers shocked the community, giving it the name "Village of Widows" (Markey, 2005; Environment Canada, 2007).

Looking to the future, the involvement of Aboriginal organizations in co-management of lands in their settlement areas may also promote further local involvement in the design, operation, and ownership of resource projects. One example in the southern edge of the Subarctic in Manitoba involves the Nisichawayasihk Cree Nation, which was involved in the construction of the Wuskwatim hydroelectric project in northern Manitoba. The relationship between Manitoba Hydro and the Nisichawayasihk Cree First Nation is discussed later in this chapter. A series of decisions by the Supreme Court of Canada provides another path to the involvement of Aboriginal organizations on traditional lands. The Supreme Court of Canada in 2004 (Haida Nation v. British Columbia) made it clear that the duty to consult exists when a development project is proposed for the so-called traditional lands that surround reserves and for all lands not yet under treaty. But there are no guarantees that the outcome will favour Aboriginal claimants (Newman, 2014). In 2013, the Grassy Narrows First Nation of Treaty 3 argued before the Supreme Court that Ontario's legal right to authorize clear-cut logging on their traditional lands was illegal. The Supreme Court disagreed because the Indians of northwest Ontario had surrendered their rights to those lands at the time of the signing of Treaty 3 (Supreme Court of Canada, 2014), thus indicating that the duty to consult does not imply a veto held by Aboriginal peoples over development projects. Still, the Court reiterated that the province has to uphold the honour of the Crown and encouraged Ontario to ensure that First Nations are consulted before logging leases are assigned to forest companies. Clearly, the Court's hope is that the two sides will find common ground through negotiations.

Protecting the Environment: Environmental Impact Assessments

Without a doubt, Canadians want to protect their environment. The problem is that the federal government also wants resource development and that form of economic activity places a heavy imprint on the environment. The Canadian Environmental Assessment Act, which was overhauled by the Conservative government in 2012 to streamline the assessment process and to limit those projects that must undergo **environmental impact assessment** (EIA), represents the key legislative measure designed to protect the environment. With governments making the final decision of approving (or rejecting) proposed resource projects, the EIA process is a critical element. The **Canadian Environmental Assessment Agency** (2011) states that its "role is to provide Canadians with high-quality environmental assessments that contribute to informed decision-making, in support of sustainable development." In its decision-making over oil sands projects and protecting the environment, however, the federal government is leaning over backward to support oil sands development because of its positive impact on the Canadian economy. At this point it is worth noting a political distinction in terms. **Tar sands** is the term originally applied to the deposits in Alberta, as it is in Venezuela's Orinoco tar sands belt, because the material is heavy and sticky like tar or pitch. But in the 1960s, the Alberta government began to use "oil sands" rather than "tar sands" to lessen the image of "dirty oil," and the oil industry quickly followed the Alberta lead. Opponents of oil sands development generally continue to use the term "tar sands" because it sounds messier (which it is) than "oil sands."

Public pressure, often expressed through environmental groups, has put the spotlight on resource development. Resource companies have responded and are more environmentally responsible than in the past. Even giant oil companies accept that global warming is real and that industry and governments must find a way to reduce greenhouse gas (GHG) emissions (Shell, 2014). Yet, despite its public statements in support of corporate environmental responsibility, Shell is pressing ahead with its controversial Jackpine mine near Fort McKay that will have a heavy imprint on the local terrain and will emit large amounts of greenhouse gases. This contradiction is not surprising. On the one hand, Shell and other oil companies operating in the oil sands first of all seek profits. On the other hand, they want the public to perceive them as responsible companies that respect the environment.

The same public pressure resulted in Canadian governments passing environmental legislation to minimize the impact of resource projects on the environment— and to keep a level playing field for resource companies. For example, all resource companies must undertake at their expense an environmental review known as an EIA if the proposed project is of a size or type to require such a review. If the resource proposal is approved, as the vast majority are, the company must ensure that it has the resources to clean up the project site once the resource is depleted and the project shuts down.

The objective of such reviews is not to stop mining undertakings, for example, but to ensure that their impacts on the environment and local populations are acceptable, that is, that the environmental and social impacts are minimized to an arbitrary "acceptable level" and that remediation of the site takes place at the company's expense. Only a few times have resource projects been rejected outright; more often, they are required to adjust details in their plans to lessen environmental and/or social impacts. One of the few rejected mining proposals was the Windy Craggy copper mine in northern British Columbia. This wilderness area along the Tatshenshini River is now a World Heritage Site.[6] In the case of the oil sands, all projects have affected the environment, but so far the environmental review process has not rejected a single proposal. For instance, the most recent approved proposal submitted by Shell Oil, the Jackpine mine expansion, will have a very heavy imprint on the environment, including the permanent loss of 185,872 hectares of wetlands used by the Athabasca Chipewyan First Nation (Sterritt, 2013). The federal government position was that the economic benefits outweighed the environmental costs (Canadian Environmental Assessment Agency, 2013):

> The Jackpine Mine Expansion Project has undergone a rigorous review by a federal–provincial panel. The federal Minister of the Environment has concluded that the project is likely to cause significant adverse environmental effects as defined in CEAA 2012 and the **Governor-in-Council** [cabinet] has determined that those effects are justified in the circumstances. Therefore, the project may proceed in accordance with conditions set out in the Decision Statement issued by the Minister.

Types of Environmental Impacts

Resource projects have several types of spatial impacts on the landscape. Each varies by duration and magnitude, with megaprojects having a longer construction period and a great capital investment. The direct impacts are expressed on the landscape as linear, areal, and accumulated impacts. Indirect or secondary effects refer to related developments that support the project, such as a paved highway to a new school, but that are not part of the project's construction budget.

Linear effects are associated with projects such as highways, seismic lines, and pipelines. While small in total area, these changes to the landscape have a greater effect than their areal extent would suggest. For example, wildlife is affected by such developments, especially migrating animals such as caribou that are attracted to these pathways cut in the bush/forest. Hunters, too, take advantage of resource roads by driving their trucks along these pathways in search of game.

Areal effects are associated with industrial developments that affect huge geographic areas. A hydroelectric project, for example, can have an impact on entire river basins by diverting rivers, creating giant reservoirs that flood extensive areas, and reversing the seasonal peak flows of the rivers, thus transforming the natural landscape into an industrial one. Hydroelectric development in northern Quebec has had a massive impact on La Grande Rivière Basin—some describe the project as an ecological disaster (Berkes, 1982, 1998; Peters, 1999; Roebuck, 1999; Rosenberg et al., 1987). Similar impacts have occurred in northern British Columbia, Manitoba, and Labrador at large-scale hydroelectric sites.

Cumulative impacts are caused by more than one project, usually in relative proximity, which results in a combination of impacts over time and, therefore, greater damage to the environment than a single project would cause (Noble, 2010: 133). In the 1980s, the proposed Alberta-Pacific pulp mill represented the first Canadian environmental impact assessment that examined the discharge of toxic chemicals from all pulp mills along a river, in this case the Athabasca River, which flows northward to the Aboriginal community of Fort Chipewyan (O'Reilly, 2005). The concern was that all toxic chemicals in the river water affect the fish and wildlife and, therefore, through the process of **biomagnification**, the health of the Dene and Métis who consume large quantities fish and game. The frequent failure to require assessment of cumulative effects is a special problem in areas, such as the oil sands, where the impacts of numerous projects, taken individually, might be manageable and remediable, but considered as a whole they can have devastating effects on ecosystems and human communities. Other issues related to cumulative effects arise when proponents are permitted, without environmental review, to expand previously approved projects, and, as has occurred especially with hydroelectric projects, when proponents (often government entities) are allowed to submit and gain separate approvals for various stages or aspects of an overall project so that the project as a whole is never examined for its total environmental and social impacts.

Co-management of the Environment

Aboriginal participation in the management of the lands, waters, and wildlife in their settlement areas represents a major step forward. Such co-management is a by-product of comprehensive land claim agreements. The purpose of co-management boards is to share the approval process for industrial proposals in land claim settlement areas. A process of cross-cultural exposure and learning by the members of these boards is an important spinoff, and these inevitable cultural compromises gradually permeate the broader populations, thus bringing a greater mutual appreciation and respect between Western and Aboriginal cultures.

Co-management environment impact assessment boards consist of members of the land claim agreement and the federal/territorial governments. These co-managed

---

Vignette 6.4 The Origin and Structure of Co-management Environmental Boards

Two bodies—screening committees and review boards—assess environmental impacts in settlement areas defined in each comprehensive land claim agreement. The screening committee within a particular jurisdiction examines all industrial proposals, to approve those with acceptable impacts and to recommend those with potential significant impacts to a review board. Co-management review boards have exercised considerable power over the resource development process in settlement areas of comprehensive agreements.

Co-management boards first appeared in 1984 when the transfer of the environmental impact assessment administration from Ottawa to Inuvik took place under the Inuvialuit Final Agreement. Out of this realignment, a new co-management arrangement for environmental impact assessment in the Inuvialuit Settlement Region was achieved. The co-managed agencies (Environmental Impact Screening Committee and the Environmental Impact Review Board) have equal Inuvialuit and non-Inuvialuit representatives. The Inuvialuit Game Council nominates the Inuvialuit members while the federal and territorial authorities select their representatives. Unlike the Ottawa-based agency, these two co-managed boards place much more emphasis on possible negative impacts of proposed industrial projects on the local environment and wildlife. Holding the hearings in Inuvik rather than Ottawa or Yellowknife reinforces local interests and symbolizes the power shift. This "northern face" to the environmental process reflects the Inuvialuit culture and the importance in that culture of wildlife harvesting and the consumption of country food. The first example of the power shift took place in 1991 when concerns over a potential oil spill from a proposed offshore oil drilling project led to rejection of the proposal by the Inuvialuit Environmental Impact Review Board (Vignette 6.5).

In 2014, another step towards provincial-like status in the North seemed to occur when the Northwest Territories government gained control over land, water, and resources through the Northwest Territories Devolution Act. However, the downside of that Act is that the five regional co-management boards associated with the Mackenzie Valley Resource Management Board—which does not include the Inuvialuit environment regulatory regime—were consolidated into one board with only 11 members (Power and Petz, 2014). In doing this, the federal government has made it so that only one Aboriginal representative comes from each of the former regional boards for a total of five, thus significantly diminishing local control over decision-making and virtually eliminating local citizens' input.

boards have strengthened the Aboriginal hand in the review process for industrial proposals. In particular, the Aboriginal voice on these boards has strengthened concerns over local issues, including the impact on wildlife in their land claim settlement areas. As a consequence, the focus of the assessment process has shifted from national issues to local concerns. National issues tended to reflect on larger matters, such as the net economic benefit of the proposed resource project to the country or province or territory, and economics have tended to trump environmental concerns; local concerns look at how the impacts will affect the environment, wildlife, and the harvesters of those resources. Co-management boards consist of appointed members from the beneficiaries of land claim agreements and members of the general public selected by the federal, territorial, or provincial governments. These public boards, funded by governments, have two functions: to manage the environment and wildlife and to make decisions related to the use of resources.

Without a doubt, co-management of the environment and wildlife represented an enormous step forward for the beneficiaries of comprehensive land claim agreements. Now, only the Inuvialuit enjoy a co-management system that permits a large measure of control over resource development, land-use management, and wildlife harvesting. Co-management means that governments and Aboriginal organizations manage the environment together. This power was watered down under the 2014 **Northwest Territories Devolution Act**. The net result has fundamental implications—the Northwest Territories has undergone a realignment from a decentralized system to a centralized one at a cost of local input and control, which raises the question: Is this any better than the *advisory* role of the **Beverly-Qamanirjuaq Caribou Management Board**, which encompasses Aboriginal, territorial, and provincial jurisdictions? (Kendrick, 2000).

Vignette 6.5   Gulf Canada's Kulluk Drilling Proposal

Gulf Canada had drilled many wells in the Beaufort Sea without encountering problems. In 1990, Gulf proposed to drill four new wells in the offshore waters of the Beaufort. The objective was to add to the known oil reserves of the Amauligak field. The proposed well sites were located in the ice transition zone, i.e., between the normal extent of land-fast ice and the polar ice pack. For the first time, Gulf Canada's proposal was assessed by a co-management board consisting of Inuvialuit and non-Inuvialuit members in Inuvik rather than by the Canadian Environment Assessment Agency based in Ottawa. Local members of the Environmental Impact Review Board (EIRB) were particularly concerned about how Gulf would deal with an oil spill caused by a blowout of a well. Gulf calculated that, in the worst-case scenario, 2.6 million barrels of oil would escape into the Beaufort Sea. Gulf estimated that cleanup costs would reach $400 million. The EIRB learned that the maximum liability under federal regulations for such a blowout was $40 million. Gulf's proposal was rejected because of the potential damage of a blowout to the marine environment, wildlife, and the harvesting of marine country food. The Board also faulted the federal regulatory agency, the Canada Oil and Gas Lands Administration, for failing to develop adequate oil spill contingency plans, including a much higher liability for oil companies.

Source: Bone (2003: 390–1).

Another Step Forward: Cleanup and Decommissioning

Since the late 1990s, governments have required companies to restore the natural habitat found at their closed mine sites. Now, all mining proposals require closure plans that must be approved by the appropriated public regulatory agency. The final closure plan may not be exactly the same as returning the site to the original conditions, but the aim is for a minimal environmental footprint.

The closing of uranium mines, including decommissioning, is a special case and falls under the Nuclear Safety and Control Act, which places the responsibility on Ottawa and the Canadian Nuclear Safety Commission. Not surprisingly, the closure of uranium mines is much more demanding and involves considerable decommissioning costs. Detailed decommission plans are embedded in their original proposals. Often, these decommission plans take the form of environmental management agreements between the company undertaking a resource project and the appropriate territorial and/or provincial government. Such an agreement between the Saskatchewan government and Areva, a French-owned uranium mining company, resulted in the successful decommissioning of the Cluff Lake mine in northwest Saskatchewan. After 22 years of production, the Cluff Lake mine ceased operations in 2002 when the ore reserves were depleted. From 2004 to 2006, the company's decommissioning program took place. The first priority was to ensure that radiation exposure hazards were neutralized so that hunters passing through this area will not exceed the regulatory radiation exposure limit. The next priority was to remove all signs of mining activities, including demolishing the buildings, filling the open-pit mine, and removing the waste rock piles. From 2006 to 2012, a small group of post-decommissioning monitoring staff remained at the Cluff Lake site. The staff conducted environmental activities focusing on long-term groundwater and surface water monitoring to ensure that radiation levels met the national standards. Since 2013, automatic monitoring, coupled with periodic visits, takes place. By June 2015, radiation levels remained above the acceptable national levels and the transfer of the site to the province remains in the future (Areva, 2012; *Regina Leader-Post*, 2014).

The Berger Inquiry Sets the Standard

The **Berger Inquiry** (1974–7) into the proposed **Mackenzie Valley Gas Pipeline Project** was the first environmental and social assessment of a major resource project. Justice Thomas Berger broke new ground and set the standard for future inquiries. Most importantly, Berger moved the hearings from the boardroom to Aboriginal communities and the central discussion from technical aspects of the project to local concerns about the potential impact of the project on wildlife and Aboriginal culture.[7]

Four elements of the Berger Inquiry are now found in environmental impact assessments and reports:

- Industrial proposals are more carefully scrutinized by the federal/provincial/territorial co-managed environmental assessment and review inquiries.
- The role of public participation has been greatly enlarged.

- Hearings are held in the communities affected by proposed projects as well as larger cities.
- Environmental issues affecting wildlife/country food now form a prominent part of inquiries.

Environmental Impact of Pipelines

In the second half of the twentieth century, several large pipeline proposals arose, but some of these pipelines failed to reach the construction stage because market conditions deteriorated.[8] Three—Alaska Gas Pipeline, Mackenzie Valley Gas Pipeline, and Polar Gas Pipeline—were designed to transport oil or gas from the Arctic to southern markets. From the Subarctic, the much smaller Norman Wells pipeline did come into play and most environmental concerns were focused on possible leaks due in part to its construction in a permafrost zone. In the 1980s, little was known about constructing and operating an oil pipeline in permafrost (Smith and Burgess, 2004). Added to this challenge, the Norman Wells pipeline route would cross the three permafrost zones—continuous, discontinuous, and sporadic. As a result, the pipe would be buried in frozen and unfrozen ground where ground temperatures would vary from colder than $-3°C$ to warmer than $-2°C$, thus complicating the setting of the temperature of the oil—too warm would melt the permafrost and too cold could reduce its viscosity and thus slow the rate of flow. In 1985, the temperature of the oil was set at $0°C$, which, it was hoped, would minimize the possibility of the pipe either thawing or freezing the ground around the pipe (Duffy, 1981: 33–4). In fact, oil was initially chilled to $-2°C$ before entering the pipe at Norman Wells to minimize ground subsidence caused by the melting of ground ice (Burgess and Riseborough, 1989; MacInnes et al., 1989). After 1993, the company requested and was allowed to adjust to a summer/winter temperature regime with the oil chilled to $-4°C$ in winter and raised to $12°C$ in summer (Burgess et al., 1998: 95). Until 2011, pipeline failures had not occurred even though significant subsidence was noted, placing pressure on pipeline joints (Figure 6.3). However, a more serious problem emerged in 2011: oil seepage was attributed to leaks caused by pipeline corrosion. Water seeping through the pipe insulation had reached the bare steel, thus initiating the process of corrosion that eventually caused a rupture. By 2013, concern over the integrity of the entire line was expressed and the **National Energy Board** (NEB) reacted by ordering the company to conduct a full engineering assessment and submit a plan to mitigate future spills (National Energy Board, 2013). From the company's perspective, replacing the pipeline is far too expensive. Enbridge prefers to react to specific leaks by placing containment sleeves around the sections of leaking pipe. So the regulator, the NEB, must determine—if oil production continues to 2020, as predicted—whether this aging pipeline can carry the product to market without a major spill or if it should order the pipeline shut down. Enbridge's experience with an oil spill in a river in Michigan, discussed below, indicates the difficulty of a cleanup.

In addition to these northern pipelines, the Trans Mountain pipeline that stretches from Edmonton to Burnaby, BC, was completed in 1953. At that time, the public mood was strongly supportive of megaprojects. In fact, environment impact assessment legislation did not exist until some 20 years later.

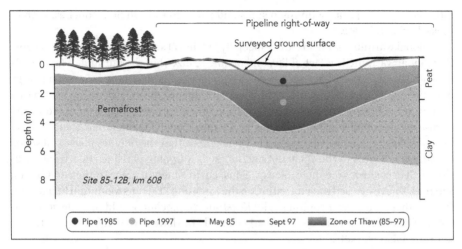

Figure 6.3 Geological Cross-Section of the Norman Wells Pipeline near Fort Simpson, NWT

This geological cross-section, based on the Geological Survey of Canada's monitoring program, illustrates the retreat of permafrost beneath the Norman Wells pipeline at a site near Fort Simpson between 1985 and 1997 (Burgess et al., 1998). As a result of ice melt, the ground subsided nearly four metres and the pipe over two metres.

Source: Adapted from Natural Resources Canada (2006).

By the twenty-first century, although the public was behind the environmental movement, the lure of Asian markets called for oil and gas pipelines to the west coast. Two proposed oil pipeline proposals are under considerable attack from environmental groups. While Ottawa supports oil pipelines to the west coast, the BC public and its government have not taken well to the notion of bitumen pipelines, especially since the risk is largely that of BC and the gain is that of Alberta and the oil multinationals. The first oil pipeline project, the Northern Gateway Project proposed by Enbridge, has received approval of the National Energy Board and the federal government but it continues to face strong opposition. Its proposed route begins in Bruderhein, Alberta, and reaches the Pacific coast at Kitimat, BC (Figure 6.4). The second oil pipeline project, the Trans Mountain Expansion, consists of twinning the existing one. On 16 December 2013, Kinder Morgan submitted its proposal to the National Energy Board. The hearings were scheduled for August 2015 in Calgary and the following month in Burnaby, BC. Strong opposition to the proposal is expected at the interveners' September meeting in Burnaby.

The BC government and its citizens welcome gas pipeline proposals but have opposed proposals related to oil pipelines. Since the gas deposits lie within the province, the government would receive the full benefit from gas royalties. Not surprisingly, then, the province believes that its economic future hinges on natural gas development and export to Asian countries. Even First Nations situated along the route and at the LNG facilities seem pleased with the gas project while they are fiercely opposed to the oil pipelines. LNG terminals may be constructed at Prince Rupert, Kitimat, and Squamish. In the public's mind, a ruptured gas pipeline or the sinking of an LNG tanker would cause relatively little environmental damage compared to a spill of diluted bitumen—gas evaporates; bitumen sinks and sticks to everything it touches, such as rocks, birds, seals, etc.

Global warming is a wild card, especially for the oil sands. The world now accepts that global warming is a fact of life, and people realize that the burning of fossil fuels contributes much of the greenhouse gases responsible for absorbing more solar energy into the atmosphere. Some international environment groups, especially those based in the United States, are dedicated to halting energy projects that would add to the release of greenhouse gases and thus accelerate global warming. For them, the oil sands expansion is a perfect target. They recognized that the Achilles heel to oil sands expansion is access to the US and Asian markets. Stopping pipelines that would provide such access became their mantra. Thus, environmental organizations in Canada and the US have focused their attention on the Keystone XL, Trans Mountain Expansion, and Northern Gateway proposals. The Keystone XL pipeline would provide access to the Gulf coast refineries and world markets, while the latter two would connect the oil sands to Asian markets.

Northern Gateway Project

The Northern Gateway Project consists of two pipelines: one to bring Alberta bitumen to Kitimat for shipment overseas and the other to carry a chemical and oil condensate, used for thinning bitumen so it can be transported by pipeline. Since its approval by the National Energy Board in December 2013, the Northern Gateway Project has aroused fierce opposition from many corners of BC, including the Sierra Club, First Nations, and the general public. The Sierra Club (2013b) has led the way, arguing that:

> The National Energy Board Joint Review Panel's recommendation to the federal government, that Enbridge's proposed Northern Gateway pipeline and tankers project be approved with conditions, flies in the face of overwhelming evidence presented to the panel and is a message of disrespect to British Columbians on the eve of the holiday season.

The following June, the federal government announced a conditional approval of the $7 billion Northern Gateway pipeline, while the opposition parties all opposed the project. With a federal election looming in 2015, Northern Gateway became a wedge issue, thus pushing the project deeper into federal political waters. In addition, Premier Rachel Notley of the newly elected NDP government in Alberta has stated that the Northern Gateway Project "is not the right decision" (CBC News, 2015).

Among British Columbia residents, opposition to the Northern Gateway Project has taken on almost mythical proportions. For a few, stopping the project is the last chance to save the planet from global warming. Most, however, focus on the potential impact on their province. While the prospects for an oil spill are considered by some to be low, fears are that such a spill would have devastating effects on the **Great Bear Rainforest** and would cause a catastrophe in the Douglas Channel and/ or the surrounding coastline, an area of difficult navigation and frequent extreme weather (see Figure 6.4). In the minds of many British Columbians, this risk is simply

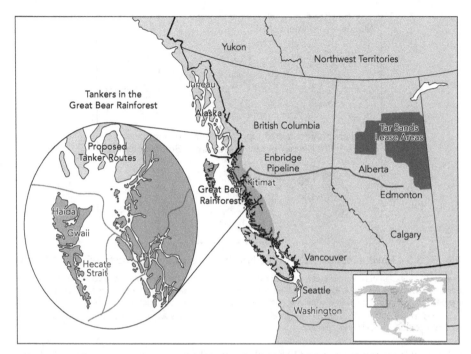

Figure 6.4   Northern Gateway: The End of the Line

For BC, the environmental risks of the Northern Gateway Project far outweigh those inherent in similar natural gas projects. The BC government opposed the Northern Gateway pipeline proposal at the Joint Review Board, and has stated that it will not approve this oil project until five conditions are met (BC Ministry of Environment, 2012). One condition was met with the National Energy Board's approval of the project, but the most contentious condition is that the province expects to receive a fair share of the financial benefits of such a project that would reflect the risk to the province and its inhabitants.

Source: Sierra Club BC (2013b).

unacceptable. The question of shipping **dilbit**, a diluted form of bitumen, across the land and sea raises many questions regarding spills fouling the land, lakes, rivers, and coastal waters of British Columbia. At the Northern Gateway hearings, insufficient scientific evidence was presented for a definitive statement on dilbit sinking in water, which would complicate any cleanup efforts (National Energy Board, 2014). The only example of a dilbit spill was in fresh water, when an Enbridge pipeline ruptured and spilled an estimated 843,000 gallons of oil into a tributary of the Kalamazoo River near Marshall, Michigan, in July 2010. Under these circumstances, dilbit did sink. The remediation to remove the dilbit from the bottom of the river took four years and the final riverbank restoration was completed by November 2014 (Parker, 2014).

Environmental Impact of Mining

Mining, by its very nature, has a harsh impact on the local landscape. The upside is its contribution to economic growth and revenue for provincial/territorial governments, as exemplified by Alberta receiving over $3.5 billion in oil sands royalties in 2012–13

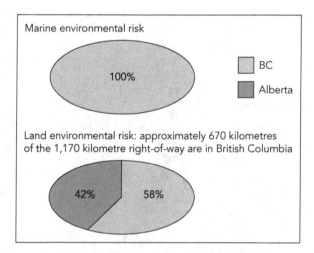

Figure 6.5   Environmental Risks of the Northern Gateway Pipeline to BC and Alberta

Premier Christy Clark has argued that BC would be taking most of the risk but would receive little of the benefit of the Northern Gateway pipeline. One of her conditions is that BC must receive its fair share of the benefits, which can only mean receiving some of the anticipated revenue flowing to Alberta.

Source: BC Ministry of Environment (2012). Copyright © Province of British Columbia. All rights reserved. Reproduced with permission of the Province of British Columbia.

(Alberta, 2014a). But at what cost? The full extent of the past effects of oil sands projects on the environment is not known because of limited monitoring by the industry-funded Regional Aquatics Monitoring Program (Kelly et al., 2010; Schindler, 2010, 2014; Dowdeswell et al., 2010; Parajulee and Wania, 2014). What is known is that the massive size of these operations creates an extensive geographic area of impact. In the case of open-pit mines, the industrial pollution ranges from scarring of the landscape to water contamination, emissions of greenhouse gases, and seepage from the giant tailings ponds (Figure 6.6).

Environmental bodies from Greenpeace to the Sierra Club have identified the tar sands of Alberta as a flash point in global warming. The political rhetoric between opponents and supporters of oil sands development has reached extremes. On the one side, "**dirty oil**" comes from the tar sands; on the other side, oil sands developments produce "**ethical oil**." On the local scene, environmental monitoring of Alberta's oils sands turned into a battleground between industry and environmentalists. Reports from Dr David Schindler of the University of Alberta that fish from the Athabasca River had tumours, deformities, and signs of disease were disturbing and suggested that the level of pollution from the oil sands was much higher than previously claimed. By 2010, confidence in the industry-funded Regional Aquatics Monitoring Program sank so low within the scientific community that the former federal Minister of the Environment, Jim Prentice, appointed a panel of independent scientists to quickly assess the situation and determine if a first-class, state-of-the-art monitoring system is in place. Their answer was a resounding no (Dowdeswell et al., 2010). Two years after the Dowdeswell Report, Alberta and Canada put in place a more extensive monitoring system. In fact, Alberta only moved towards creating a "first-class" monitoring system under the 2012 Joint Canada/Alberta Implementation Plan for Oil Sands Monitoring (Vignette 6.6). A clearer picture of

Figure 6.6 Tailings Ponds just above the Athabasca River

Suncor's plant is situated along the bank of the Athabasca River. Its huge tailings ponds are adjacent to the Athabasca River, which is shown on the right side of the photograph. The tailings ponds in the Alberta oil sands region, which store a chemical-laden mix of water, residual oil, and clay after oil sands processing, cover an area of 176 km²—an area close to 25 per cent of the size of Edmonton. Seepage into the groundwater and the river is a relatively "minor" issue compared to the unlikely but still possible breach in the wall separating the tailings pond from the Athabasca River. Could it happen? Probably not, but until it happened no one fore-saw the disastrous breach in the Mount Polley tailings pond on 4 August 2014 that sent toxic waste into the Quesnel and Cariboo River systems and possibly to the salmon-bearing Fraser River (CBC News, 2014). The breach of the tailings pond at the Mount Polley mine—which is owned by Imperial Metals Corporation—sent millions of cubic metres of waste into central BC waterways.

Source: Daniel Barnes/istockphoto

pollution from the oil sands will take time because the first step is to collect the data and then the data can be analyzed. At that point, the stage will be set for more fully identifying local and regional impacts of oil sands operations. This joint plan enlarges the geographic area under study by including the downstream waters of the Athabasca River and ana-lyzing lake sediments contaminated by airborne polycyclic aromatic compounds (PACs). In the past, reports of toxic substances in the Athabasca River causing abnor-malities in wildlife were dismissed as being without scientific proof by industry and political leaders. Now the Joint Canada/Alberta monitoring program should provide a "scientific" answer to those reported abnormalities as well as a better understanding of the effects, if any, on residents of Fort Chipewyan, many of whom consume country food on a daily basis.

Vignette 6.6 Joint Canada/Alberta Implementation Plan for Oil Sands Monitoring

|||||||||||||||||||||||||||||||||||||||||||||||||||||||||||||||||||||||||||||||||||||||||||||||||||||||||||||||||||||||||||||||||||||||||||

The importance of the joint monitoring plan cannot be underestimated. Monitoring of the impact of the oil sands on the air, land, and waters of northern Alberta has been inadequate for years. After considerable public pressure and various scientific reports, in 2012 the governments of Canada and Alberta committed to implement scientifically rigorous monitoring of the oil sands region to ensure this important resource is developed in a responsible way—which means developing the oil sands but trying to minimize the environmental impacts. In the past, the oil companies funded most monitoring, and these efforts were narrowly based in terms of subject matter and spatial range. In the new monitoring plan, the two governments will significantly expand the subject matter and geographic study area. Equally important, they will directly manage all aspects of the monitoring program, including the oil sands monitoring activities currently managed by independent organizations funded by the oil sands companies. One important outcome is that the scope of the monitoring will extend over much of northern Alberta, thus allowing for studies of airborne pollutants and their possible impact on wildlife.

Source: Canada and Alberta (2012).

Since each mine has passed an environmental assessment, why have the oil sands become the symbol for environmental degradation and a critical source of greenhouse gases? The answers are found in the vastness of this industrial undertaking and its powerful impact on those lands. Open-pit mining results in an ugly landscape. Yet, such a landscape is inevitable due to the system of extraction. The problem confronting industry is that bitumen is difficult to access and turn into heavy oil without great expenditure of water and energy and huge environmental impacts. Simply stated, oil sands cannot be pumped from the ground in a natural state but must be mined or extracted by underground heating (steam injection) and then subjected to processing by upgraders to produce a synthetic **heavy oil**. This poses a dilemma. The black gold from the oil sands is driving the Canadian economy, but its extraction and processing ruin the environment by impacting the landscape, the water regime, and the atmosphere—and the health of those who consume regular quantities of fish and game.

Key questions are:

- Has industry done all it can to curtail the spewing of gases into the atmosphere, the release of toxic water into tailings ponds, the integrity of the ponds, and the scarring of the landscape?
- Have the federal and provincial governments been on top of the situation?
- What are the current state of the environment, the quality of the monitoring system, and the prospects for restoration?

Ultimately, as with all development projects, the basic questions remain: *Can the environmental consequences be limited to an acceptable level? And who gets to decide*

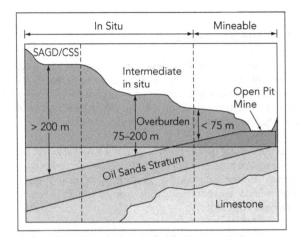

Figure 6.7 Schematic Cross-Section of the Oil Sands

The much-desired bitumen lies in the oil sands stratum. As this diagram indicates, only a small portion (20 per cent) is located sufficiently close to the surface to allow open-pit mining. When the overburden, which is mainly glacial deposits and bedrock, reaches a depth of 75 metres or more, the cost of removing the surface material becomes too expensive and the two drilling techniques, steam-assisted gravity drainage and cyclic steam stimulation, are employed to extract the bitumen. The upside to drilling is a smaller environmental footprint; the downside is that only a portion of the deposit can reach the surface.

Source: University of Alberta (2014).

what an "acceptable level" really is—industry, politicians, the scientific community, the First Nations people at Fort Chipewyan? At present, while an "acceptable level" is determined by the government regulator, i.e., the National Energy Board, the federal cabinet ultimately approves (or, theoretically, turns down) proposed projects. Perhaps the answer lies in the two types of extraction—open-pit mining or steam-assisted gravity drainage and **cyclic steam stimulation** (see Figure 6.7). Open-pit mining has limited scope for future mines for three reasons. First, the best sites where the oil sands stratum extends near the surface already have mines. Second, the cost of construction of new open-pit mines is much higher than for steam-injected mining. Third, the impact on the environment of open-pit mining means that remediation costs will be extremely high. Suncor's Pond 1 has demonstrated that reclamation is possible, but at a cost in the billions. On the other hand, steam-injected mining has relatively little surface impact and remediation costs should be much lower than those associated with open-pit mines. Other factors favouring more steam-assisted gravity drainage extraction are its ability to access the deep strata of the oil sands, its lower demand for water, and its ability to expand incrementally.

While all projects have been approved by environmental assessments, environmental groups believe the bar is set too low (Vignette 6.7). Both the federal and provincial governments express concern about the level of pollution, especially about the toxic tailings ponds and the fouling of the Athabasca River and its reduced streamflow, which is a result of the huge amounts of water drawn from the river for processing, but neither government wants to see production curtailed. In fact, production was forecasted to grow from its 2013 level of 1.98 million barrels per day to 3.7 million barrels

---

Vignette 6.7   Is the Bar Too Low?

Since 2011, federal–provincial **joint review panels** have approved two more open-pit mining operations, the Joslyn mine in 2011 and the Jackpine mine in 2013. In both cases, the review panels recommended a number of required conditions that must be addressed before construction begins. Nonetheless, John Bennett, Sierra Club Canada executive director, decried the Joslyn mine decision:

> Alberta and the federal government have not seen a tar sands project that they didn't love. We have no faith in this process whatsoever, because it's a system geared to encouraging the creation of more and more of these things, rather than to monitoring and controlling their environmental impact. (Vanderklippe, 2011)

Simon Dyer, policy director at the Pembina Institute, expressed his disappointment at the approval of Shell's Jackpine project, saying that "Until oil sands projects are required to meet the environmental standards that are meant to govern this sector, responsible oil sands development will be little more than a slogan" (Dyer, 2013). Perhaps high construction costs combined with low oil prices are the saviour for the environment as construction of both the Joslyn and Jackpine proposed open-pit mines has been put on hold (Healing, 2015; Tait, 2014).

---

per day by 2020 and 5.2 million barrels per day by 2030 (Alberta Energy, 2015). However, the sharp downturn in the price of oil has put a damper on construction projects, as well as on those proposals still on the drawing board (Cattaneo, 2015). While existing plants are expected to continue to produce bitumen, the level of production is likely to remain around 2 million barrels per day. Almost all production goes to the United States, either the Midwest refineries or the Gulf coast refineries. The three proposed pipelines would open markets to Europe and Asia, but the odds are low for all three reaching the construction phase and the potential sales to Asian countries.

As discussed above, the extraction of bitumen involves two methods: open-pit mining and steam injection. The physical nature of the oil sands helps explain why pollution and its impacts are so great. The first impact of open-pit mines involves the removal of the surface material to access the oil sands. Open-pit mines are only possible in a relatively small geographic area where the overburden is less than 75 metres. Beyond that depth, steam injection wells are used. The first step in surface mining is to drain and remove the muskeg and scrape off the overburden. The exposed oil sands are then scooped up, crushed, and transported, via conveyor belt or truck, to the extraction plant.

The second step associated with open-pit mining is to remove the bitumen from the sand, clay, and water. This process takes place in an upgrading plant where the oil in bitumen is separated from sand and water and fine particles are removed to produce a synthetic crude oil. Hot water is used to separate the oil from the sand. The effluent, a mixture of water, clay, sand, and residual bitumen, flows into a tailings pond. Huge tailings ponds are designed to recycle water in the mining operation, but

Figure 6.8   Suncor Energy's Pond 1 in June 2010

By 2014, Suncor's Pond 1 was well on its way to revegetation but still some distance from reaching the state of a sustainable landscape (Suncor, 2014). As Suncor stated: "Over the next two decades, we will closely monitor progress on Wapisiw Lookout, including the growth of 630,000 shrubs and trees planted in 2010. Ongoing soil, water and vegetation assessments will help ensure this site is on course for return to a self-sustaining ecosystem."

Source: Suncor Energy Inc.

their main purpose is to allow the toxic particles in the water to settle to the bottom of the pond. Under the terms of approved projects, companies must restore the tailings ponds to a "sustainable landscape." These ponds, in total, extend over 176 km² and their number is increasing as more mines come into production. So far, only Suncor has changed a tailings pond (Pond 1) into a surface solid enough to be actively revegetated (Figure 6.8).[9]

Is restoration possible? The time frame for restoration is long and its cost is high. So far, the Wapisiw Lookout project (Pond 1) has cost just over $1 billion and it represents only 1 per cent of the area of all tailing ponds. Other oil sands companies may follow Suncor's reclamation system, which has the following steps: draining the pond, burying the toxic waste at the bottom of the pond, reshaping the topography, and then covering with topsoil. Trees, grasses, and shrubs are then planted. When the former tailings pond becomes a sustainable landscape, the Alberta government will release the company from its obligation and the land will be returned to the province.[10]

Environmental Impact of Hydroelectric Projects

Each resource project has a distinct environmental footprint. Mining sites are highly localized. Hydroelectric projects, however, have a regional impact on river basins. Dams flood vast areas of land, alter river courses and streamflows, cause shoreline erosion, and increase the mercury content of the water in reservoirs. In the past, the cost of pre-flood clearing of trees was considered too expensive, and anyone who has visited such a reservoir is struck by the extent of distorted shoreline where slumping has taken

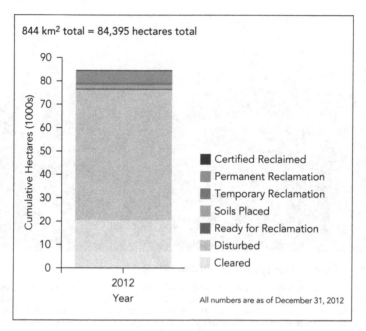

Figure 6.9 Alberta's Oil Sands Reclamation Status, 2012

The area of Alberta's oil sands exceeds that of the three Maritime provinces (142,200 km² to 133,852 km²). However, surface mining is limited to sites where the bitumen deposits lie close to the surface. These lands, some 4,800 km², make up less than 4 per cent of the total area of the oil sands, but by 2013 open-pit mining had only taken up 844 km² of land. While an estimated 20 per cent could be suitable for open-pit mining, construction costs are very high and the cost of returning the land to a "sustainable" level will be equally high. A more likely approach for new projects is steam injection extraction. While companies with open-pit mines are not required to begin reclamation operations until their commercial operations are completed, a few reclamation efforts have taken place; still, only 104 hectares have reached a sustainable level.

Source: Adapted from Alberta (2014b).

place and dead trees poke above the water level. Unfortunately, nature does not correct such man-made problems quickly. Some 80 years after the construction of the Island Falls Dam on the Churchill River in Saskatchewan, standing snags (dead trees) lie just below the surface of the reservoir (Sokatisewin Lake), while along its shoreline trees have a tilted or "drunk" appearance.

Hydroelectric projects have an enormous impact on the environment. Crown corporations are the leading agencies in the design, construction, and operation of northern hydroelectric projects. As discussed in Chapters 3 and 5, provincial power companies have built large-scale hydroelectric projects across the Provincial North, and these projects have profoundly altered the physical character of river basins and the seasonal rhythm of streamflow in British Columbia, Manitoba, and Quebec. In all cases, hydroelectric development is a critical element of the provincial economy. The purpose of energy development is to supply low-cost electrical power to industrial users and to export surplus power to other provinces or to large US markets. The physical transformation of the landscape and its impact on wildlife and Aboriginal peoples were considered acceptable trade-offs to governments. Hydroelectric developments in Manitoba and Quebec provide two case studies.

In Manitoba, the construction of hydroelectric projects on the Saskatchewan, Churchill, and Nelson rivers resulted in flooding of lands and diversion of water, both of which transformed the river landscape from wilderness to industrial and had a negative impact on First Nations inhabitants of northern Manitoba. In 1960, construction began on the first northern power station, Grand Rapids, on the Saskatchewan River as it enters Lake Winnipeg. In the following years, six hydroelectric projects were completed, starting in 1961 with the Kelsey hydroelectric dam and ending with the Wuskwatim dam in 2013. Four other dams are: Jenpeg, Kettle Rapids, Long Spruce, and Limestone.

In the early 1970s, the diversion of water from the Churchill River to the Nelson River added a new dimension to the environmental impact on the two river systems. A control dam was built at the outlet of Southern Indian Lake to prevent water from continuing to flow to the mouth of the Churchill River, and a channel was dug between the southern edge of the lake and the Rat River, which flows into the Nelson River. The water level on Southern Indian Lake rose by three metres and in 1976 water began to flow into the Nelson Basin. While the amount of electric power generated by existing power stations on the Nelson River increased, the diversion of water from the Churchill River to the Nelson River is seen as an ecological disaster (Rosenberg et al., 1987: 81). Submerged vegetation still chokes the lake, causing oxygen depletion during the winter, and mercury levels in fish have risen, making frequent consumption unwise. Changes in the depth of water in the lake have greatly diminished the whitefish population, resulting in the collapse of the commercial whitefish fishery. Those fishers living in the Indian village of South Indian Lake received special compensation through the Northern Flood Agreement with Manitoba Hydro (Rosenberg et al., 1987: 83). Overall, five First Nations in northern Manitoba were adversely affected by flooding from hydroelectric projects on the Nelson and Churchill rivers, and they were to receive compensation from Manitoba Hydro through the Northern Flood Agreement. However, this agreement proved difficult to implement. In the 1990s, Tataskweyak Cree Nation, York Factory Cree Nation, Nisichawayasihk Cree Nation, and Norway House Cree Nation signed comprehensive implementation agreements that, with more precise terms, are proving easier to implement. As well, resource co-management boards (Manitoba and these First Nations) have been established as part of the agreements. Most importantly, positive partnerships between Manitoba's First Nations and Manitoba Hydro are now a fact of life. For instance, the Nisichawayasihk Cree Nation and Manitoba Hydro formed a partnership in 2006, the Wuskwatim Power Limited Partnership (see Chapter 5). One outcome has been the inclusion of Aboriginal traditional knowledge in its environmental assessment (Wuskwatim Power Limited Partnership, 2010).

The scope of the James Bay Hydroelectric Project is staggering. Like the Manitoba hydroelectric development, the James Bay Project has involved the transformation of the northern landscape in Quebec and the establishment of a "settlement" lifestyle for the Cree and Inuit living in this north. The aim of the project is to eventually harness the energy of all the rivers flowing through northern Quebec. This massive project involves some 20 rivers and it represents about one-fifth of the total area of Quebec. The original intent was to develop three major river basins: La Grande, the first phase of which was completed in 1985; Nottaway-Eastmain-Rupert, completed in 2006; and Great Whale, which remains on hold (see Chapter 5 for more details).

With their loss of traditional hunting grounds, the James Bay and Northern Quebec Agreement sought to soften the social impact on Quebec's northern Aboriginal peoples of this massive development. Harnessing La Grande River in northern Quebec resulted in extensive flooding of river valleys to create giant reservoirs, and the diversion of waters from one river to another maximized the water flow through the power stations. Environmental impacts caused by damming La Grande River are well documented (Gill and Cooke, 1975; Berkes, 1982; Rosenberg et al., 1987; Cloutier, 1987; Gorrie, 1990; Hornig, 1999). Coppinger and Ryan (1999: 69) played down the direct ecological impact of the project and suggested, rather, that newly constructed roads in the James Bay area have a much greater impact on the natural vegetation and wildlife because they increase the number of southern visitors and expand the hunting range of local people. In their words:

> As we have pointed out, the La Grande complex is not going to be an ecological disaster. But the roads leading to it and the increasing human population could be. People can have a very real impact on productivity, diversity, and species survival. The increasing number of people, be they miners, foresters, developers, electric workers, vacationers, campers, hunters, anglers, or residents, all change the nature of the region. While habitat loss through construction and development reduces the region's bioproductivity, the real problem is that greater access to harvest individual species because of their commercial, ritual, or recreational value results in another round of damage through reduced productivity and diversity.

Fikret Berkes (1998: 106) substantiates the threat of road access. He reports that the Chisasibi Cree in 1983–4 took advantage of access to an area near the LG-4 dam where a herd of caribou had congregated: "Large numbers were taken (the actual kill was unknown), even though the caribou stayed in the area only for a month or so. Chisasibi hunters used the road, bringing back truckloads of caribou. There was so much meat that, according to one hunter, 'people overdosed on caribou.'" This hunt involved "shooting wildly, killing more than they could carry, and not disposing of wastes properly." Like other caribou-hunting peoples, the Quebec Cree strongly believe that caribou must be respected, meaning that hunters must only take what they need. This concept of a mutual relationship between the Cree and the caribou is strongly etched into Cree spirituality. Cree hunting leader Robbie Matthew translated this cultural tradition into a simple but foretelling phrase: "show no respect and the game will retaliate" (Berkes, 1998: 106). The following winter, 1984–5, no caribou appeared on the road where thousands were seen the previous year. The Elders recalled that a similar disaster took place at the turn of the century when hunters with repeating rifles lost all self-control and slaughtered the caribou at a crossing point along the Caniapiscau River. Then, too, the caribou disappeared. Elders, who know that change occurs in cycles, said that the caribou would return but that the hunters must take only what they need. When the caribou returned in 1985–6, the Chisasibi hunt was conducted in a controlled and responsible manner, illustrating that the Cree can exercise self-management.

Environmental Impact of the Forest Industry

Preserving forests, particularly primary and old-growth forests, is key to preserving biodiversity and the boreal forest ecosystem, as well as for maintaining a suitable environment for traditional activities of First Nations hunters and trappers. While this principle of sustainability is widely accepted in environmental quarters, forest companies depend on harvesting trees by clear-cut methods. In Canada, over 90 per cent of logging is clear-cutting (Conference Board of Canada, 2014). Clear-cutting is a controversial subject. On the one hand, environmentalists consider this a threat to the ecosystem of the boreal forest while First Nations see it as ruining their hunting grounds. On the other hand, logging firms argue that this is the most efficient way of harvesting trees.

While the two sides argued, a compromise was in the works. In 2010, the Canadian Boreal Forest Agreement came into force, helping chart a new path for the forest industry and for the conservation of the boreal forest and its wildlife. The incentive for the forest companies was threefold. First, no industry is more vulnerable to public pressures than the forest industry because it depends on Crown-owned timber leases for its raw material. The forest industry realized that it must be more environmentally sensitive or else it could be denied timber leases, particularly in the more fragile northern environments. Second, the negative publicity about its questionable logging practices distributed to key European politicians, largely by Greenpeace, had made an impact on sales. Before the 2010 agreement, European "greens" had urged the European Parliament to pass legislation boycotting Canadian forest products. Third, while Ontario has the right to issue logging permits on traditional lands, companies should consult with affected First Nations on where to clear-cut before commencing operations (Supreme Court of Canada, 2014).

The Canadian Boreal Forest Agreement, signed by nine environmental organizations and 21 of Canada's forestry companies, covers the 72 million hectares licensed to the Forest Products Association of Canada, with 30 million hectares of woodland caribou habitat now off-limits to road-building and logging (Boychuk, 2011: 44). The goal of the environmental organizations is to ensure the natural ecology of the boreal forest by having forest companies harvest trees to minimize disturbance to ecosystems. As a major concession to the forest industry, environmental groups (Canopy, David Suzuki Foundation, ForestEthics, and Greenpeace) suspended their "do-not-buy" campaigns in foreign markets and adopted a pro-buy campaign of "green" forest products. In sum, the heart of the agreement calls for four major developments (Canopy, 2011):

- committing to comprehensive land-use planning, which will reinforce the ecological-driven harvesting plan outlined in the agreement;
- pursuing more sustainable forest practices, including full utilization of trees into commercial products;

- greening the marketplace by convincing buyers to purchase higher-priced "green" forest products because of the ecological soundness of their harvesting;
- increasing the number of government-legislated protected areas.

Will this truce between formerly bitter rivals hold? One weakness in the decision-making process for the original agreement was the absence of First Nations representatives at the negotiating table. Also, logging arrangements were not as airtight as the environmental groups wanted. In 2012, Greenpeace accused the forest companies of violating the agreement and pulled out of the agreement. Yet, the arrangement, while not perfect, is holding and new regional agreements are dealing with previous shortcomings. Perhaps the initial agreements were too ambitious? As Chanda Hunnie stated (Los, 2014), "This job takes time if you want to do it right." Success is slow, but across the boreal forest a series of agreements remain in force. Some are more robust than others. One positive development is that the Canadian Boreal Forest Association has more than 120 scientists, foresters, and administrators from Newfoundland and Labrador to British Columbia working together to create regional agreements that specifically deal with local concerns. Forest science and planning play a key role. Forest Management Licence-2 in Manitoba offers the best opportunity to test forest science and planning in a form of "active adaptive management" (Los, 2014). At nine million hectares, Forest Management Licence-2 in Manitoba represents the largest forest tenure in the world where foresters, scientists, and Indigenous peoples are working together to plan, log, and regenerate the forest. Within this region of forest, six caribou ranges, 10 First Nations communities, and vast stretches of old-growth forest exist. Out of this experience, the stakes are high for a positive outcome that may serve as a model for other agreements.

Conclusion

Rachel Carson's *Silent Spring*, published in 1962, changed the dynamics of public opinion about the impact of industrial activities on the environment. NGOs quickly picked up the challenge, but Canadian governments reacted slowly and sometimes reluctantly before passing environmental legislation and approving of environmental regulatory offices. Why? Because governments want industrial development to fuel the economy. From their perspective, the goal of environmental legislation and regulatory offices is not to halt development but to ensure that the environmental impacts are "minimized."

One positive sign is that governments now require resource companies to return their industrial lands to a satisfactory condition. The federal government has even stepped forward with the Federal Contaminated Sites Action Plan. These changes represent an enormous step forward. One indication of public support for the environment comes from the large number of Canadian-based environmental organizations. They range from local ones such as the Pembina Institute located in Alberta to international ones such as the Sierra Club. Finally, even though Canadian governments have created environmental impact assessment programs to "minimize" damage, their mandate is to facilitate economic growth, and that includes resource projects. On the bad news side of events, the Northwest Territories Devolution Act of 2014 affects the management of lands, waters, and resources by reducing the role of Aboriginal people in the co-management of resources in the Territorial North.

Governments do face a dilemma—how to stimulate the economy and not damage the environment. This dilemma calls for trade-offs. Tipping the scale towards industry came in 2012 with modifications to the original Canadian Environmental Assessment Act. Yet, another more powerful force, public opinion, can sway the balance. For instance, in the same year, the Alberta and Canadian governments announced a much more rigorous monitoring program for the Alberta oil sands with the Joint Canada/Alberta Implementation Plan for Oil Sands Monitoring. Nonetheless, governments can clash over resource development. The Northern Gateway Project provides a case in point. Ottawa considers building this pipeline to the west coast in the "national interest," but with the loss of support from Alberta, stiff demands from BC, and strong NGO and First Nations opposition, the Northern Gateway pipeline as the spigot for substantial royalties may well be an illusion.

In the decades to come, one question will continue to confront the North: How will its governments balance their reliance on resource development to support their economies, boost employment, and generate public revenues and still maintain a sustainable environment? Another question involves oil patch firms that, down the road, will face huge reclamation costs. Are they committing enough to Alberta's Environmental Protection Security Fund to cover future reclamation costs? (Alberta, 2014c; Rubin, 2015).

Figure 6.10   Clear-Cutting in the Boreal Forest

Clear-cutting, while the most efficient form of logging, makes the landscape useless for hunters and trappers. Such logging practice also strips away the forest, leaving the exposed land vulnerable to various types of erosion. As shown in the photograph, clear-cut logging is prohibited near water bodies because in times of heavy rainfall sediments and other debris will be carried into the stream or lake, causing both severe gullying and heavy deposition. The previously logged area left of the road indicates the slow nature of forest regeneration.

Source: Greenpeace. Photo by J. Henry Fair.

Perhaps Peter Drucker's concept of corporate responsibility has fallen on deaf ears, but since he wrote of the need for such responsibility a more powerful force has entered the arena—a host of non-governmental organizations, concerned citizens, and First Nations that can influence public opinion, change governments, and thereby shift the fate of a resource proposal as well as force companies to pay for the environmental pollution they create. The Northern Gateway Project and the oil sands may be two such examples.

Challenge Questions

1. Why is the North so vulnerable to industrial impacts?
2. In your opinion, has environmental assessment of megaprojects in the Alberta oil sands been vigorous enough? Or do you think that John Bennett of the Sierra Club Canada is correct in claiming that the environmental bar for approving oil sands projects is too low?
3. In the oil sands, companies are required to restore their industrial sites. Does the experience of Suncor's Pond 1 make you comfortable that companies will have the technology and funds to restore all the land affected by their activities?
4. Did the oil industry want the 2012 Joint Canada/Alberta Implementation Plan for Oil Sands Monitoring put in place years sooner?
5. How is co-management in the Northwest Territories affected by the Northwest Territories Devolution Act of 2014? If you were the Premier of the NWT would you be pleased with this new federal legislation that, ostensibly, has given more power to your government? Explain.
6. In the 1990s, the original plan for storing the toxic arsenic waste in the mine shafts at the former Giant gold mine near Yellowknife was based on the assumption that permafrost in the mine shaft would freeze the waste, thus keeping it secure. What was the error in this assumption?
7. Clear-cutting of our forests has its supporters and detractors. If you were a supporter, what positive point(s) would you make?
8. How have large hydroelectric projects in Manitoba and Quebec transformed northern wilderness areas into industrial landscapes?
9. Premier Christy Clark argues that BC would be taking most of the risk involved with the construction and operation of the proposed Northern Gateway Project but would receive little of the benefit. Explain her position. Do you agree with it? How, if at all, does the "national interest" factor into your thinking?
10. While northern resource development is a fact of life, the idea of "minimizing" industrial impacts on the environment seems ineffective. Should governments up the ante for corporations that violate their environmental responsibilities by introducing larger fines and even jail sentences for CEOs?

Notes

1. In 1973, Canada moved towards protecting the environment with the federal **Environmental Assessment and Review Process** (EARP) and the **Federal Environmental Assessment Review Office** (FEARO). However, EARP was not supported by legislation. Twenty years passed before the Canadian Environmental Assessment Act was enacted in 1992—but it applied only to federal lands. During the same period, provinces passed

similar but not identical legislation. Saskatchewan, for example, approved its environmental legislation (Environmental Assessment Act) in 1979–80, while Yukon's Environmental and Socio-economic Assessment Act was enacted in 2003 (Sinclair and Doelle, 2010: Table 17.2). Except for the three Maritime provinces, all of the provinces have northern lands that fall under their respective provincial environmental statutes and assessment processes. The territories have a more complicated environmental assessment system that includes direct involvement and partial control by Aboriginal organizations under co-management arrangements with the federal government as a result of the inclusion of environmental matters within comprehensive land claim agreements.

2. Ottawa continues to increase the pressure on mining companies to pay for the decommissioning of mine sites. Much progress has been made and most companies comply. However, a few still slip through the regulations. Even those required to provide financial assurance in the form of a deposit or bond to governments can dodge the bullet. Their legal trick is bankruptcy.

3. Between 1962 and 1975 Dryden Chemicals Ltd, a subsidiary of Reid Paper Ltd, produced chlorine and other chemicals used as bleach in the pulp and paper mill of Reid Paper at Dryden, Ontario. The mill flushed its waste products into the Wabigoon River. The mill effluent contained a relatively high level of mercury, which worked its way into the aquatic food chain of the river system. In 1970, the Ontario government discovered that the level of mercury found in fish in a 500-km stretch downstream from the pulp and paper mill was dangerous to health, and advised the Ojibway communities at Grassy Narrows and Whitedog reserves not to eat fish from these rivers. It also banned commercial fishing on all lakes and tributaries of the English and Wabigoon rivers. The impact on the Ojibway was staggering. First, 90 members from the Whitedog and Grassy Narrows reserves exhibited serious neurological symptoms characteristic of methyl mercury poisoning (Roebuck, 1999: 80). Second, the Ojibway lost their economic base with the collapse of commercial fishing and guiding income. Third, they could no longer trust their environment because one of their traditional sources of food, fish, had caused their illness. This loss of trust in a basic tenet of their culture was psychologically devastating.

4. Modernity has imposed its way on traditional ways, but it has not destroyed its basic values. Caribou hunters, for example, have taken advantage of modern methods to access caribou herds, by small aircraft and by trucks along resource company access roads.

5. Until the Port Radium mine was closed in 1960, Sahtu Dene workers were hired to load and unload 100-pound sacks of radioactive ore from barge to truck to barge as the ore was transported around the rapids on the Bear River. When cancer cases started showing up among the former workers, the community became alarmed. Deline, NWT, became known as the "Village of Widows." The Sahtu Dene brought their concerns to the public and the federal government decided to support a joint investigation into the possible impacts of the Port Radium mine operation on the local people and their environment. According to the 2005 report on possible impacts on the health of Sahtu Dene men who handled the sacks of ore, "it is not possible to know for certain if the illness or death of any individual Deline ore carrier was directly caused by radiation exposure" (Markey, 2005). The following year, the federal government approved funds for the cleanup of the mine site at Port Radium. The cleanup includes sealing mine openings, safely disposing of equipment and demolishing structures on the site, dealing with exposed tailings, and continued environmental monitoring (Environment Canada, 2007).

6. The pristine Tatshenshini River nearly became the site of a copper mine, but instead this river basin was named a World Heritage Site. This marks one of the few times that an industrial project has been rejected for environmental reasons. In the late 1980s, Geddes Resources Limited, a Toronto-based mining company, wanted to develop the huge Windy

Craggy copper/gold deposit in the mountainous corner of northwest British Columbia near Alaska, only some 80 km downstream from Glacier Bay National Park and Preserve and 32 km from Kluane National Park. This rich copper/gold deposit was valued at $8.5 billion. Geddes applied to the British Columbia environmental agency for permission to proceed with development. The original submission was unacceptable and the company withdrew it. A major flaw in the initial proposal was the danger from storage and disposal of sulphide-bearing waste rock, which posed serious complications due to the potential for acid rock drainage of sulphuric acid from the sulphide-rich tailings and the open-pit mine into the Tatshenshini River. The revised mine plan, released in January 1991, was also rejected. At this point the company abandoned its plans. Efforts by environmental groups such as the Western Canada Wilderness Committee, World Wildlife Fund, Sierra Club, National Audubon Society, and Tatshenshini Wild played a critical role in shaping the final outcome. In 1993, the Tatshenshini–Alsek Wilderness Park was created. The following year the park was designated as a World Heritage Site. An interesting side note: only two months before the termination of the Windy Craggy project, Royal Oak Mines bought 39 per cent of Geddes Resources, thereby becoming its controlling shareholder. Royal Oak Mines claimed millions of dollars to recover exploration expenditures and lost mining revenues. The province settled by paying Royal Oak Mines $167 million (Walters et al., 2007: 10). Royal Oak Mines is the same company that left behind deadly arsenic tailings at its Giant mine near Yellowknife. The federal government is spending over a $1 billion to clean up this mess.

7. The community hearings were closely followed by the national media and in this way the Canadian public became aware of the issues surrounding the pipeline proposal. The media interest resulted from Berger's strategy of shifting the focus from technical issues to a much broader examination of the consequences of industrial development on the environment and the culture of Aboriginal peoples. In a sense, Berger was educating the Canadian public about northern development, the environment, and Aboriginal peoples. The outcome was a remarkable shift in public attitude towards industrial projects—a shift from acceptance of the inevitable to one of sharp questioning about potential impacts.

8. Arctic pipeline proposals and inquiries certainly did not end with the **Berger Report**. In 1977, Foothills proposed an alternate route for Alaskan gas—the Alaska Highway Gas Pipeline Project. This project was approved in 1982 but has not yet been built because the cost of delivering natural gas from Prudhoe Bay to the Chicago market is still not commercially attractive (though this project is currently being reviewed and may receive financial assistance from the US government). Next in line was a joint project by Esso Resources Canada Ltd and Interprovincial Pipe Lines (NW) Ltd to construct a pipeline from Norman Wells to Zama, Alberta. In this case, the proposed project was approved and completed in 1985. The oil industry saw this project as a test case for northern pipelines and, if successful, as a model for much larger pipeline projects to bring Mackenzie Delta gas and Beaufort Sea oil to North American markets. So far, the Norman Wells pipeline has been successful.

9. Gateway Hill, a former Syncrude surface mine, began its restoration in 1983. In 2010, Alberta Environment certified Gateway Hill restored to the equivalent of its pre-mine landscape. However, this restoration was not a tailings pond but a shallow open-pit mine. For more on this, see Testa (2010).

10. Suncor's first tailings pond (Pond 1), built in the late 1960s, was in operation for 40 years. In 2006, Pond 1 began the long process of reclamation into a sustainable landscape. First, excess water and fine tailings were removed from the bottom of the pond and transferred to an active tailings pond. Then, the pond was filled with 30 million tonnes of reclaimed tailings sand. At that point, drainage systems were established and low tracts of land

were used to manage water runoff. Landscaping then began with 1.2 million cubic metres of topsoil placed over the surface, to a depth of 50 centimetres. Trees, shrubs, and grasses were planted, allowing Suncor to reach the first step in its restoration commitment. A full account is found at Suncor, "Wapisiw Lookout Reclamation," http://www.suncor.com/en/responsible/3708.aspx.

References and Selected Reading

Aboriginal Affairs and Northern Development Canada. 2010. *Giant Mine Remediation Project: Developer's Assessment Report*. 8 July. http://www.reviewboard.ca/upload/project_document/EA0809-001_Giant_DAR_1288220431.PDF.
————. 2013. "Remediation Plan." 24 July. https://www.aadnc-aandc.gc.ca/eng/1100100027395/1100100027396.
————. 2014. "Giant Mine Remediation Project." 8 July. https://www.aadnc-aandc.gc.ca/eng/1100100027364/1100100027365.
Alberta. 2014a. "Alberta's Oil Sands, Economic Benefits." http://oilsands.alberta.ca/economicinvestment.html.
————. 2014b. "Alberta's Oil Sands, Reclamation." http://www.oilsands.alberta.ca/reclamation.html.
————. 2014c. *Environmental Protection Security Fund Annual Report*. 1 Apr. 2013–31 Mar. 2014. http://esrd.alberta.ca/lands-forests/land-industrial/programs-and-services/reclamation-and-remediation/documents/SecurityFundReport-Jan22-2015.pdf.
Alberta Energy. 2015. "Facts and Statistics." http://www.energy.alberta.ca/oilsands/791.asp.
Areva. 2012. *Cluff Lake Decommissioning Project*. http://kiggavik.ca/wp-content/uploads/2013/04/Cluff_DoubleSided_Factsheet_Sept2012_xsm.pdf.
Armitage, D.R. 2005. "Environmental Impact Assessment in Canada's Northwest Territories: Integration, Collaboration and the Mackenzie Valley Resource Management Act." In Hanna (2005: 185–211).
————, Fikret Berkes, and Nancy Doubleday, eds. 2007. *Adaptive Co-Management: Collaboration, Learning and Multi-Level Governance*. Vancouver: University of British Columbia Press.
Berger, Thomas R. 1977. *Northern Frontier, Northern Homeland: The Report of the Mackenzie Valley Pipeline Inquiry*. Ottawa: Department of Supply and Services.
Berkes, Fikret. 1982. "Preliminary Impacts of the James Bay Hydroelectric Project, Quebec, on Estuarine Fish and Fisheries." *Arctic* 35, 4: 524–30.
————. 1998. "Indigenous Knowledge and Resource Management Systems in the Canadian Subarctic." In Berkes and Folke (1998: 98–129).
————. 2003. "Inuit Observations of Climate Change: Designing a Collaborative Project." Public lecture at the University of Saskatchewan, 9 Jan.
———— and Carl Folke, eds. 1998. *Linking Social and Ecological Systems: Management Practices and Social Mechanisms for Building Resilience*. Cambridge: Cambridge University Press.
Bone, Robert M. 2003. "Power Shifts in the Canadian North: A Case Study of the Inuvialuit Final Agreement." In Robert B. Anderson and Robert M. Bone, *Natural Resources and Aboriginal People in Canada: Readings, Cases and Commentary*, 382–93. Concord, Ont.: Captus Press.
Boychuk, Rick. 2011. "The Boreal Handshake." *Canadian Geographic* 131, 1: 30–43.
BC Ministry of Environment. 2012. "British Columbia Outlines Requirements for Heavy Oil Pipeline Consideration." 23 July. http://www.newsroom.gov.bc.ca/2012/07/british-columbia-outlines-requirements-for-heavy-oil-pipeline-consideration.html.
Bronson, J.E., and B.F. Noble. 2006. "Health Determinants in Canadian Northern Environmental Impact Assessment." *Polar Record* 42, 4: 1–10.
Burgess, Margaret M., and Daniel W. Riseborough. 1989. *Measurement Frequency Requirements for Permafrost Ground Temperature Monitoring: Analysis of Norman Wells Pipeline Data, Northwest Territories and Alberta*. Geological Survey of Canada, Paper 89-1D. Ottawa: Energy, Mines and Resources Canada.
————, J.F. Nixon, and D.E. Lawrence. 1998. "Seasonal Pipe Movement in Permafrost Terrain, KP2 Study Site, Norman Wells Pipeline." In A.G. Lewkowicz and M. Allard, eds, *Proceedings, Seventh International Conference on Permafrost*, 95–100. Ste Foye, Que.: Centre d'étude nordiques, Université Laval.

Canada. 2013. "High Priority Projects." *Federal Contaminated Sites Portal.* 24 July. http://www.
federalcontaminatedsites.gc.ca/default.asp?lang=En&n=9009D23F-1.
Canada and Alberta. 2012. Joint Canada/Alberta Implementation Plan for Oil Sands Monitoring. http://
www.ec.gc.ca/pollution/EACB8951-1ED0-4CBB-A6C9-84EE3467B211/Final%20OS%20Plan.pdf.
Canada and the Northwest Territories. 2009. *Governments of Canada & of the Northwest Territories: Final
Response to the Joint Review Panel Report for the Proposed Mackenzie Gas Project.* Nov. www.ceaa.
gc.ca/Content/1/5/5/155701CE-6B5C-4F54-84E3-5D9B8297CD15/MGP_Final_Response.pdf.
Canadian Environmental Assessment Agency. 2009. "Sustainable Development Strategy: 2007–2009: 2.1
Role of Environmental Assessment in SD." 10 Mar. http://www.ceaa.gc.ca/EB9E8950-F6DC-4D8A-
9CA6-0520435C89AA/0709SED_e.pdf
————. 2011. "Canadian Environmental Assessment Agency." 19 Jan. www.ceaa.gc.ca/default.
asp?lang=En.
————. 2013. "Jackpine Mine Expansion Project—Release of Environmental Assessment Decision
Statement." http://www.ceaa.gc.ca/050/details-eng.cfm?evaluation=59540.
Canadian Environmental Law Association. 2014. "Canadian Environmental Assessment Act." http://
www.cela.ca/collections/justice/canadian-environmental-assessment-act.
Canopy. 2011. *Canadian Boreal Forest Agreement.* canopyplanet.org/uploads/Agreement101ENG-lr.pdf.
Cantox Environmental. 2006. "Oil Sands." www.cantoxenvironmental.com/sectors/oilgas/oilsands/.
Carroll, A.L., J. Régnière, J.A. Logan, S.W. Taylor, B. Bentz, and J.A. Powell. 2006. *Impacts of Climate Change
on Range Expansion by the Mountain Pine Beetle.* Mountain Pine Beetle Initiative Working Paper
2006–14. Victoria, BC: Natural Resources Canada, Canadian Forest Service, Pacific Forestry Centre.
Carson, Rachel. 1962. *Silent Spring.* Boston: Houghton Mifflin.
Cattaneo, Claudia. 2015. "Canada's Oil Industry to Slash Spending, Reduce Production as Prices Sink."
National Post, 21 Jan. http://business.financialpost.com/news/energy/capp-cuts-western-canadian-
oil-output-forecast-warns-of-more-revisions?__lsa=1c56-f199.
Caulfield, Richard A. 2000. "The Political Economy of Renewable Resource Management in the Arctic." In
Mark Nuttall and Terry V. Callaghan, eds, *The Arctic: Environment, People, Policy,* ch. 17. Amsterdam:
Harwood Academic Publishers.
CBC News. 2007. "Governments Spending $25M to Clean Up Uranium Mines." 3 Apr. http://www.cbc.ca/
news/canada/saskatchewan/governments-spending-25m-to-clean-up-uranium-mines-1.666559.
————. 2010. "Dewline Clean-up Continues in Nunavut." 29 Nov. www.cbc.ca/canada/north/
story/2010/12/29/dew-line.html.
————. 2011. "Toxic Chemicals Released by Melting Arctic Ice." 25 July. www.cbc.ca/news/canada/
north/story/2011/07/25/science-climate-arctic-ice.html.
————. 2014. "Mount Polley Tailings Pond Breach Called Environmental Disaster." 4 Aug. http://
www.cbc.ca/news/canada/british-columbia/mount-polley-mine-tailings-pond-breach-
called-environmental-disaster-1.2727171.
————. 2015. "Rachel Notley's NDP Win Puts Future of Enbridge Northern Gateway Pipeline in
Question." 6 May. http://www.cbc.ca/news/canada/british-columbia/
rachel-notley-s-ndp-win-puts-future-of-enbridge-northern-gateway-pipeline-in-question-1.3063190.
Christensen, J., and M. Grant. 2007. "How Political Change Paved the Way for Indigenous Knowledge: The
Mackenzie Valley Resource Management Act." *Arctic* 60, 2: 115–23.
Cloutier, Luce, 1987. "Quand le mercures élevé trop haut Ã la Baie James . . ." *Acta Borealia* 4, 1 and 2: 5–23.
Conference Board of Canada. 2014. "Use of Forest Resources." http://www.conferenceboard.ca/hcp/
details/environment/use-of-forest-resources.aspx.
Coppinger, Raymond, and Will Ryan. 1999. "James Bay: Environmental Considerations for Building Large
Hydroelectric Dams and Reservoirs in Quebec." In Hornig (1999: 41–72).
Damsell, Keith. 1999. "Mining's Toxic Orphans Come of Age." *National Post,* 18 Sept., FP1.
Delormier, Treena, and Harriet V. Kuhnlein. 1999. "Dietary Characteristics of Eastern James Bay Cree
Women." *Arctic* 52, 2: 182–7.
Department of Justice. 2007. Mackenzie Valley Resource Management Act (1998, c. 25). http://laws.
justice.gc.ca/PDF/M-0.2.pdf.
Dowdeswell, Liz (Chairperson), Peter Dillon, Subhasis Ghoshal, Andrew Miall, Joseph Rasmussen, and
John Smol. 2010. *A Foundation for the Future: Building an Environmental Monitoring System for the Oil
Sands.* Report of the Oil Sands Advisory Panel. Environment Canada, Dec. http://publications.gc.ca/
collections/collection_2011/ec/En4-148-2010-eng.pdf.

Draper, Dianne. 2002. *Our Environment: A Canadian Perspective*, 2nd edn. Toronto: Nelson.

Duffy, Patrick. 1981. *Norman Wells Oilfield Development and Pipeline Project*. Report of the Environmental Assessment Panel. Ottawa: Minister of Supply and Services.

Duhaime, Gérard, Nick Bernard, and Robert Comtois. 2005. "An Inventory of Abandoned Mining Exploration Sites in Nunavik, Canada." *Canadian Geographer* 49, 3: 260–71.

Dyer, Simon. 2013. "Pembina Reacts to Decision on Shell Jackpine Oil Sands Mine Expansion." Pembina Institute. 9 July. http://www.pembina.org/media-release/2462.

Ellis, S.C. 2005. "Meaningful Consideration? A Review of Traditional Knowledge in Environmental Decision Making." *Arctic* 58, 1: 66–7.

Environment Canada. 2007. "Federal Contaminated Sites Receive Funding." www.ec.gc.ca/default. asp?lang=En&xml=81941DCD-F8FA-4012-9266-CEC4F186B0F7.

———. 2009. "Governments of Canada and Northwest Territories Release the Final Response to the Joint Review Panel's Report for the Mackenzie Gas Project." 15 Nov. ec.gc.ca/default. asp?lang=En&n=714D9AAE-1&news=5FC1AA1E-5DE5-4A50-9C19-E67FA8379A5A.

———. 2011. "Canada's Environment Minister Announces an Integrated Plan for Oil Sands Monitoring." 21 July. www.ec.gc.ca/default. asp?lang=En&n=714D9AAE-1&news=DA1E8CBC-D0A6-4304-A1DD-A9206D0818AB.

Fowlie, Jonathan. 2010. "Ottawa Halts $815-Million Prosperity Mine." *Vancouver Sun*, 3 Nov. www2.canada.com/vancouversun/news/story.html?id=19531422-0a35-4e37-8ac9-9f3c1c950e11.

Furgal, C.M., T.D. Garvin, and C.G. Jardine. 2010. "Trends in the Study of Aboriginal Health Risks in Canada." *International Journal of Circumpolar Health* 69, 4: 322–32.

———, S. Powell, and H. Myers. 2005. "Digesting the Message about Contaminants and Country Foods in the Canadian North: A Review and Recommendations for Future Research and Action." *Arctic* 58: 103–14.

Galbraith, Lindsay, Ben Bradshaw, and Murray B. Rutherford. 2007. "Towards a New Super-Regulatory Approach to Environmental Assessment in Northern Canada." *Impact Assessment and Project Appraisal* 25, 1: 27–41.

Geddes Resources Limited. 1990. *Windy Craggy Project: Stage 1: Environmental and Socioeconomic Impact Assessment*. Toronto: Geddes Resources.

Gilbert, Richard. 2011. "Kearle Oilsands Project Faces Protest." *Daily Commercial News and Construction Record*, 11 Apr. dcnonl.com/article/id43847.

Gill, Don, and Alan D. Cooke. 1975. "Hydroelectric Developments in Northern Canada: A Comparison with the Churchill River Project in Saskatchewan." *The Musk-Ox* 15: 53–6.

Gorrie, Peter. 1990. "The James Bay Power Project." *Canadian Geographic* 110, 1: 21–31.

Grant, Jennifer, Simon Dyer, and Dan Woynillowicz. 2008. *Fact or Fiction: Oil Sands Reclamation*. Pembina Institute. pubs.pembina.org/reports/fact-or-fiction-report-rev-dec08.pdf.

Hanna, Kevin S., ed. 2009. *Environmental Impact Assessment: Practice and Participation*, 2nd edn. Toronto: Oxford University Press.

Healing, Dan. 2015. "Shell Canada Withdraws Oilsands Mine Application." *Calgary Herald*, 23 Feb. http://calgaryherald.com/business/energy/shell-canada-withdraws-oilsands-mine-application.

Hornig, James F., ed. 1999. *Social and Environmental Impacts of the James Bay Hydroelectric Project*. Montreal and Kingston: McGill-Queen's University Press.

Indian and Northern Affairs Canada. 2010. "Giant Mine." 12 Aug. www.ainc-inac.gc.ca/ai/scr/nt/cnt/gm/ab/index-eng.asp.

Itaakati. 2014. "Traditional Knowledge Festival." Yellowknife, 21–2 Sept. http://www.traditionalknowledge.ca.

James Bay Energy Corporation. 1988. *La Grande Rivière: A Development in Accord with Its Environment*. Montreal: James Bay Energy Corporation.

Joint Review Panel. 2009. *Foundation for a Sustainable Northern Future: Report of the Joint Review Panel for the Mackenzie Gas Project*. Dec. www.ngps.nt.ca/report.html.

Keeling, Arn, and John Sandlos. 2012. *Giant Mine: Historical Summary. Project Report*. Abandoned Mines in Northern Canada Project. 8 Aug. http://research.library.mun.ca/638/.

Kelly, Erin N., David W. Schindler, Peter V. Hodson, Jeffrey W. Short, Roseanna Radmanovich, and Charlene C. Nielsen. 2010. "Oil Sands Development Contributes Elements Toxic at Low Concentrations to the Athabasca River and Its Tributaries." *Proceedings, National Academy of Sciences*. www.pnas.org/content/early/2010/08/24/1008754107.full.pdf+html?sid=33a2f9a7-e7ba-42ee-b6f4-d41203d748a7.

————, Jeffrey W. Short, David W. Schindler, Peter V. Hodson, M. Ma, A.K. Kwan, and B.L. Fortin. 2009. "Oil Sands Development Contributes Polycyclic Aromatic Compounds to the Athabasca River and Its Tributaries." *Proceedings, National Academy of Sciences* 106, 52: 22346–51.

Kendrick, Anne. 2000. "Community Perceptions of the Beverly-Qamanirjuaq Caribou Management Board." *Canadian Journal of Native Studies* 20, 1: 1–33.

Los, Fraser. 2014. "Boreal Truce." *Canadian Geographic* (Jan.–Feb.). http://www.canadiangeographic.ca/magazine/jf14/canadian-boreal-forest-agreement.asp.

McCarthy, Shawn. 2010. "Ottawa, Alberta Blamed for Lax Oil-Sands Oversight." *Globe and Mail*, 21 Dec., B1.

MacInnes, K.L., M.M. Burgess, D.G. Harry, and T.H.W. Baker. 1989. *Permafrost and Terrain Research and Monitoring: Norman Wells Pipeline.* Environmental Studies No. 64, vols 1 and 2. Ottawa: Minister of Supply and Services.

Markey, Andrea. 2005. "Deline Uranium Report Released." *Northern News Service.* www.nnsl.com/frames/newspapers/2005-09/sep12_05rad.html.

Marslen, William. 2008. *Stupid to the Last Drop: How Alberta Is Bringing Environmental Armageddon to Canada (and Doesn't Seem to Care).* Toronto: Vintage Canada.

Morgan, Shauna, 2015. "The True Price of a Resource Economy in Canada's North." Pembina Institute. http://www.pembina.org/op-ed/the-true-price-of-a-resource-economy-in-canadas-north.

Myers, H., and C. Furgal. 2006. "Long-Range Transport of Information: Are Arctic Residents Getting the Message about Contaminants?" *Arctic* 59, 1: 47–60.

Natcher, Paul. 2007. "The Gift in the Animal: The Ontology of Hunting and Human–Animal Sociality." *American Ethnologist* 34, 1: 25–43.

National Defence and the Canadian Armed Forces. 2014. "The Distant Early Warning (DEW) Line Remediation Project." 7 Mar. http://www.forces.gc.ca/en/news/article.page?doc=the-distant-early-warning-dew-line-remediation-project/hgq87xvs.

National Energy Board. 2010. *Mackenzie Gas Project: Reasons for Decision.* Apr. www.neb.gc.ca/clf-nsi/rthnb/pplctnsbfrthnb/mcknzgsprjct/rfd/rfd-eng.html.

————. 2013. "Letter and Order CO-001-2013 to Imperial Oil Resources Ltd.—Norman Wells." 19 July. https://www.neb-one.gc.ca/sftnvrnmnt/cmplnc/brdrdr/mprll/2013/co-001-eng.html.

————. 2014. "Environmental Behaviour of Products to Be Transported by the Project." Enbridge Northern Gateway Project Joint Review Panel. 16 Jan. http://gatewaypanel.review-examen.gc.ca/clf-nsi/dcmnt/rcmndtnsrprt/rcmndtnsrprtvlm2chp6-eng.html.

Natural Resources Canada. 2006. "Norman Wells Pipeline Research." Geological Survey of Canada. gsc.nrcan.gc.ca/permafrost/pipeline_e.php.

————. 2007a. "The Mountain Pine Beetle Program." mpb.cfs.nrcan.gc.ca/biology/index_e.html.

————. 2007b. "Statistics and Facts on Forestry." www.nrcan.gc.ca/statistics/forestry/default.html.

Newman, Dwight. 2014. *The Rule and Role of Law: The Duty to Consult, Aboriginal Communities, and the Canadian Natural Resource Sector.* Aboriginal Canada and the Natural Resource Economic Series. Macdonald/Laurier Institute. http://www.macdonaldlaurier.ca/files/pdf/DutyToConsult-Final.pdf.

Nikiforuk, Andrew. 2008. *Tar Sands: Dirty Oil and the Future of a Continent.* Vancouver: Greystone Books.

Noble, Bram F. 2000. "Strengthening EIA through Adaptive Management: A Systems Perspective." *Environmental Impact Assessment Review* 20: 97–111.

————. 2010. *Introduction to Environmental Impact Assessment: A Guide to Principles and Practice,* 2nd edn. Toronto: Oxford University Press.

North American Council for Environmental Cooperation. 2007. "Ontario Logging." www.cec.org/trio/stories/index.cfm?varlan=english&ed=20&ID=216.

O'Reilly, Kevin. 2005. "New Maps Show Impact of NWT Gas Development." CARC news release. 72.14.253.104/search?q=cache:BfHT084xNdEJ:www.carc.org/2005/news%2520release%2520backgrounder.doc+Cumulative+environmental+impacts+of+the+Mackenzie+Delta+Project&hl=en&ct=clnk&cd=1&gl=ca&client=firefox-a.

Parajulee, Abha, and Frank Wania. 2014. "Evaluating Officially Reported Polycyclic Aromatic Hydrocarbon Emissions in the Athabasca Oil Sands Region with a Multimedia Fate Model." *Proceedings of the National Academy of Sciences* 111, 9: 3344–9. doi:10.1073/pnas.1319780111.

Parker, Rosemary. 2014. "Enbridge Wrapping Up Kalamazoo River Oil Spill Cleanup and Restoration, on Target for Fall Completion." *Kalamazoo Gazette*, 11 July. http://www.mlive.com/news/kalamazoo/index.ssf/2014/07/hold_the_dam_is_gone_cerescos.html.

Pearson, Meghan. 2014. "Yellowknife Is Sitting on Enough Arsenic to Kill Every Human on Earth." *Vice News*, 15 Apr. http://www.vice.com/en_ca/read/yellowknife-is- sitting-on-enough-arsenic-to-kill-every-human-on-earth.

Peters, Evelyn J. 1999. "Native People and the Environmental Regime in the James Bay and Northern Quebec Agreement." *Arctic* 52, 4: 395–410.

Pitt, Doug. 2009. "Harvesting Practices in Canada's Boreal Forest." Natural Resources Canada. canadaforests.nrcan.gc.ca/article/video/boreal/harvesting/trscpt.

Poirier, Sylvie, and Lorraine Brooke. 2000. "Inuit Perceptions of Contaminants and Environmental Knowledge in Salluit, Nunavik." *Arctic Anthropology* 37, 2: 78–91.

Power, Samantha and Sarah Petz. 2014. "Northwest Territories Devolution Act Causes Controversy." *Capital News Online*, 14 Feb. http://www.capitalnews.ca/archive/index.php/news/controversial-legislation-changes.

Regina Leader-Post. 2014. "The Return of Cluff Lake Today's Mine Decommissioning Practice." 24 May. http://www.leaderpost.com/business/return+Cluff+Lake+today+mine+decommissioning+practice/9873145/story.html.

Robinson, Allan. 1991. "Exploring the Risks at Windy Craggy." *Globe and Mail*, 19 Jan., B1.

Roebuck, B.D. 1999. "Elevated Mercury in Fish as a Result of the James Bay Hydroelectric Development: Perception and Reality." In Hornig (1999: ch. 4).

Rosenberg, D.M., R.A. Bodaly, R.E. Hecky, and R.W. Newbury. 1987. "The Environmental Assessment of Hydroelectric Impoundments and Diversions in Canada." In M.C. Healey and R.R. Wallace, eds, *Canadian Aquatic Resources*, 71–104. Ottawa: Department of Fisheries and Oceans.

Rubin, Jeff. 2015. "Money Pit: Who's on the Hook If Low Energy Prices Shut Down the Oil Sands?" *The Walrus* 12, 6: 19–21.

Saskatchewan Environment. 2002. *An Assessment of Abandoned Mines in Northern Saskatchewan*. File R3160. Regina: Clifton Associates Ltd.

Schindler, D. 2010. "Tar Sands Need Solid Science." *Nature* 468, 7323: 499–501.

————. 2014. "Unravelling the Complexity of Pollution by the Oil Sands Industry." *Proceedings of the National Academy of Science* 111, 9: 3209–10.

Searle, Rick. 1991. "Journey to the Ice Age: Rafting the Tatshenshini, North America's Wildest and Most Endangered River." *Equinox* 55, 1: 24–35.

Shell Global. 2014. "Climate Change." http://www.shell.com/global/environment-society/environment/climate-change.html.

Shkilnyk, Anastasia M. 1985. *A Poison Stronger Than Love: The Destruction of an Ojibwa Community*. New Haven: Yale University Press.

Sierra Club BC. 2013a. "Tar Sands Pipeline and Tanker Traffic." http://www.sierraclub.bc.ca/our-work/gbr/issues/tar-sands-pipeline-and-tanker-traffic.

————. 2013b. "Panel Fails to Listen to British Columbians." 19 Dec. http://www.sierraclub.bc.ca/media-centre/press-releases/panel-fails-to-listen-to-british-columbians.

————. 2014. "Who We Are." http://www.sierraclub.org/about.

Sinclair, John, and Meinhard Doelle. 2010. "Environmental Assessment in Canada: Encouraging Decisions for Sustainability." In Bruce Mitchell, ed., *Resource and Environmental Management in Canada: Addressing Conflict and Uncertainty*, 462–94. Toronto: Oxford University Press.

Sinclair, William F. 1990. *Controlling Pollution from Canadian Pulp and Paper Manufacturers: A Federal Perspective*. Ottawa: Environment Canada/Minister of Supply and Services.

Smith, Sharon L., and Margo M. Burgess. 2004. *Sensitivity of Permafrost to Climate Warming in Canada*. GSC Bulletin 579. Ottawa: Government Printing.

Spears, Tom. 2002. "Arctic Pollution Causing Natives Serious Health Problems: Study." *National Post*, 19 Aug., A13.

Sterritt, Angela. 2013. "'Jackpine Mine Will Destroy Wetlands and Wildlife, First Nations Say." CBC News. 9 July. http://www.cbc.ca/news/aboriginal/jackpine-mine-will-destroy-wetlands-and-wildlife-first-nations-say-1.2455963.

Stout, R., Dionne Stout, and R. Harp. 2009. *Maternal and Infant Health and the Physical Environment of First Nations and Inuit Communities: A Summary Review*. Project #183 of the Prairie Women's Health Centre of Excellence, Winnipeg, Apr. www.pwhce.ca/pdf/AborigMaternal_environment.pdf.

Suncor Energy. 2011. "Pond 1 Reclamation." Jan. sustainability.suncor.com/2010/en/responsible/3508.aspx.

———. 2014. "Wapisiw Lookout Reclamation." http://www.suncor.com/en/responsible/3708.aspx.

Supreme Court of Canada. 2014. *Grassy Narrows First Nation v. Ontario (Natural Resources)*. http://scc-csc.lexum.com/scc-csc/scc-csc/en/item/14274/index.do.

Tait, Carrie. 2014. "Total Shelves $11-billion Alberta Oil Sands Mine." *Globe and Mail*, 29 May. http://www.theglobeandmail.com/report-on-business/joslyn/article18914681/.

Tait, Carrie, Nathan VanderKlippe, and Josh Wingrove. 2011. "Alberta Conservation Plan Stuns Oil Patch." *Globe and Mail*, 5 Apr., B1.

Tait, Heather. 2007. *Harvesting and Country Food: Fact Sheet*. Statistics Canada 2001 Aboriginal Peoples Survey, Catalogue no. 89-627-XIE, Apr. www.statcan.gc.ca/pub/89-627-x/89-627-x2007001-eng.pdf.

Tesar, Clive. 2000. "POPs: What They Are; How They Are Used; How They Are Transported." *Northern Perspective* 26, 1: 2–5.

Testa, Bridget Mintz. 2010. "Reclaiming Alberta's Oil Sands Mines: Forest, Wetlands and a Lake Suggests a Boreal Ecosystem Can Be Rebuilt." *Earth*. www.earthmagazine.org/earth/article/30d-7da-2-16.

Tyrrell, M. 2006. "Making Sense of Contaminants: A Case Study of Arviat, Nunavut." *Arctic* 59, 4: 370–80.

University of Alberta. 2014. "Oil Sands Overview." http://www.oilsands.ualberta.ca/wqm/?page_id=230.

Usher, Peter. 2000. "Traditional Ecological Knowledge in Environmental Assessment and Management." *Arctic* 53, 2: 183–93.

Vanderklippe, Nathan. 2011. "Total's Joslyn Oil Sands Mine Approved." *Globe and Mail*, 27 Jan., B1.

Walters, Joshua, Shin-Ling Hsu, and Gareth Duncan. 2007. "Windy Craggy: Mining in British Columbia." UBC Law School. http://www.arc.law.ubc.ca/files/pdf/enlaw/windycraggy_05_15_09.pdf.

Watt-Cloutier, Sheila. 2002. "Speech Delivered by Sheila Watt-Cloutier, President ICC Canada and Vice-President ICC International, at the 9th General Assembly and 25th Anniversary of the Inuit Circumpolar Conference." *Silarjualiriniq: Inuit in Global Issues* 12 and 13: 1–8.

Weber, Bob. 2007. "Contaminant Levels Dropping among Arctic Mothers, Blood Studies Show." *Globe and Mail*, 29 Oct., A10.

———. 2009. "Arctic Pipeline Environmental Study Expected This Week." *Globe and Mail*, 27 Dec., B1.

———. 2010. "Mutant Fish Lead to Calls for Ottawa to Monitor Oil Sands." *Globe and Mail*, 17 Sept., B1.

———. 2013. "Giant Mine's High Cleanup Bill Shakes Up Policy on Toxic Sites." *Globe and Mail*, 1 Apr. http://www.theglobeandmail.com/news/national/giant-mines-high-cleanup-bill-shakes-up-policy-on-toxic-sites/article10659731/.

Wenzel, George W. 1999. "Traditional Ecological Knowledge and Inuit: Reflections on TEK Research and Ethics." *Arctic* 52, 2: 113–24.

White, Graham. 2006. "Cultures in Collision: Traditional Knowledge and Euro-Canadian Governance." *Arctic* 59, 4: 401–14.

Woo, M.-K., et al. 2007. "Science Meets Traditional Knowledge: Water and Climate in the Sahtu (Great Bear Lake) Region, Northwest Territories, Canada." *Arctic* 60, 1: 37–46.

World Commission on Environment and Development (WCED). 1987. *Our Common Future*. Oxford: Oxford University Press.

Wuskwatim Power Limited Partnership. 2010. "About the Wuskwatim Generating Station Project." www.wuskwatim.ca/project.html.

- 7 -

Aboriginal Economy and Society

Great Strides and Defining Changes

Over the past four decades, **Aboriginal peoples** have taken giant strides towards securing a new place in Canadian society. In that time, they have moved from a restrictive colonial world to a more open one where they have gained a measure of control over their lives. Nunavut is one example of political progress: in the 1970s, the Berger Report awakened the Canadian public to the impact of resource projects on Aboriginal peoples; 20 years later Nunavut represented a previously unthinkable political advance. In a different manner, the **Truth and Reconciliation Commission**, which completed its work in May 2015, marked an important social turning point in this continuing journey. And this growing empowerment, while not yet complete, continues to gain momentum.

Empowerment came neither from Canadian governments nor from an upwelling of support from the general public. Instead, empowerment has come mainly from the efforts of the affected peoples and their handful of supporters who used our legal system to address their grievances. Comprehensive land claim agreements and decisions by the Supreme Court of Canada have played a critical part in this transformation by allowing a duality—participation in both the market economy and their traditional economy. In fact, the 1973 *Calder* **decision** of the Supreme Court pushed the federal government to accept the legal concept of Aboriginal title, which, in turn, led to comprehensive land claim negotiations. The first comprehensive land claim agreement was the Inuvialuit Final Agreement, signed in 1984. An essential element in that agreement recognized the duality of the modern Aboriginal world whereby the Inuvialuit Regional Corporation functions in the market economy while the Inuvialuit Game Council maintains their interest in their hunting economy.

In this chapter, our attention is focused on defining social and political change:

- The emergence of an understanding of the importance of Aboriginal **homelands**.
- The recognition of Aboriginal title and the achievement of comprehensive land claim agreements.
- The role of Aboriginal-controlled enterprises in the northern resource economy.

Solving the Puzzle

While the northern resource economy forms only one piece of a much larger puzzle, the creation of Aboriginal-controlled enterprises and their participation in the resource economy mark a quantum leap into the capitalistic world by Aboriginal peoples. At the same time, the Aboriginal peoples living in small northern communities want to retain their close connection with their traditional roots—the land, their languages, and their traditional way of life, including the concept of sharing. The challenge is a formidable one—how to have one foot in the market economy and the other foot in a twenty-first-century version of their traditional economy. With the broad parameters of their journey through the remainder of the twenty-first century already set through land claim agreements and Supreme Court of Canada decisions, their relationship with the Crown is changing and the new relationship will shape their place in Canadian society. Coates and Crowley (2013: 1) describe this new relationship:

> The emergence of development corporations—Aboriginal-run, community-based, and collectively-owned commercial enterprises—is perhaps the most significant development in the field. Many have received funds from modern treaties, legal settlements, and revenue from resource activity, and are already significant players in Indigenous economic development.

Financial independence is a worthy goal and has implications for other aspects of the Aboriginal world. The search for Aboriginal self-government, for instance, loses much of its meaning if this political gain is not accompanied by similar economic gains. Resource-sharing is another avenue, and a number of Aboriginal governments are benefiting from the sharing of resources.

Our discussion about these pressing issues hinges on two past events that began to reshape the Aboriginal world in the last half of the twentieth century:

• A geographic shift occurred with the relocation of Aboriginal peoples from the land to settlements.
• A political shift began to sweep through Indigenous country with the growing recognition and significance of Aboriginal title.

Geographic Shift

By the 1950s, the centuries-old practice of living off the land continued for northern Aboriginal peoples. Yet, times were changing. Several incidents of inland Inuit dying from starvation struck a chord in the media, forcing the federal government to act (Williamson, 1974). In the late 1950s, Ottawa came to the conclusion that the best solution was to relocate northern Aboriginal peoples into settlements where food security and basic medical services would be available. This decision solved the immediate problem, but it opened the door to a host of other issues. One was the daunting cultural leap from a quasi-nomadic lifestyle to a modern and more sedentary one. Another was the absence of settlement economy strong enough to support the newcomers.

As an outcome of the relocation strategy, approximately 200 "relocation" settlements exist across the North. For the most part, these Aboriginal communities, like Clyde River on Baffin Island, are struggling to survive (Vignette 4.5). Under normal market circumstances, urban places that lose their economic function soon die, whether they are single-industry towns or rural communities. Native settlements, however, do not follow this pattern of urban evolution. In fact, almost all have increased in population size, thus placing pressure on limited public resources. In Nunavut, for example, the Inuit population nearly doubled from 15,850 in 1986 to 27,360 in 2011 (Appendix IV). The reasons for Nunavut's population increase are due to two demographic factors: natural increase is high and relocation to other parts of Canada is low. The basic reason for low Inuit mobility is their desire to remain in their cultural homeland. Here, most speak Inuktitut and practise Inuit culture. Another factor is that access to traditional lands for hunting and recreational purposes is possible, especially in the smaller communities. While few jobs are available, public services, including housing as well as various forms of income support, serve to keep people in their communities. Still, economic opportunities are limited, leaving the growing number of young Inuit with few options. The high rate of suicide among young men/teenagers in Nunavut is a sign of discontent with their lives. In any event, a small but growing stream of Aboriginal families and individuals are relocating to southern cities. In 2006, 11,005 Inuit or 21.8 per cent resided outside of their traditional Arctic homeland of Nunangat; five years later, this number had increased to 15,985 or 26.9 per cent (Table 4.3; Statistics Canada, 2014a). The good news is that some find their footing in this urban environment while others return home with degrees and improved health. The bad news is that some fall between the cracks, adding to homeless populations in some southern Canadian cities.

The Political Shift

Shortly after World War II a series of world events caught the eye of Aboriginal leaders, political activists, and intellectuals. First, colonial empires were replaced by independent states in Africa and Asia. Second, the American civil rights movement was challenging segregation by taking to the streets. Third, separatist ethno-nationalist groups in some European countries, and in Quebec, were seeking political independence, and their activities often led to violent acts. Lastly, relocation of scattered extended hunting families into settlements provided the catalyst for the emergence of a tribal consciousness, creating a form of Aboriginal **ethnocentrism**. This newly found unity served as a tipping point for engaging in political change. Thus, one unintended consequence of the relocation strategy was the emergence of a sense of regional consciousness. The most striking examples occurred in the Territorial North with **Nunavut** and the unsuccessful attempt to create a Dene territory of **Denendeh**, and in northern Quebec with its Cree Regional Government in **Eeyou Istchee** and an Inuit-dominant government in **Nunavik**.

By the 1970s, the Dene tribes in the Mackenzie Basin were moving towards what then seemed to Ottawa as a troubling and even threatening political position. In 1975, their chiefs announced the **Dene Declaration**, which for the time was a radical political statement. This declaration called for political independence within Canada

(Erasmus, 1977: 3–4). In their land claim negotiations with Ottawa, the Dene demanded their own separate territory and government in the western half of the Northwest Territories. This new state would be called Denendeh.

Ottawa was taken aback by such a bold request for an ethnic form of government, which, it seemed, could lead to a series of semi-autonomous Aboriginal states within Canada and perhaps even tip the balance in favour of Quebec independence. By 1990, the Dene land claim had reached the advanced stage of a final agreement with Ottawa, but the draft agreement required approval from the Dene/Métis assemblies. The majority voted against the agreement because it did not deal with self-government. Ottawa discontinued further discussions with the Dene/Métis and offered to begin the comprehensive land claims negotiations with individual tribes—what might be seen as a divide-and-conquer strategy. Two northern groups, the Gwich'in and Sahtu Dene/ Métis, accepted Ottawa's offer. As a result, the proposed political entity of Denendeh unravelled, putting an end to the political dream of a Dene state within Canada (Vignette 7.1).

For the Inuit of the eastern Arctic, however, their political dream was translated into reality with the formation of Nunavut. Though the Inuit had originally proposed an ethnic state at the same time as the Dene, they were able to adjust to the political realities of the day—and, significantly, in their proposed territory they had a larger majority than did the Dene. In their revised version, the Inuit called for a public government rather than an ethnic one, i.e., a government that in practice would be under Inuit control because of their large numerical majority but where all citizens, regardless of ethnicity, would be treated equally. In 1999, the Inuit wish for a separate territory was fulfilled with the creation of Nunavut.

Vignette 7.1 Denendeh, Self-Government, and Comprehensive
Land Claim Negotiations

In 1975, the concept of Denendeh represented a bold breakthrough in Aboriginal political thinking. Dene/Métis negotiators were focused on self-government while Ottawa was looking at a land settlement. In this way, Denendeh foreshadowed the acceptance by Canada of a more vigorous version of Aboriginal self-government. Unfortunately for the Dene/Métis, the time was not right because the federal government of the day was strongly opposed to such a radical idea. Even the first comprehensive land claim negotiations, which led to agreements in 1984, 1992, and 1993, excluded any discussion of Aboriginal self-government. Ottawa was not ready for such discussion.

A major shift in federal policy towards Aboriginal self-government took place in the Charlottetown Accord. While the Accord was rejected in a national plebiscite in 1992, it called for entrenchment in the Constitution of the inherent right of Aboriginal peoples to self-government. The rejection of the Charlottetown Accord meant that prospects for constitutional change were no longer in the cards. In 1995, the federal Liberal government recognized the inherent right of self-government as an existing right within section 35 of the Constitution Act, 1982. Since 1995, self-government has become part of comprehensive land claim negotiations.

Emergence of Aboriginal Homelands

While the Dene first conceived of a political homeland, the Inuit were first able to achieve this. In 1999, Nunavut was formally carved out of the Northwest Territories. Most forms of Aboriginal self-government take the form of "ethnic" governments and involve relatively small territories, but Nunavut does not. Nunavut has a "public" government but in fact represents a de facto ethnic state within Canada that "fits comfortably within established traditions of mainstream Canadian governance" (Hicks and White, 2000: 31). For the Inuit, accepting a public form of government was not a problem because Nunavut has very few non-Inuit residents.

Nunavut

In the late twentieth century, Nunavut was the most innovative political development to appear on the northern horizon. In 1976, Inuit Tapirisat of Canada (now called **Inuit Tapiriit Kanatami**) proposed creating an Arctic territory called Nunavut ("our land" in Inuktitut) that would represent the political and cultural interests of Canada's Inuit. It took nearly 25 years, but on 1 April 1999, following a settlement agreement reached six years earlier, Nunavut became a new political entity (see Table 7.1).

As the largest political unit in Canada, Nunavut extends over 2 million km². At the same time, it has the smallest population—31,910 inhabitants in 2011, but most (85 per

Table 7.1 The Road to Nunavut: A Chronological History

Year	Event
1973	Inuit Tapirisat of Canada (ITC) documents its traditional lands by mapping traditional land use and occupancy under the direction of Professor Milton Freeman.
1976	ITC proposes the creation of a separate Arctic territory for the Inuit of the Northwest Territories; the Inuit of the western Arctic (the Inuvialuit) opt to undertake their own land claim, thus dividing the Inuit land claim and territory into two parts.
1990	Following years of negotiations, the **Tungavik Federation of Nunavut**, the Northwest Territories, and the federal government sign an agreement-in-principle that includes the formation of a separate territory for the Inuit of the central and eastern Arctic subject to the approval of the division of the Northwest Territories by means of a plebiscite.
1992	The residents of the Northwest Territories approve of its division into two separate territories; The Tungavik Federation of Nunavut and the federal government sign the Nunavut Political Accord; and the Inuit of Nunavut ratify the Nunavut Land Claims Agreement.
1993	Parliament passes the Nunavut Land Claims Agreement Act and the Nunavut Act; Nunavut Tunngavik Inc. replaces the Tungavik Federation of Nunavut.
1997	The Office of the Interim Commissioner is established to help prepare for the creation of Nunavut.
1999	The Territory of Nunavut is formed, and its government begins to function; the **Clyde River Protocol** establishes the norms that will govern the relationship between Nunavut Tunngavik Inc. and the new government of Nunavut.

Vignette 7.2   Royalty Payments in Nunavut

Of all the provinces and territories, Nunavut represents the last place without an agreement with Ottawa to receive royalties from mining operations on Crown lands. While the Nunavut Land Claims Agreement contains provision for sharing royalty revenue from mining on Crown lands within the Nunavut Settlement Area, Crown lands outside of the settlement area do not have such an arrangement and all royalties flow to Ottawa. Where subsurface rights within the Settlement Area were assigned to the Inuit, all royalties go to **Nunavut Tunngavik Inc.** (NTI), which holds title to the minerals on Inuit-owned lands. In 2011, the Meadowbank gold mine was the first mining operation to generate royalties for NTI. These and future royalty funds will be placed in the NTI Resource Revenue Trust.

---

cent) are **Nunavummiut.** Its population, however, is growing rapidly, at an annual rate of 1.6 per cent, due to its high birth rate (the highest in Canada). Fuelled by an exploding population, its demographic increase has outpaced that of most developing countries—in 1961, its population was just over 4,000, but this had climbed to 32,000 by 2011 (Statistics Canada. 2012a). The same increase is found in Nunavut communities, with Iqaluit leading the way: in 1961, Iqaluit (then Frobisher Bay) had a population of 512; by 2011, this number had jumped to 6,699 (Krolewski, 1973; Statistics Canada, 2012b). From 2006 to 2011, the total population of Nunavut, as well as that of Iqaluit, increased by 8.3 per cent (Statistics Canada, 2014b). Among all Aboriginal peoples, but particularly the Inuit, an extremely high rate of natural increase adds significantly to the demand for public services, especially public housing. The relatively large number of people under the age of 15 pushes the population into a position where the **age-dependency ratio** is high so that more public resources are required to respond to the need for public institutions and services designed for infants, children, and teenagers.

The people of Nunavut are scattered across the Arctic in 26 small, isolated settlements without a series of highways connecting these places (Table 7.2). Given its extremely weak economy, Nunavut is heavily dependent on transfers from Ottawa. Transfer payments amount to over 90 per cent of Nunavut's annual budget. Beyond transfer payments, the mining industry provides some economic growth, but the problem with the mining industry is the relatively short lifespan of mines, limited employment of Inuit workers, and the collection of royalties not by Iqaluit but by Ottawa. Most Inuit families rely on a combination of local employment, social welfare payments, and hunting for country food. A sign of the dysfunctional nature of the Nunavut economy is the high level of unemployed and underemployed. While its population growth shows no signs of slowing, the number of Inuit residing in southern cities remains small but is on the rise.

Along with a shortage of jobs in Inuit communities, many people depend on welfare, reside in overcrowded housing, and face a bleak future. Colonialism and now neocolonialism help to explain the roots of this social malaise—the intrusion of Western culture destroyed the Inuit cultural anchors, leaving the people in a state of cultural drift. Not surprisingly, suicides occur much more frequently in Nunavut than in southern Canada. Suicides are just one indicator of the level of distress in communities (Hicks, 2007). In 2013, the number of suicides in Nunavut reached a record high of 45, making it by far the highest rate in Canada (CBC News, 2014).

Table 7.2 Population Counts: Urban Centres in Nunavut, 1961, 2006, and 2011

Centre	2011	2006	% Change	1961
Iqaluit	6,699	6,184	8.3	512
Arviat	2,318	2,060	12.5	168
Rankin Inlet	2,266	2,358	–3.9	586
Baker Lake	1,872	1,728	8.3	386
Cambridge Bay	1,608	1,477	8.9	531
Pond Inlet	1,549	1,315	17.8	53
Igloolik	1,454	1,538	–5.5	133
Coppermine/Kugluktuk	1,450	1,302	11.4	230
Pangnirtung	1,425	1,325	7.5	114
Cape Dorset	1,363	1,236	10.3	161
Gjoa Haven	1,279	1,064	20.2	98
Repulse Bay	945	748	26.3	116
Clyde River	934	820	13.9	40
Spence Bay/Taloyoak	899	809	11.1	124
Coral Harbour	834	769	8.5	117
Arctic Bay	823	690	19.3	49
Sanikiluaq	812	744	9.1	169
Pelly Bay/Kugaaruk	771	688	12.1	94
Hall Beach	546	654	–16.5	
Broughton Island/Qikiqtarjuaq	520	473	9.9	70
Lake Harbour/Kimmirut	455	411	10.7	90
Whale Cove	407	353	15.3	125
Chesterfield Inlet	313	332	–5.7	146
Resolute	214	229	–6.6	116
Grise Fjord	130	141	–7.8	98

Source: Nunavut Bureau of Statistics (2012); Krolewski (1973).

Nunavut Tunngavik Inc., the Inuit land claim organization, has taken on the role to ensure that Nunavut is not only a homeland for the Inuit but also an embodiment of Inuit culture (Légaré, 2000), and is sometimes in conflict with the Nunavut government because Inuit culture and participating within Canadian society do not always mesh well. The challenge for both the government and Nunavut Tunngavik is to ensure that government policies and operations reflect the Inuit way of thinking and living (**Inuit Qaujimajatuqangit**).

An acute shortage of qualified Inuit for jobs within the public service signals a critical education issue—not enough Inuit are completing high school and post-secondary education. Fortunately, this gap is narrowing as the percentage of Inuit in

Table 7.3 Inuit/Qallunaat Employees in the Government of Nunavut, December 2014

Position	Number of Inuit Employees	Inuit Percentage of Total Employment	Number of Positions Filled in Category
Executive	13	38	34
Senior management	29	21	141
Middle management	96	24	396
Professional	327	27	1,202
Paraprofessional	824	72	1,096
Administrative support	418	88	427
Total	1,707	50	3,309

Note: Qallunaat refers to non-Inuit employees; more specifically, Qallunaat means "white people."
Source: Nunavut, Finance (2015: 5).

the territorial civil service has climbed from 44 per cent in 1999 to 50 per cent in 2013 (Nunavut, Finance, 2015:19). Still, the distribution of Inuit employees remains heavily weighted towards the lower-skilled and lower-paying jobs. For example, Table 7.3 reveals that, in 2013, 91 per cent of public service workers in the lowest category, administrative support, were Inuit, while Inuit employees comprised only 20 per cent of senior management.

Vignette 7.3 Protecting Inuktitut but Not Willing to Pay the Price

Language is a sensitive issue. Nunavut's government has introduced legislation to protect Inuktitut. The two bills—Bill 6, the Official Languages Act, and Bill 7, the Inuit Languages Protection Act—became law in 2008. Nunavut Tunngavik says the bills are "a good start" but that the government should do more to protect Inuit language rights, especially in education, where most of the education program should be delivered in Inuktitut as soon as possible (Bell, 2007). The Qikiqtani Inuit Association, which represents the interests of the Inuit of the Baffin Region, the High Arctic, and the Belcher Islands, strongly opposes these two bills because it feels they are "too weak" and because the Inuit language requires the same status within Nunavut that the French language enjoys in Quebec under that province's famous Bill 101 language law (Bell, 2007). Berger (2006: 1–67) addressed this contentious language issue and recommended that Inuktitut become the language of instruction from kindergarten to Grade 4, with English during these years as a second-language program; commencing in Grade 5, instruction would be in both Inuktitut and English. This proposed bilingual program, according to Légaré (2008a: 365), had three flaws. First, Ottawa was not prepared to fund the $20 million required to support the program. Second, an insufficient number of teachers currently speak Inuktitut. Third, resource materials in Inuktitut are lacking, especially beyond Grade 4. Of course, the last two flaws could be tackled and resolved with sufficient funds—which, as noted, is the central flaw in such a plan. As pointed out in Chapter 6, the paradigm followed by the federal government is more about promoting economic development than about addressing environmental and social issues.

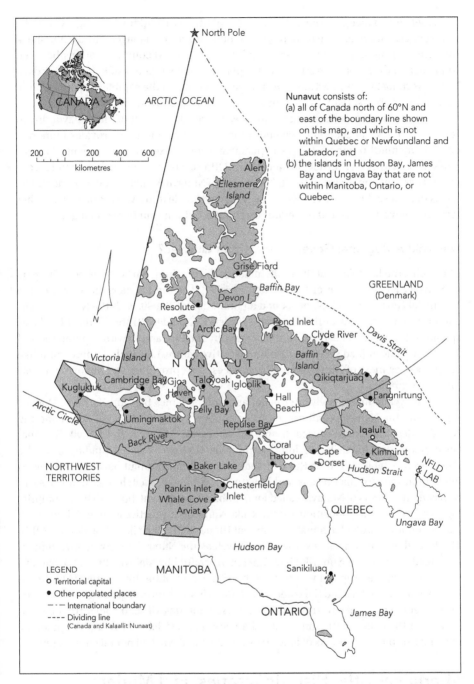

Nunavut consists of:
(a) all of Canada north of 60°N and east of the boundary line shown on this map, and which is not within Quebec or Newfoundland and Labrador; and
(b) the islands in Hudson Bay, James Bay and Ungava Bay that are not within Manitoba, Ontario, or Quebec.

LEGEND
O Territorial capital
● Other populated places
— ·— International boundary
- - - - Dividing line
(Canada and Kalaallit Nunaat)

Figure 7.1   Nunavut: Canada's Newest Territory
Source: Adapted from *The Atlas of Canada* (2006).

A second challenge is the lack of use of Inuktitut as a working language within the civil service. The main reasons are (1) the bulk of senior management do not speak Inuktitut; (2) the Inuit civil servants are bilingual; and (3) communications between Nunavut and the federal government take place in English. As a result, English tends to be the working language within government, especially in the capital of Iqaluit. On the other hand, public servants speak Inuktitut more frequently in smaller centres.

A third challenge is fiscal independence. Political fulfillment of the Nunavut dream is unlikely unless its economy becomes more diversified and thus provides more revenues to Iqaluit through business and personal income taxes. At the moment, Nunavut relies most heavily on transfer payments from Ottawa, and such financial dependence—over 90 per cent—may curtail the full flowering of Nunavut's political development. The fear is that these three barriers will cause Nunavut to slide into a version of the two other territorial governments and therefore run counter to Inuit Qaujimajatuqangit.

Nunavik: A Regional Government within Quebec?

The Arctic, the homeland of the Inuit, includes the northern part of Quebec. With part of Nunavik located north of the sixtieth parallel, it is the northernmost area of the Provincial North. In 2011, Nunavik had a population of 12,090, with the Nunavimmiut composing 89 per cent of this number (Statistics Canada, 2013b). Of the 14 villages spread out along its coastline, Kuujjuaq was by far the largest with a population of 2,375 in 2011. Kuujjuaq serves as its administrative centre. Like Iqaluit, this capital-like centre houses most of Nunavik's public services, thus providing an employment base for the community.

In the years immediately after negotiation of the James Bay and Northern Quebec Agreement, the Nunavimmiut assumed control over the Kativik School Board and the Nunavik Regional Board of Health and Social Services. In 1978, the Kativik Regional Government and Makivik, the Inuit corporation that oversees the funding from the JBNQA, were established. With these institutions, the Nunavimmiut have learned how to function in modern institutions and how to shape them to match their culture. As in Nunavut, the leaders of Nunavik called for a "public" government, but with Inuit comprising 85 per cent of the population, Nunavik, like Nunavut, in practice is an Inuit homeland with an Inuit-controlled regional government (Bone, 2001; Makivik, 2011a; Rodon, 2014). The federal and provincial governments, as well as the Nunavik Regional government, were ready to sign the Nunavik Final Agreement crafted by Makivik. In a 2011 referendum, however, the proposal was rejected by 70 per cent of the Inuit voters (CBC News, 2011a). The question is: Did the voters reject Makivik's version of regional government or the concept of regional government? The next step for Makivik is to hold further discussions with the communities and make the necessary revisions. At that point, another referendum could be held. As of June 2015, however, Makivik had not taken this step.

Aboriginal Title, Historic Treaties, and Modern Land Claim Agreements

Aboriginal title refers to those lands traditionally used by Aboriginal peoples. The British recognized that Aboriginal peoples in North America had some form of title to their land. Treaties provided a means of the Crown to obtain title to those lands. Over time,

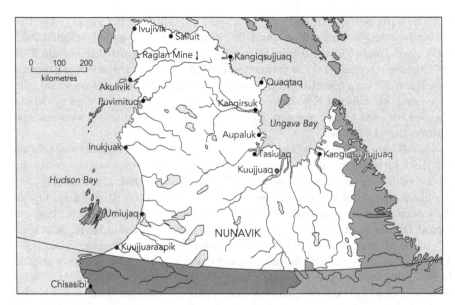

Figure 7.2 Nunavik

Quebec, unlike other provinces, contains a large portion of Canada's Arctic—nearly 18 per cent. This Arctic region is known as Nunavik. With Hudson Strait forming its northern border and the fifty-fifth parallel its southern border, Nunavik, the homeland of Quebec's Inuit, extends over 443,685 km². The 14 isolated villages along the coasts of Hudson Bay, Hudson Strait, and Ungava Bay depend on sea transportation for basic food and construction supplies while an air service links the 14 communities and, most importantly, provides a connection with Montreal. Air Inuit, owned by the Inuit corporation Makivik, provides scheduled and emergency air service.

Source: Nunavimmiut Kavamanga (2007).

the purpose of treaties changed. For the British and later the Canadian government, treaties allowed the Crown to obtain title to the land and thus open the lands for settlement and resource development. At first, Indian tribes saw them as peace and friendship treaties. Later, Indians sought treaties for protection from the newcomers, relief for their lost hunting grounds, and reserves to ensure at least some land base. The first such treaties took place on southern Vancouver Island (the Douglas treaties) and in Upper Canada (the Upper Canada land surrenders) and for large tracts east of Lake Huron and north and east of Lake Superior (the Robinson treaties). The so-called numbered treaties were designed to secure title to lands in western Canada for agricultural settlement. In addition, the Williams treaties involved lands "missed" in pre-Confederation Ontario land surrenders and were concluded in 1923.

For the federal government, the need for treaty negotiations, so necessary to deal with Aboriginal title, was determined by the imminence of agricultural settlement and resource development. At the end of the nineteenth century, the North held little prospects for such settlement or development. Despite requests for treaty negotiations from the North, Ottawa took no action and continued to support the Hudson's Bay Company to provide "relief" when necessary, thus minimizing public expenditures. The federal government believed that northern **Native peoples** "were best left as subsistence harvesters and saw no justification for any systematic attempts to restructure their lives" (Coates and Morrison, 2010). In 1920, the discovery of oil at

Norman Wells changed Ottawa's position and it pressed hard for a treaty to include the lands of the Mackenzie River Basin. By 1923, the last tribe in the Mackenzie Basin had signed Treaty 11, which thus extinguished their Aboriginal title—or did it? In 1973, the Dene argued in the Supreme Court of the NWT that, based on their oral history, Treaty 11 was a friendship and peace agreement. Justice Morrow concluded that Aboriginal title may not have been extinguished in Treaty 11 as claimed by Ottawa. Coupled with the *Calder* decision, comprehensive land negotiations were initiated with the Dene by Ottawa.

In 1973, the federal government accepted the legal concept of Aboriginal title and the claim to lands not under treaty. In Ottawa's comprehensive claims policy, the ground rules were established to resolve such claims by negotiation. Such negotiations would result in a legal agreement between a First Nation or other Aboriginal group and the federal government to extinguish their Aboriginal title in exchange for land and capital. Ottawa commonly cedes a portion of the land identified as having Aboriginal title to the Aboriginal claimants. The percentage ceded varies from one agreement to the next, possibly accounting for the "productive value" of the land. At the low end, the 14 First Nations in Yukon claimed virtually all the territory of Yukon but received ownership of 41,440 km^2, which represents 8.6 per cent of the land mass of Yukon. At the high end, the Inuit of Nunavut obtained 18 per cent of the total land mass of Nunavut or 350,000 km^2.

Comprehensive land claim agreements have had a profound impact on Aboriginal peoples, their homelands, and their participation in Canadian society. However, the first agreements remained silent on the matter of self-government. The first exception took place in 1993 when the Nunavut Land Claim Agreement was ratified, setting in motion the formal establishment of the new territory six years later. While the Inuit had to struggle to have a form of self-government included in the negotiations, the

---

Vignette 7.4 The *Calder* Decision and Aboriginal Title

The 1973 *Calder* decision opened up the debate on Aboriginal title. Prior to 1973, the British Columbia government recognized the 14 treaties signed before Confederation but it claimed that, on the remaining lands, Aboriginal title had been extinguished by the act of joining Confederation. The *Calder* decision threw doubt on that assumption. In the 1973 Supreme Court case, six of the seven judges stated that the Nisga'a still had some kind of an Aboriginal right to their traditional lands (although the Nisga'a lost this particular decision by 4–3 on a technicality). Since most northern Aboriginal peoples had not signed treaties, the ruling implied that they still had a claim to those lands. Consequently, Ottawa decided that modern treaties were necessary to resolve this unfinished business. Thus, the federal government announced its plans for comprehensive land claim negotiations with those Aboriginal peoples without treaty. In another part of the country, the James Bay and Northern Quebec Agreement (1975) represented another form of modern treaties.

negotiations towards the 2005 Tlicho Agreement for part of the Northwest Territories combined the land claim and self-government from the start (INAC, 2008). Since 2005, comprehensive land claim negotiations automatically consider both land claims and self-government.

For Aboriginal peoples living in the Territorial North, the impact of modern treaties has been revolutionary. Comprehensive agreements have transformed their economic and political lives because they provide the means and structure to participate in Canada's economy and society and yet retain a presence in their traditional economy and society. In conjunction with modern treaties, favourable court decisions have recognized Aboriginal rights to natural resources (Bartlett, 1991; Usher et al., 1992). Most significantly, modern land claim agreements involve approximately 4 million km^2 of Canada's northern lands, but only just over 600,000 km^2 or 15 per cent resulted in outright Aboriginal ownership (AANDC, 2014). For instance, Nunavut encompasses 1.9 million km^2, but, as noted above, the agreement assigned the Inuit clear title to 350,000 km^2 (AANDC, 2010).

Aboriginal peoples in the Provincial North who had signed treaties in the late nineteenth and early twentieth centuries (Figure 7.4) could only turn to specific land claims to seek redress for past violations of their treaties, such as the improper loss of reserve land.[1] Others without treaty were eligible for comprehensive land claims. By 1970, the treaty process had made little progress in much of the North. The Inuit and Cree of northern Quebec, First Nations in northern British Columbia, and the Inuit and Innu in Labrador had not yet settled their land claims. Since then, many have negotiated modern land claim agreements. By 2015, 24 such agreements had been concluded (Table 7.4). Except for three in BC—the Nisga'a (2000), the Tsawwassen (2009), and the Maa-nulth (2011)—all took place in the Canadian North.

Modern land claim agreements are not all the same. While a modern treaty, the JBNQA took place before the first federal comprehensive land claim agreement with the Inuvialuit was concluded in 1984. For that reason, the JBNQA differs in several ways from comprehensive land claim agreements that focus on engaging in the market economy. Rodon (2014) states that the JBNQA was modelled after the Alaskan Land Claim Settlement Act of 1971, which emphasized cash settlement over land (Chance, n.d.). The Inuvialuit led the way by demanding subsurface rights to lands for which they gained full title—hoping to select land with oil potential. Back in the 1970s, the Quebec Cree and Inuit were primarily concerned about protecting their traditional lifestyle and therefore sought to obtain as much hunting lands as possible. In the James Bay and Northern Quebec Agreement, the Quebec government interpreted the Indigenous goal as wishing to continue to live on the land:

The needs and interests of the native peoples are closely tied to their lands; their lands are the very centre of their existence. That is why in this Agreement we have devoted ourselves especially to the establishment of a land regime that will satisfy the needs both of the native peoples and of Quebec. (Ciaccia, 1975: "Philosophy of the Agreement")

The resulting land regime consists of different categories of land that allow the Cree and Inuit to continue to harvest country food and furs. Category I, by far the smallest, surrounds settlements; Category II provides hunting lands exclusively for Cree and Inuit; and a third category allows use by both Aboriginal and non-Aboriginal hunters (see Table 7.4). Twenty years later, both the Quebec Cree and Inuit accepted the concept of sharing the wealth generated by resource development as described in the Paix des Braves and the Sanarrutik agreements (Vignette 7.9). This change of strategy marks a partnership between those two Indigenous peoples and the Quebec government and an acceptance by these groups of the inevitability of resource development.

Quebec's support for a "stay-on-the-land program" through its Income Security Program for hunters and trapping families falls outside of the JBNQA. Still, it was a product of those negotiations. Quebec officials recognized that the social costs of moving the Cree from a land-based culture to a welfare/settlement lifestyle would place demands on the Quebec treasury. The officials believed that the cost of the program would be offset by reduced welfare payments and lower social costs. Their position proved correct, as so many Aboriginal communities across the Canadian North faced a host of social ills, ranging from family violence to drug abuse. The sorry tale of Davis Inlet proves this point (Vignette 7.5).

Professor Richard Salisbury (1986) documented the initial impact of cash from the program that allowed for greater use of modern hunting equipment and air trips to hunting grounds.[2] Some 40 years since its initiation, this program remains popular with older traditionalist residents and also provides a safety value for unemployed and

---

Vignette 7.5   From Davis Inlet to Natuashish

It may not be the worst example of the federal government's relocation program—that dubious distinction might be reserved for the removal of Inuit, beginning in the 1950s, from Arctic Quebec to Grise Fiord on Ellesmere Island to create a Canadian presence in the Far North—but Davis Inlet certainly is close to the worst case of social engineering by the Canadian state. Like other northern peoples, the Innu of Labrador were nomadic hunters until the 1960s, when they were relocated into settlements, principally Sheshatshiu and Davis Inlet.

Davis Inlet was a social disaster. This dysfunctional village represented an Aboriginal community in crisis because the old ways of hunting caribou were gone and the new way was dependency and welfare. In the case of Davis Inlet, not only did the settlement remove the Innu from their homeland, but it left them isolated on an island—making access to traditional caribou hunting grounds impossible. In the early 1990s, the Canadian public saw terrible television images of Davis Inlet children sniffing gasoline. Not surprisingly, the community had "one of the highest suicide rates in the world" (MacDonald, 2001), and alcoholism and drug abuse among parents, accidental death, physical abuse, and family dysfunction were common features of the Mushuau Innu community. A new settlement, Natuashish, was completed in 2003 on the Labrador mainland, at an estimated cost of $165 million, and the nearly 700 Davis Inlet Innu left behind their squalid village for a new town with a sewer and water system, a well-designed school, and a band building. Yet, the physical move, while a great improvement, has not resolved the deep-rooted social problems. The healing process for the next generation has just started and its chances of success will only be known in the decades to come.

Figure 7.3   Cree Hunting Camp in Northern Quebec

Modern hunting camps allow for more comfortable and secure conditions for Cree families living in the bush.

Source: Cree Hunters and Trappers Income Security Board (2014).

young people who seek relief from village life by spending time in the bush. In 2012–13, this income security program made payments of just over $23 million to 2,675 people for a per capita payment of $8,600. A family of three, for instance, could receive nearly $26,000. The program is attracting more participants for several reasons. First, the Quebec Cree population has increased over this time period. Second, the hardship of living on the land has eased greatly (Figure 7.3) and quick access to local communities

---

Vignette 7.6   Importance of Land-Use Studies for Aboriginal Land Claims

Geographers have played a key role in land-use studies that provided the evidence of Aboriginal title and thus for land claims. Freeman (1976) and Usher (1976), for example, helped to develop a system of mapping Inuit traditional land use. Land selection is a critical factor in the claims process, and it involves geographic concerns such as location and quality of land. Frank Duerden (1990: 35–6), a geographer who worked extensively with Yukon First Nations, outlined four spatial components of the land claims process to ensure maximum benefit for the Yukon First Nations (see also Duerden and Keller, 1992; Duerden and Kuhn, 1996, 1998; Duerden, 2004):

- mapping areas of use and occupancy in order to develop a case for legitimacy of a claim;
- preparation of maps of land-use potential to identify "optimum" areas to be retained as an outcome of negotiations;
- evaluation of land selection positions as negotiation proceeds to establish the extent to which they satisfy the goals of the process;
- evaluation of non-ownership agreements regarding land to ascertain the effectiveness of the control they offer and their long-term impact on land-use patterns.

Table 7.4   Modern Land Settlements in Canada's North

Agreement	Date	Beneficaries*	Land (km²)/ per Capita	Cash Payment/ per Capita
Cree: James Bay and Northern Quebec	1975	6,650	70,677/10.6**	$135,000,000/$29,300
Inuit: James Bay and Northern Quebec	1975	4,390	98,756/22.5**	$90,000,000/$20,501
Naskapis: James Bay and Northern Quebec	1978	330	4,471/13.6**	$9,000,000/$23,136
Inuvialuit	1984	2,500	91,000/3.3	$55,000,000/$22,000
Yukon Indians	1990	6,500	41,440/6.4	$242,700,000/$37,339
Gwich'in	1992	2,300	23,976/10.4***	$141,000,000/$61,304
Nunavut	1993	17,500	318,084/18.2	$1,170,000,000/$66,857
Sahtu Dene and Métis	1993	2,428	41,437/17.1	$75,000,000/$30,890
Tlicho	2003	2,020	39,000/19.3	$152,000,000/$75,248
Nunatsiavut	2005	2,160	15,799/7.3	$140,000,000/$66,667

*Population at the time of signing.

**Land settlement for the James Bay and Northern Quebec Agreement (JBNQA) differs from comprehensive agreements. The JBNQA was divided into three categories. In Table 7.4, Column 4 refers to Category I lands (5,589 km² for Cree and 8,106 km² for Inuit) and Category II lands (65,086 km² for Cree and 90,650 km² for Inuit). Category I includes the land surrounding Cree and Inuit communities and is for their exclusive use. Category II (a total of 155,736 km²) is reserved exclusively for Inuit and Cree hunters. Category III (980,000 km²) is open to all hunters. Mining, forestry, and tourism can occur in Categories II and III. In 2008 and 2010, the Inuit and Cree signed agreements regarding offshore islands.

***Gwich'in lands in Yukon are 1,554 km² and 22,422 km² in the Northwest Territories.

Sources: Crowe (1990); Bone (1992: 236); Land Claims Agreements Coalition (2014).

provides a safety net for those injured or ill. Third, the program provides an option for those who become unemployed. The concluding statement from its 2012–13 report provides a similar analysis:

> The participation to the Cree Hunters and Trappers Income Security Program has increased steadily since 2008–2009. Possibly, the completion of major developments and the associated decrease in employment opportunities, combined with an emerging trend for new retirees to join the income security program, may be explanations for this trend. (Cree Hunters and Trappers Income Security Board, 2013: 34)

Impact and Benefits Agreements

Impact and benefits agreements (IBAs) are formal, written agreements between development companies and Aboriginal organizations that are designed to compensate Aboriginal peoples for the predicted impacts associated with an industrial development and to permit the development to proceed (Kennett, 1999). Geographers are playing a leading role in the field by studying and assessing IBAs (Galbraith et al., 2007; Prno and Bradshaw, 2007; Noble, 2010). While details of IBAs are confidential, they do provide specific business, employment, and training opportunities for Aboriginal groups with a traditional land claim to the area proposed for industrial development. IBAs vary considerably in their scope and complexity, depending on the scale and nature of the project and the issues identified by the negotiating parties involved. However, most agreements address such topics as:

- environmental protection, including special concerns about wildlife;
- protection of Aboriginal social and cultural values;
- education, training, and employment;
- health and safety;
- business opportunities;
- Aboriginal access to the project site;
- financial compensation, including royalties;
- dispute resolution mechanism.

Impact and benefits agreements vary from one group to the next depending on the political strength and negotiating skills of each group. Under most circumstances, the company has an advantage because its negotiators are more experienced and have the legal and other resources of the company at their disposal. The Aboriginal group has one important advantage—Aboriginal title. The company must reach an agreement in order to proceed with the project. However, not all Aboriginal groups are sufficiently aware of the stakes and some have little experience with such negotiations. Two cases are considered here: Victor diamond deposit in northern Ontario and the Ekati diamond deposit in the Northwest Territories (the most recent IBA involved Meadowbank gold mine in Nunavut; see Chapter 5). The IBAs for the Victor and Ekati mines were signed in the last decade of the twentieth century and "broke" the ground for resource developments far from settlements but on traditional lands. In both cases, local residents and Aboriginal officials grasped the significance of an IBA and learned from the experience of the negotiations, but they were, in a sense, learning on the run. At the same time, the resource corporations involved in these projects learned, too, and ultimately recognized that a partnership approach would achieve the best results for both the local residents and their companies.

In the early 1990s, the Luxembourg-based diamond company, De Beers, which originated in South Africa, discovered a vast diamond deposit in the Hudson Bay Lowland of northern Ontario, approximately 90 km west of the coastal community of Attawapiskat. The company undertook an environmental assessment and, over a period of several years, negotiated an IBA with the Attawapiskat First Nation since this diamond deposit lies in the traditional land claimed by the Attawapiskat. Both the EIA and the IBA were fraught with problems—largely because First Nations officials were inexperienced and unsure of their grounds (Whitelaw and McCarthy, 2010: 473–4). The Victor mine lies close to two other First Nations—the Moose Cree First Nation and the Taykwa Tagamou Nation, so these groups also were involved. Negotiations began but were complicated. Unlike First Nations that had achieved a regional sense of belonging and therefore had gained a level of sophistication in dealing with the

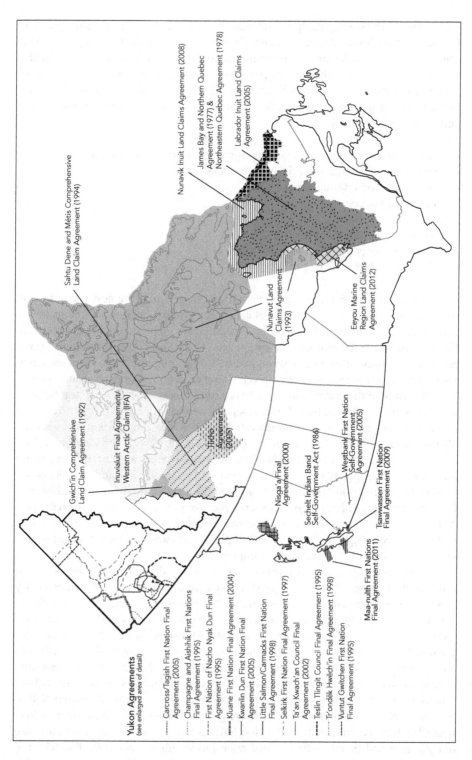

Yukon Agreements
(see enlarged area of detail)

---- Carcross/Tagish First Nation Final Agreement (2005)
.... Champagne and Aishihik First Nations Final Agreement (1995)
——— First Nation of Nacho Nyak Dun Final Agreement (1995)
——— Kluane First Nation Final Agreement (2004)
——— Kwanlin Dun First Nation Final Agreement (2005)
——— Little Salmon/Carmacks First Nation Final Agreement (1998)
– – – Selkirk First Nation Final Agreement (1997)
——— Ta'an Kwach'an Council Final Agreement (2002)
■■■■ Teslin Tlingit Council Final Agreement (1995)
■■■■ Tr'ondëk Hwëch'in Final Agreement (1998)
——— Vuntut Gwitchen First Nation Final Agreement (1995)

Sahtu Dene and Métis Comprehensive Land Claim Agreement (1994)

Nunavik Inuit Land Claims Agreement (2008)

James Bay and Northern Quebec Agreement (1977) & Northeastern Quebec Agreement (1978)

Labrador Inuit Land Claims Agreement (2005)

Eeyou Marine Region Land Claims Agreement (2012)

Nunavut Land Claims Agreement (1993)

Gwich'in Comprehensive Land Claim Agreement (1992)

Inuvialuit Final Agreement/ Western Arctic Claim (IFA)

Tlicho Agreement (2005)

Westbank First Nation Self-Government Agreement (2005)

Nisga'a Final Agreement (2000)

Sechelt Indian Band Self-Government Act (1986)

Tsawwassen First Nation Final Agreement (2009)

Maa-nulth First Nations Final Agreement (2011)

Vignette 7.7   A Foot in Both Cultures

From an Aboriginal perspective, comprehensive land claim agreements provide a legal and financial foundation that blends the land-based **subsistence economy** with the commercial economy, thus allowing Aboriginal peoples a foot in two cultures. The first agreement, the Inuvialuit Final Agreement of 1984, set the pattern for future negotiations. In the case of this agreement, the Inuvialuit Regional Corporation represents the business structure while the Inuvialuit Game Council is concerned about the environment and wildlife. Co-management of the environment and wildlife, involving the Aboriginal group and government officials, flowed out of the agreement. The underlying premises are: (1) that the land-based economy will continue to attract young members and thus remain a viable option; and (2) that Aboriginal culture, language, and traditions will thrive among members of the land-based economy. But are these assumptions valid? The most successful program, the Quebec **Cree Hunters and Trappers Income Security Program**, involves substantial subsidy from the Quebec government.

When resource developments are proposed within the **settlement area** of an Aboriginal group, the developer must negotiate an impact and benefits agreement. Such agreements are between the Aboriginal group and the company. As such, they are treated as confidential documents. In 1990, the first IBA, the Raglan Agreement, was concluded between Makivik Corporation (representing the Inuit of Nunavik) and Falconbridge Corporation (purchased in 2006 by the Swiss mining giant, Xstrata). Since the mine site and access route to an ocean port are located in Nunavik, lands controlled by the Inuit of northern Quebec, Falconbridge was compelled to negotiate an IBA. Falconbridge gained Inuit approval for an underground nickel/copper mine and an access route to Deception Bay on Hudson Strait. According to O'Reilly and Eacott (1998), the Raglan Agreement contains the following items:

- The Inuit receive $14 million plus 4.5 per cent of mine profits, estimated at $60 million over 15 years.
- Detailed project descriptions trigger renegotiation if the project deviates from original specifications.
- An implementation committee consists of three Falconbridge representatives and three Inuit. Makivik also got agreement to have a representative appointed to the mine's board of directors.
- A joint committee oversees training programs.
- Inuit enterprises, such as Inuit Air, are given preference for contracts and Inuit communities or individuals have entered into joint ventures with companies to seek contracts with the mine.

◄ **Figure 7.4**   Map of Modern Treaties

Signed in 1975, the James Bay and Northern Quebec Agreement was the first modern treaty and land claim settlement. Since then, modern claims have fallen into two categories: comprehensive and specific. The first comprehensive land claim agreement was signed in 1984 between the Inuvialuit and the federal government. By 2014, two claims in the Territorial North remained outstanding, both in the Northwest Territories: the Deh Cho claim lying west of Great Slave Lake and the Slavey claim east of that lake. The main areas without treaties are found in British Columbia and the Northwest Territories, and portions of Quebec and Labrador.

Source: Land Claims Agreements Coalition (2014).

outside world, the First Nations of northern Ontario tended to act independently and, some might say, lacked the capacity to negotiate on an equal footing with De Beers.

The first step saw De Beers and Attawapiskat sign a memorandum of understanding (MOU) in November 1999. After more consideration, in July 2002, the Attawapiskat First Nation declared that it was terminating the MOU, but would like to begin negotiation towards an impact and benefits agreement. During negotiations, Attawapiskat demanded that work halt for at least two months. The reason for the demand was simple: the First Nation leaders were unsure of the consequences of the project and needed time to review technical documents and to consult with band members and advisers. They also believed that their interests would be better served by an IBA, but to achieve such an agreement requires negotiating experience and skills that the First Nation may lack to strike a "good" deal (Hasselback, 2002). In 2005, an IBA was concluded between the Attawapiskat First Nation and De Beers. In 2007, the Moose Cree First Nation and De Beers announced that an IBA was signed. Like the Attawapiskat First Nation, the Moose Cree First Nation voted in favour (85 per cent) of accepting the IBA. This agreement is the third one involving the Victor diamond mine. The Taykwa Tagamou Nation did not sign an IBA but, rather, a working relations agreement (an MOU). These agreements commit De Beers to provide education, training, compensation, and business and other initiatives to maximize opportunities from the Victor mine and to minimize impacts arising from construction and operation. Chief Patricia Faaries-Akiwenzie of the Moose Cree stated that:

> It's important for the Moose Cree citizens to be entering into this IBA with De Beers because it shows respect for our Homeland. Our people have been here since time immemorial and will continue to live in and off this land. Now that De Beers is extracting our resources it is only fair that we have reached an agreement that provides benefits for our First Nation. This Agreement represents a major step forward in our relationship with outside resource users and breaks the trail for other companies who want to utilize our resources in the Moose Cree Homeland. (Beaton, 2007)

Broken Hill Proprietary Company (BHP), an Australian multinational, negotiated IBAs with each of the four Aboriginal groups—North Slave Métis Alliance, Akaitcho Treaty 8, Kitikmeot Inuit Association with the Hamlet of Kugluktuk, and Dogrib Treaty 11—whose traditional lands are potentially affected by the Ekati diamond mine. The agreements ensure that these Aboriginal groups will not interfere with the project. In return, the company agrees (1) to ensure that training, employment, and business opportunities are made available to these people, and (2) to minimize any potential adverse environmental and social impacts of the project. Other objectives are to respect local cultures and the land-based economy, to provide a working relationship between the company and the Aboriginal groups, and to determine how the local communities will be involved in employment, training, and business opportunities. Since impact and benefits agreements are private agreements that include a confidentiality clause, exact details, including any cash compensation, are not part of the public record (personal communication with Denise Burlingame, Senior Public Affairs Officer, Ekati Diamond Mine, 2 Aug. 2002; Noble, 2010: 174–5).

Resource-Sharing

Over the next decade, the prospects for resource megaprojects in the Canadian North are bright. World demand is driving these projects. Northern Aboriginal governments will share in these projects (Irlbacher-Fox and Mills, 2008). Moving from Aboriginal title to a land claim agreement and then to resource-sharing is a long journey (Hurley, 2000; Coates, 2015). Along these lines, McNeil (1998: 29) argued:

> Despite its shortcomings, the ***Delgamuukw*** **decision** could usher in a new era for Aboriginal rights in Canada. For the first time, the right of Aboriginal peoples to participate as equal partners in resource development on Aboriginal lands has been acknowledged.

This right to share resource revenue is most compelling in northern Canada where comprehensive land claim agreements define the sharing of royalties between Aboriginal peoples and Canadian governments (Table 7.5). Such sharing also takes place on lands outside of the settlement areas associated with comprehensive land claim agreements. The Mary River iron ore project on northern Baffin Inland provides

Table 7.5 Agreements Related to the Sharing of Royalties between Aboriginal Groups and Canadian Governments

Agreement	Initial Share of Royalties of the First $2 Million (%)	Secondary Share of Royalties (%)
Yukon Umbrella Final Agreement	50.0	10.0
Gwich'in and Sahtu Final Agreements	7.5	1.5
Tlicho Final Agreement	10.429	2.086
Labrador, Nunavut, and Nunavik Inuit Final Agreements	10.429	5.0
Deh Cho First Nations*	12.15	2.45
Quebec Cree Nation**		
Nunatsiavut Inuit and Innu (Voisey's Bay)	5.0	5.0
BC First Nations**	?	5.0
Northwest Territories	25.0	25.0
Ontario***	?	?
Alberta and Saskatchewan****	0.0	0.0

*Interim Resource Development Agreement, 2003.
**Varies by project.
***Ontario is considering sharing resource royalties.
****Alberta and Saskatchewan have said they will not share resource royalties.
Source: Adapted from Simeone (2014). Reproduced with the permission of the Library of Parliament, 2015.

an example where the royalties are defined in an impact and benefits agreement between Nunavut Inuit and the Baffinland Iron Mines Corporation. Like all of these agreements, the financial arrangements of the Mary River agreement are not disclosed. For Aboriginal settlement areas associated with comprehensive land claim agreements, resource-sharing is defined within each agreement. The sharing of royalties on non-settlement land and on lands associated with historic treaties is negotiable between the resource corporation and the particular Aboriginal group(s).

The Dynamic Nature of Culture

All cultures adjust to new circumstances. Since contact, Indian and Inuit cultures have evolved to adjust to new circumstances. In fact, the Métis culture is a product of early contact between fur traders and Aboriginal women. Culture, being a dynamic phenomenon, constantly changes, though it must retain at least some of its core elements, such as language, belief system, and way of living. This paradox—changing but holding on to core elements—is particularly noticeable in Aboriginal cultures because of the rapid rate of change caused by the intense contact with Western culture and efforts by the settlers to assimilate the Indigenous peoples, thus wiping out their languages and culture. In Canada, assimilation, now referred to as cultural genocide, took the form of the Indian residential schools. The outcome was a disaster, as revealed by the findings of the Truth and Reconciliation Commission. For northerners, in addition to their children being forced to endure residential schools, their relocation to settlements cut the link to the land and their culture. In this sense, settlement life presented a more subtle form of assimilation.

Without a doubt, the relocation strategy to remove Aboriginal peoples from the threat of starvation succeeded, but it did not provide a road to the future. Instead, it replaced food insecurity with high unemployment rates, housing shortages, and a host of social problems. The toll has been particularly hard on young people, who, facing terrible circumstances, are prone to take their lives. Suicide rates for Aboriginal peoples are double those of the rest of the population (Health Canada, 2005; Laliberté and Tousignant, 2009) and in some communities have reached epidemic proportions. In the community of Sandy Bay in northern Saskatchewan, suicides occurred in a cluster in early 2007, with five suicides and 12 attempted suicides. During that time, one day just before noon a 15-year-old Métis girl used the school computer to type a poem, written by an American teen (Warick, 2007). During the lunch break, she hung herself at home. The poem, "A Simple Smile, a Scream Inside" (Schultz, 2001), was found later, and included the following lines:

I'm screaming on the inside
But a smile is what you see
I'm not content
With this person I seem to be

Another drawback of settlement life is the deterioration of Aboriginal people's health. Diabetes, for instance, has affected an unusually high proportion of Aboriginal people, including those under 30 years of age. Many are overweight. Health officials point to inactivity and diet as two causes for the poor state of their health. Prior to the 1950s, Aboriginal people were extremely active as hunters, fishers, and food gatherers. Their diet was based on fresh meat and fish (country food) that supplied them with most of their caloric and nutritional needs. As well, country food was an important cultural anchor providing both an economic and spiritual link with the land and wildlife. While tea, flour, sugar, and salt were purchased at the local trading store, it was not until the Indigenous peoples moved into settlements that a wide range of foods with high levels of sugar and starch were purchased and consumed. As Collings, Wenzel, and Condon (1998: 303) reported, "This shift from a high-protein, polyunsaturated fat diet to one high in carbohydrates and saturated fats, especially among the generation(s) raised in the settlement, has not been without adverse health effects."

In spite of the high rate of suicides, alcohol and substance abuse, and spouse battering, these communities offer a cultural home where individuals can reformulate their personal lives and their Aboriginal culture. Swidrovich (2001: 99) maintains that "Aboriginal peoples have, as part of the strength of their traditions, the ability to continually adapt and change as their circumstances demand without losing their sense of cultural integrity." Swidrovich concludes that "the adoption of European principles or beliefs should not, in itself, signal a sudden or wholesale change as to who they are as peoples." David Newhouse (2000: 408), an Onondaga from the Six Nations of the Grand River and professor of Native Studies at Trent University, sees the development of modern Aboriginal societies as a blending of Aboriginal and Canadian cultures:

> The idea of one citizen, one vote gives to an individual a role and power that may be at odds with the roles of individuals within traditional societies. These modern notions are being combined with Aboriginal notions of collectivism and the fundamental values of respect, kindness, honesty, sharing, and caring. Yet, despite these combinations, we still recognize the governments as Aboriginal and the people as Aboriginal. What has happened is that our notion of "Aboriginal" has changed and now incorporates features of both the traditional and modern worlds.

Aboriginal Corporations

Are Aboriginal corporations a business example of responding to new circumstances? The fur trade represented a response to new circumstances, so why not see Aboriginal corporations in the same light? Some are products of comprehensive land claim agreements, which in total have involved the cash transfer of $3.2 billion (AANDC, 2014), and this infusion of capital has led to the creation of business corporations, such as the Inuvialuit Regional Corporation, which has various subsidiary business ventures. In addition, both Makivik Corporation and the Quebec Cree reached agreements with Quebec, adding another $2 billion, and Makivik, like the Inuvialuit Regional Corporation and the Nunavut Development Corporation, invests the settlement funds, supports regional enterprise, and is involved in various businesses. Such Aboriginal business entities are a significant force in the northern economy. First Nations also have Aboriginal corporations, but, under historic treaties, they did not benefit from large cash settlements. Instead, the most successful ones are a product of an arrangement with natural resource companies, encouraged by provincial governments, to demonstrate "northern benefits." For example, the proximity of the Fort McKay First Nation to the Alberta oil sands created the opportunity for servicing the nearby operations through its Fort McKay Group of companies and from a series of joint ventures with private companies. The advantages of joint ventures are: (1) they provide the Aboriginal partner with a strong business culture and a profitable operation; and (2) the resource companies can point to "northern benefits." Kitsaki, the business corporation of the Lac La Ronge Indian Band in northern Saskatchewan, followed a similar path with uranium mining companies. One of its first businesses took the form of a partnership with a successful private trucking firm that transports supplies and equipment to the mine and trucks its concentrated uranium powder, known as yellowcake, to the United States.

Now operating within the market economy, Native-owned corporations, like other enterprises, are confronted with many issues, including finding qualified (Aboriginal) staff, especially senior managers, ensuring that these operations are profitable (which may mean laying off workers in slow times), protecting the environment, and, above all, avoiding overexploitation of renewable resources (Vignette 7.8).

While Aboriginal corporations cannot solve the deep-rooted economic, employment, and social challenges single-handedly, they do represent a start in that direction. More importantly, Aboriginal corporations are making progress in breaking the dependency on government, though they often receive financial backing from Ottawa and/or provinces. Such backing, it must be emphasized, is hardly any different from that received by large corporations through their elaborate lobbying efforts.

Just as the Aboriginal political movement generates political leaders, Aboriginal corporations are drawing Aboriginals into the business world, and, by doing so, are making a new social class within the Aboriginal community. Aboriginal title, however,

---

Vignette 7.8   The Gap between Aboriginal Education and Employment

Canada faces a shortage of skilled workers. Traditionally, Canada has sought skilled workers from other countries. John Kim Bell, president of the National Aboriginal Achievement Foundation, sees another solution. The demographic explosion among the Aboriginal population has created a large and growing young "potential" labour force—one that is growing more than twice as fast as the non-Aboriginal workforce. The problem is that most young Aboriginal people are not graduating from high school and, therefore, cannot enter into trade schools or other post-secondary institutions. In the short run, Bell (2002) calls for private industry to draw more Aboriginal workers into their companies by providing on-the-job training. As well, flying in skilled workers from the south is both expensive and fails to recirculate their wages in the local community. Makivik Construction approached this problem by working closely with the Kativik School Board and the Kativik Regional Government to develop a number of on-the-job carpenter training programs for local students who spend their earnings in their own communities (Makivik, 2011b).

is an important factor in allowing such businesses to come into being. The Innu of Labrador provide one example. Aboriginal title has favoured the Innu, who have a claim to land around the Muskrat Falls hydroelectric project on the Lower Churchill River, 225 km downstream from Churchill Falls and approximately 300 km from the Labrador–Quebec border. The province hopes to reap major economic rewards, but the province needed and got the approval of the Innu. Under the New Dawn Agreement, ratified by the Innu in 2011, the economic benefits and royalties from the proposed project, plus cash compensation ($2 million annually) for the Churchill Falls dam built nearly 50 years ago (1967–71), were just too attractive for the Innu to reject. On the other hand, Elders such as Elizabeth Penashue (Tshaukuesh, 2010) were more skeptical about the benefits and strongly critical of the impact of the proposed dam on the Lower Churchill River and Innu traditions. Peter Penashue, Elizabeth's son and an Innu leader and former minister in the Harper government, said that "while there are diverging opinions, the deal is in the long-term interests of the Innu communities" (CBC News, 2011b). For its part, the province was delighted with the New Dawn Agreement: former Premier Kathy Dunderdale called it a landmark for the province and its goal of exporting power to the US. While the economics of the Muskrat Falls project is based on the sale of huge exports of power to New England via subsea transmission lines from Labrador to the island of Newfoundland and from Newfoundland to Nova Scotia, the potential high costs have put a damper on this project, placing it on the shelf for now. Then, too, the provincial government was less upbeat about the land claim aspect of the New Dawn Agreement, considering it "non-binding" and stating that it "will form the basis for negotiating a final land claims agreement or treaty" that "will define Innu treaty rights and where those rights will apply in Labrador" (CBC News, 2011c).

Aboriginal title alone provides an opportunity for Native peoples to participate in resource developments, but comprehensive land claim agreements supply the beneficiaries with a business structure as well as a large cash settlement. The cash settlements obtained by Aboriginal groups through comprehensive agreements in the Territorial North, as noted above, have enabled various business ventures. In the Provincial North, cash transfers to Quebec Cree and Inuit came from two sources: the James Bay and Northern Quebec Agreement and a series of agreements with the Quebec government; in the rest of the Provincial North, most cash transfers have come from impact and benefits agreements, such as those for the Victor diamond mine in Ontario and the Voisey's Bay nickel mine in Labrador.

Northern Quebec leads the Provincial North in terms of modern treaties and related agreements and benefits. The 1975 JBNQA transferred a total of $234 million to the Cree and Inuit organizations. With what has happened since and will happen in the future, while not exactly "pocket money," this amount pales with that from later agreements. In 2002, the Paix des Braves and Sanarrutik agreements opened the way for the development of resources within a partnership context. For example, the Inuit will receive 1.25 per cent of the annual value of electricity produced from future hydroelectric developments in Nunavik. Over the next 50 years, the total value of the Cree agreement with the province is $3.6 billion. These funds are committed to economic and community development and release Quebec from its responsibilities for such development, as outlined in the JBNQA for these two fields.

The Sanarrutik Agreement between Quebec and the Inuit of Nunavik in April 2002 was for $90 million over 25 years (George, 2002). Here, too, the agreement links northern development with community and economic projects in Nunavik. The Quebec government has approval to begin developing the hydroelectric potential of the rivers flowing into Hudson Bay and Hudson Strait (the Nastapoka, Whale, George, and Caniapiscau rivers). In June 2002, the Makivik Corporation received the first payment of $7 million.

Quebec sees these partnership agreements as "a path to the future"—with political implications that, from the province's perspective, will bind the North to the rest of Quebec. Even though the rest of the Provincial North remains far behind the partnership agreements found in Quebec, there is a growing sense in the Aboriginal community that these agreements are the wave of the future. For example, Manitoba, with similar (though smaller) northern hydroelectric developments, is finally embracing partnership agreements. But Manitoba's Cree First Nations are well behind their counterpart in Quebec. Historic treaties—the numbered treaties 1 to 6 and 10—in Manitoba did not provide anything like the same range of monetary benefits or the opportunity to form regional Aboriginal governments as was the case in northern Quebec, where no previous treaties were in place.

Aboriginal enterprises are not immune to business failure. Sometimes such failure is due to global economic conditions. For example, Kitsaki, the business arm of the Lac La Ronge Indian Band, was involved in the Wapawekka sawmill operation with Weyerhaeuser Canada, the Peter Ballantyne Cree Nation, and Montreal Lake Cree Nation. With the help of the federal government, the partners obtained $22.5 million to build a small-log sawmill just north of Prince Albert. The objective was to make better use of small logs rather than using them as pulpwood. In 1999, the market for lumber in the United States was strong and the Canadian dollar was low in comparison with the

Vignette 7.9   Paix des Braves: Shining Model?

On 7 February 2002, Quebec Premier Bernard Landry and Cree Grand Chief Ted Moses agreed to the $3.8 billion hydroelectric project involving the Eastmain and Rupert rivers in exchange for annual payments over 50 years totalling $3.6 billion. These funds are for economic and community development and will help to address the shortage of housing in the Cree villages and create construction jobs for local workers. As part of the deal the Cree dropped their $8 billion in lawsuits filed against the province of Quebec.

The agreement put an end to years of disputes over the impact of the hydroelectric projects on the Cree and their claim of unfulfilled promises from the JBNQA. It commits the traditional territory of the Cree to economic development—hydroelectric, forestry, and mining. Such an agreement represents a template for Aboriginal regional development in other areas of the Provincial North where sharing of resource development has not yet taken root. Yet, some fear that this agreement is taking the Quebec Cree in the wrong direction, and that such agreements simply snare the Quebec Cree and Inuit in a process leading to their economic assimilation into Canadian society.

Figure 7.5   The Wapawekka Sawmill, 2002

The Wapawekka mill, located in the closed boreal forest zone of northern Saskatchewan, produced 2 x 4 and 2 x 6 lumber. This forest operation was a small, marginal player, and in 2006, when the US market collapsed, the mill was forced to close.

US dollar. These economic conditions made the sawmill, which employed 45 workers, profitable. However, when the US imposed duties on Canadian lumber in 2006, the mill was no longer profitable, closed, and went into receivership. The buildings and equipment of Wapawekka Lumber Ltd were sold through an auction as part of the bankruptcy process to Millar Western Forest Products Ltd, which moved them to Alberta.

Kitsaki and Inuvialuit Regional Corporation

Circumstances have allowed some Aboriginal groups to move boldly into the resource economy. Two Aboriginal corporations, the Inuvialuit Regional Corporation (IRC) and Kitsaki Management Limited Partnership, exemplify this Native entrepreneurial spirit. The IRC is located in Inuvik, NWT, while the headquarters of Kitsaki, the business arm of the 8,000-plus Lac La Ronge Band, is at Air Ronge, Saskatchewan. While each corporation makes a profit in most years, their major impact on the economic lives of the communities is employment.

In 1984, the IRC was formed as a result of the Inuvialuit Final Agreement. Its purpose is to receive and manage the lands and cash settlement of $55 million (in 1979 dollars) flowing out of this agreement. As a company, the IRC sought to invest these funds in profitable enterprises and, by doing so, to provide both employment and dividends for its members. Its strategy has been to form a number of subsidiaries to undertake specific business activities. For example, the Inuvialuit Development Corporation (IDC) is responsible for investing in a wide range of business ventures. By 2007, IDC was

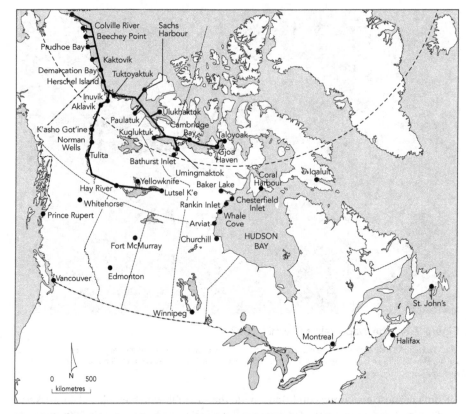

Figure 7.6 Arctic Supply Route: Northern Transportation Company

Based in Hay River, NWT, Northern Transportation Company has a fleet of 12 mainline tugs, more than 90 barges, and two Arctic Class II charter supply vessels that cover nearly 6,000 km of northern shipping routes, travelling north along the Mackenzie River nearly 1,200 km to reach Tuktoyaktuk on the shores of the Beaufort Sea. From there, the route stretches from Point Barrow, Alaska, to the west all the way to the Boothia Peninsula in Nunavut. In 2006, the company reopened its shipping route north from the port of Churchill, Manitoba, to serve the Kivalliq region in Nunavut. In the same year, the company reopened its shipping route to Fort McMurray, Alberta, from where it first started shipping to Canada's Arctic in 1934. NTCL opened an office in 2005 in Halifax to service the Atlantic coast market and the resource development market in Nunavut. Northern Transportation Company is jointly owned by Inuvialuit Development Corporation and Nunasi Corporation, which represents the Inuit of Nunavut.

Source: Northern Transportation Company (2014).

a major shareholder in over 20 enterprises (including the Northern Transportation Company along with the Nunasi Corporation of Nunavut) and joint ventures in seven sectors—transportation, petroleum services, construction/manufacturing, environmental services, northern services, real estate, and tourism/hospitality. From an initial investment of $10 million in 1987, IDC's asset base has grown to over $135 million, due in large part to its petroleum investments. IDC expanded each year until 2009 (IRC, 2006: 7). In 2008, the global economic collapse led to a difficult 2009 for IDC. According to the *Annual Report* for 2009 (2010: 7) IDC recorded a loss of $15 million. The following

Figure 7.7 Trucking Enterprise of the Lac La Ronge Band

Northern Resource Trucking grew out of a partnership between the Lac La Ronge Band and Trimac Transportation. The company was founded in 1986 to expand Aboriginal participation in northern Saskatchewan's mining industry. The primary role of Northern Resource Trucking in northern Saskatchewan is to service the two uranium companies, Cameco and Areva. Winter driving conditions along gravel roads in the boreal forest to reach remote uranium mines present a challenge.

Source: Photo by Dale Peacock.

year, IDC lost $12.7 million, indicating the risky nature of northern businesses, while the Inuvialuit Investment Corporation made $13.4 million on its stock holdings (IRC, 2012: 5). Still, in 2013 the Inuvialuit Corporate Group had net earnings of $32.1 million, the IRC made beneficiary payments to over 4,000 Inuvialuit of more than $2.4 million, and the initial cash payment from the land claim settlement of $162 million had grown to a net worth of $386 million (IRC, 2014: 2).

Kitsaki Management Limited Partnership operates at a much smaller scale and without the advantage of a modern land claim settlement with a substantial cash transfer. As the business arm of the Lac La Ronge Indian Band, Kitsaki has done well.

Kitsaki came about when the provincial government insisted that northern uranium mining companies provide employment and business opportunities for northern residents. Unlike the IRC, Kitsaki has little capital and has had to move slowly and cautiously into business opportunities in the mining and forestry industries. Often, the provincial government, as a means of ensuring northern benefits, has encouraged forestry and mining companies to offer Kitsaki business opportunities. In 1986, the Lac La Ronge Band entered into a joint venture with a private trucking firm to supply goods to the uranium mines in northern Saskatchewan (McKay, 2004: 1). Without experienced truck drivers, few band members were employed at first, but the trucking firm was willing to train and then employ people from Lac La Ronge. The chief at that time, Myles Venn, saw these new opportunities as a means to achieve "Aboriginal ownership that will enable us to possess the control we need to secure jobs for our people" (Bone, 2002: 70). By 2010, Kitsaki employed over 800 workers and up to 1,000 seasonal employees. Approximately 25 per cent of their full-time employees are band members, with wages totalling over $6 million in 2009; and almost all of the seasonal employees are young band members (Kitsaki, 2010). Kitsaki's gross revenue for the year ending 31 March 2014 was $56.2 million with a net profit of $13.3 million (Lac La Ronge Indian Band, 2014: 12). While most joint ventures are profitable, a few just break even. Occasionally, a joint venture, like the Wapawekka sawmill, fails.

Kitsaki, like the Inuvialuit Regional Corporation, has made only a small dent in addressing employment and social problems among its members. First, the magnitude of the unemployment challenge is beyond their capacity. Second, the skill set and experience found in their labour force fall short of the needs of these two business entities. This mismatch between the Aboriginal labour force and the job needs of the Aboriginal corporations extends well beyond these two groups. Consequently, the immediate impact on their respective communities has been limited. The hope is for the long term so that, for example, the youth of the Lac La Ronge Band will recognize that there is a future in wage employment and businesses. Local support for these business initiatives wavers, partly because of the limited impact on their communities, high expectations, and the investment into enterprises outside of their traditional area. The Inuvialuit Regional Corporation is required to make an annual payment to each beneficiary of at least $400. Kitsaki does not make payments to band members but it does make a donation to community projects of around $1 million annually.

The Future: Finding a Place

Over the last 50 years or so, Aboriginal peoples have made great strides in finding their place in Canadian society. While that journey is far from over, the economic and political landscape of the North has shifted with Aboriginal peoples now controlling their own governments and operating businesses. A growing self-confidence has emerged as more and more Aboriginal men and woman enter the business and political world. As Calvin Helin (2006: 238), a controversial Aboriginal author and spokesperson, puts it: "What the [Aboriginal business] models examined tell us is that where the greatest successes have been achieved, strong, ethical leadership has been critical in gaining community consensus to move forward toward economic self-reliance." Breaking the chains of financial dependency for Aboriginal governments will not be easy but it is critical for their self-determination. Resource-sharing provides one key step in that direction.

Challenges remain. One is that northern Aboriginal communities—the products of relocation—still have virtually no economic raison d'être. In addition, the continuing high rates of natural increase place enormous pressure on these communities to provide adequate housing for their people. While the relocation strategy of the 1950s has had dreadful economic and social consequences, one serendipitous outcome was the emergence of regional consciousness. That sense of togetherness forged Nunavut and led to the inclusion of self-government within comprehensive land claim negotiations. Within their newly achieved self-governance structures, the opportunity for economic advances quickly became apparent to Aboriginal groups. Of course, the learning curve is steep, and much can depend on the luck of the geographical draw: Aboriginal groups in a large portion of the Provincial North, because of treaties that long predate the modern land claim settlements, have been disadvantaged; also, the resource wealth of the Canadian North is not equally distributed among regions or groups, nor is this wealth equally accessible. Bargaining with corporations over benefits in IBAs and making critical business decisions in the marketplace have not always favoured the Aboriginal side, but these skills can only improve with experience. In that context, the key question is: *Will the next 50 years see similar advances to the last 50 years?*

Challenge Questions

1. The Paix des Braves is presented as either a shining model for Aboriginal sharing in economic development (as argued by the Quebec government) or a Trojan Horse that will destroy Aboriginal culture. Which view do you support and why?
2. Is David Newhouse referring to the dynamic nature of culture when he states that the meaning of "Aboriginal" has changed over the years and now incorporates features of both the traditional and modern worlds? Do you believe this suggested shift in meaning has been inevitable? Do you believe it is the same throughout the country?
3. What organizational structure in the 1984 Inuvialuit Final Agreement protects their environment and its wildlife; and therefore ensures their traditional way of life.
4. Why are joint ventures so vital to Aboriginal companies such as Kitsaki?
5. If the federal strategy of relocation had included support for those who want to stay on the land (perhaps along the lines of Quebec's Income Security Program), would this have eased the cultural and economic stress at places like Davis Inlet? What is the flaw in the federal argument that such a program is too expensive (hint: how did Quebec justify its program for Cree hunters and trappers, which cost $20 million in 2009–10 and $23 million in 2012–13)?
6. Living in settlements has changed the diet of Aboriginal peoples living in northern settlements. What is the connection between their high rate of diabetes and store food?
7. Resource companies must negotiate impact and benefits agreements with Aboriginal groups on whose ancestral lands proposed projects would occur. If you were the Aboriginal negotiator, what "benefits" would you try to obtain from the company?
8. How does demography help to explain why the Inuit political jurisdiction of Nunavut enshrined a public government model rather than an ethnic one?
9. Aboriginal languages are disappearing. What are the cultural consequences for Aboriginal peoples and for Canada?
10. For Inuit communities, a shortage of public housing has led to overcrowding and homelessness. As the social planner for the Inuit, how do you explain why there is a shortage of housing when the public housing stock has increased over the years?

Notes

||||||||||||||||||||||||||||||||||||||||||||||||||||||||||||||||||||||||||||||||||||||||||||||||||||||||||||||||||||||||||||||||||||||||||||||

1. Bands who claimed that they did not receive their full allotment of land at the time of signing the treaty or that they have since lost land can file for a special land claim. After documenting its case, a band is eligible to make a specific land claim. Shortfalls in full allotment of land have occurred in three ways. (1) The original land grant was correct but government officials took some land from them without proper authorization; (2) lands that should have been granted never were granted; and (3) reserve populations have increased. Most claims are based on "lost" lands. In Saskatchewan, for example, almost a third of the land placed in reserve under treaty was surrendered to the government and then sold to settlers (Brizinski, 1993: 231). In many cases, since Crown land is either not available or not suitable, land must be purchased from private landowners. Given the price of agricultural and urban land, the cost of settling specific claims may run into billions of dollars. The Saskatchewan settlement reached $500 million divided among 27 First Nations.

2. Salisbury (1986: 76–84) describes changes in hunting among the Cree of northern Quebec. Under the subsidized Income Security Program for hunters and trappers, the increased availability of cash meant that they could make greater use of modern technology and transportation. Prior to 1975, hunters took little equipment into the bush and often travelled only short distances. By 1981, the Cree hunters could use chartered aircraft to fly their families into remote bush camps, which were as comfortable as village housing. As well, hunting has become more mechanized and less physically exhausting because these Cree hunters can afford to fly a snowmobile to their winter camps. The snowmobile is the workhorse of the hunter. It allows quick inspection of a trapline and the easy hauling of large game and firewood from the bush to camp. From 1974 to 1979, Salisbury reports that there was a 20 per cent increase in the overall amount of game killed by these Cree hunters and their families (whose number increased by more than 50 per cent over the same time period).

References and Selected Reading

||||||||||||||||||||||||||||||||||||||||||||||||||||||||||||||||||||||||||||||||||||||||||||||||||||||||||||||||||||||||||||||||||||||||||||||

Aatami, Pita. 2010. *From Igloos to the Internet: Inuit in the 21st Century*. Occasional Paper Series on Canadian Studies, Canadian Studies Center, Henry M. Jackson School of International Studies, University of Washington, 10 Feb. Seattle: University of Washington Press.

Aboriginal Affairs and Northern Development Canada (AANDC). 2010. *Resolving Aboriginal Claims—A Practical Guide to Canadian Experiences*. 15 Sept. http://www.aadnc-aandc.gc.ca/eng/1100100001417 4/1100100014179.

———. 2014. *Comprehensive Claims*. 29 Aug. http://www.aadnc-aandc.gc.ca/eng/1100100030577/ 1100100030578.

Aporta, Claudio. 2010. *"Inuit Sea Ice Use and Occupancy Project (ISIUOP)."* International Polar Year Project. gcrc.carleton.ca/isiuop.

Assembly of First Nations. 2011. *Pursuing First Nation Self-Determination: Realizing Our Rights and Responsibilities*. June. www.afn.ca/uploads/files/aga/pursuing_self-determination_aga_2011_ eng%5B1%5D.pdf.

Atlas of Canada. 2006. "Discover Canada through National Maps and Facts: Nunavut." atlas.nrcan.gc.ca/ site/english/maps/reference/provinceterritories/nunavut/referencemap_image_view.

Banta, Russell. 2005. *Review of First Nation Resource Sharing*. Discussion paper prepared for the Assembly of First Nations. 2 June. http://64.26.129.156/cmslib/general/ResourceRevenueSharing-Eng.pdf.

Bartlett, Richard. 1991. *Resource Development and Aboriginal Land Rights*. Calgary: University of Calgary, Canadian Institute of Resources Law.

Beaton, Brian. 2007. "Moose Cree First Nation Signs Impact Benefit Agreement for Developing Diamond Mine." *Knet Media*. media.knet.ca/node/2996.

Bell, Jim. 2007. "QIA Wants Language Laws Dumped, Re-written." *Nunatsiaq News*, 29 June. www. nunatsiaq.com/archives/2007/706/70629/news/nunavut/70629_253.html.

Berger, T.R. 2006. *Nunavut Land Claims Agreement Implementation Contract Negotiations for the Second Planning Period 2003–2013*. Conciliator's Final Report: "The Nunavut Project." Iqaluit: Government of Nunavut.

Bone, Robert M. 1988. "Cultural Persistence and Country Food: The Case of the Norman Wells Project." *Western Canadian Anthropologist* 5: 61–79.

————. 1992. *The Geography of the Canadian North: Issues and Challenges*. Toronto: Oxford University Press.

————. 2001. *Nunavik Political Model: Is It an Appropriate Model for Other Provinces?—The Case of Northern Saskatchewan*. Report prepared for the Department of Indian and Northern Affairs. Saskatoon: Signe Research Associates.

————. 2002. "Colonialism to Post-Colonialism in Canada's Western Interior: The Case of the Lac La Ronge Indian Band." *Historical Geography* 30: 59–73.

———— and Milford B. Green. 1984. "The Northern Aboriginal Labor Force: A Disadvantaged Work Force." *Operational Geographer* 3: 12–14.

Brizinski, Peggy. 1993. *Knots in a String*. Saskatoon: University of Saskatchewan Extension Press.

Cardinal, Harold. 1969. *The Unjust Society*. Edmonton: Hurtig.

CBC News. 2006. "Researcher [Jack Hicks] Studies Inuit Suicide Rates." 13 Oct. www.cbc.ca/news/canada/ north/story/2006/10/13/suicide-inuit.html.

————. 2010. "Natuashish Booze Ban Cancelled: New Chief." 8 Mar. www.cbc.ca/news/canada/ newfoundland-labrador/story/2010/08/24/natuashish-court-challenge-innu-824.html?ref=rss.

————. 2011a. "Quebec Inuit Vote against Self-Government Plan." 29 Apr. www.cbc.ca/news/can-ada/north/story/2011/04/29/nunavik-government-referendum.html.

————. 2011b. 'Labrador Innu Vote on Contentious Land Claim Deal." 30 June. www.cbc.ca/news/ canada/newfoundland-labrador/story/2011/06/30/nl-innu-labrador-claims-churchill-vote-630.html.

————. 2011c. "Premier Celebrates Innu Backing of Hydro Deal." 1 July. www.cbc.ca/news/canada/ newfoundland-labrador/story/2011/07/01/nl-dunderdale-rxn-hydro-deal.html.

————. 2011d. "Labrador Caribou Hunt Start Delayed." 3 Aug. www.cbc.ca/news/canada/newfound-land-labrador/story/2011/08/03/nl-caribou-herd-declining-802.html.

————. 2014. "Suicide Numbers in Nunavut in 2013 a Record High." 10 Jan. http://www.cbc.ca/news/ canada/north/suicide-numbers-in-nunavut-in-2013-a-record-high-1.2491117.

Chance, Norman. n.d. "Alaska Natives and the Land Claims Settlement Act of 1971." *Arctic Circle*. http:// arcticcircle.uconn.edu/SEEJ/Landclaims/ancsa1.html.

Ciaccia, John. 1975. *The James Bay and Northern Quebec Agreement (JBNQA)*. http://www.gcc.ca/pdf/ LEG000000006.pdf.

Coates, Ken S. 2015. *Sharing the Wealth: How Resource Revenue Agreements Can Honour Treaties, Improve Communities, and Facilitate Canadian Development*. Macdonald-Laurier Institute. http://www. macdonaldlaurier.ca/files/pdf/MLIresourcerevenuesharingweb.pdf.

———— and Brian Lee Crowley. 2013. *New Beginnings: How Canada's Natural Resource Wealth Could Re-shape Relations with Aboriginal Peoples*. Macdonald-Laurier Institute, May. http://www. macdonaldlaurier.ca/files/pdf/2013.01.05-MLI-New_Beginnings_Coates_vWEB.pdf.

———— and W.R. Morrison. 2010. *Treaty Research Report—Treaty No. 11 (1921)*. 15 Oct. https://www. aadnc-aandc.gc.ca/eng/1100100028912/1100100028914#chp4.

Collings, Peter, George Wenzel, and Richard G. Condon. 1998. "Modern Food Sharing Networks and Community Integration in the Central Canadian Arctic." *Arctic* 51, 4: 301–14.

Condon, Richard G., Peter Collings, and George Wenzel. 1995. "The Best Part of Life: Subsistence Hunting, Ethnicity, and Economic Adaptation among Young Adult Inuit Males." *Arctic* 48, 1: 31–46.

Cree Hunters and Trappers Income Security Board. 2013. *Rapport Annuel 2012.2013*. http://www.osrcpc. ca/images/osrcpc/rapportannuel/2012-2013.pdf.

————. 2014. "The Program." http://www.chtisb.ca/program/.

Crowe, K.J. 1990. "Claims on the Land." *Arctic Circle* 1, 3: 1–20.

Dahl, Jens, Jack Hicks, and Peter Jull. 2000. *Inuit Regain Control of Their Lands and Their Lives*. Skive, Denmark: Centraltrykkeriet Skive A/S.

Duerden, Frank. 1990. "The Geographer and Land Claims: A Critical Appraisal." *Operational Geographer* 8, 2: 35–7.

————. 1996. "An Evaluation of the Effectiveness of First Nations Participation in the Development of Land-Use Plans in the Yukon." *Canadian Journal of Native Studies* 16: 105–24.

————. 2004. "Translating Climate Change Impacts at Ethnic Community Level." *Arctic* 57, 2: 204–13.

———— and C.P. Keller. 1992. "GIS and Land Selection for Native Claims." *Operational Geographer* 10, 4: 11–14.

———— and R.G. Kuhn. 1996. "Applications of GIS by Government and First Nations in the Canadian North." *Cartographica* 33, 2: 49–62.

———— and ————. 1998. "Scale, Content, and Application of Traditional Knowledge of the Canadian North." *Polar Record* 34, 188: 31–8.

Ellis, S.C. 2005. "Meaningful Consideration? A Review of Traditional Knowledge in Environmental Decision-making." *Arctic* 58, 1: 66–7.

Erasmus, George. 1977. "We, the Dene." In Mel Watkins, ed., *Dene Nation—The Colony Within*, 3–4. Toronto: University of Toronto Press.

Ford, J., and L. Berrang-Ford. 2009. "Food Insecurity in Igloolik, Nunavut: A Baseline Study." *Polar Record* 45, 234: 225–36

Fournier, Watson. 2005. "Presentation by Makivik Construction Division and Kativik Municipal Housing Bureau to Indian and Northern Affairs Canada." 23 Aug. Ottawa: Indian and Northern Affairs Canada.

Freeman, M.R., ed. 1976. *Inuit Land Use and Occupancy Project*, 3 vols. Ottawa: Indian and Northern Affairs.

———— and G.W. Wenzel. 2006. "The Nature and Significance of Polar Bear Conservation Hunting in the Canadian Arctic." *Arctic* 59, 1: 21–30.

Galbraith, Lindsay, Ben Bradshaw, and Murray B. Rutherford. 2007. "Towards a New Super-Regulatory Approach to Environmental Assessment in Northern Canada." *Impact Assessment and Project Appraisal* 25, 1: 27–41.

George, Jane. 2011. "Makivik Corp. Wants to Convene Special Meeting on Self-Government." *Nunatsiaq Online*, 21 June. http://www.nunatsiaqonline.ca/stories/article/210677_makivik_corp._wants_to_convene_a_special_meeting_on_self-government/.

————. 2002. "Nunavik Leaders Celebrate First Quebec Payout", *Nunatsiaq News*, 21 June. At: http://www.nunatsiaq.com/archives/nunavut020621/news/nunavik/20621_1.html

Harding, Katherine. 2007. "Visiting Premiers Will Learn of Challenges Facing Nunavut." *Globe and Mail*, 4 July. www.apathyisboring.com/en/the_facts/news/208.

Hasselback, Drew. 2002. "Natives Halt De Beers Diamond Project." *National Post*, 1 Aug., FP3.

Health Canada. 2005. *A Statistical Profile on the Health of First Nations in Canada*. Ottawa: Health Canada.

Heinrich, Jeff. 2007. "$14B Deal Gives Cree Promise of Nationhood." *National Post*, 17 July, A4.

Helin, Calvin. 2006. *Dances with Dependency: Indigenous Success through Self-Reliance*. Vancouver: Orca Spirit Publishing.

Hicks, Jack. 2007. "The Social Determinants of Elevated Rates of Suicide among Inuit Youth." *Indigenous Affairs* 4/07: 30–7. http://inuusiq.com/wp-content/uploads/2012/04/Hicks-J-2007b1.pdf.

———— and Graham White. 2000. "Nunavut: Inuit Self-Determination through a Land Claim and Public Government?" In Jens Dahl, Jack Hicks, and Peter Jull, eds, *Inuit Regain Control of Their Lands and Their Lives*, 30–117. Skive, Denmark: Centraltrykkeriet Skive A/S.

Hurley, Mary C. 2000. *Aboriginal Title: The Supreme Court of Canada Decision in Delgamuukw v. British Columbia*. Ottawa: Library of Parliament. http://www.parl.gc.ca/content/lop/researchpublications/bp459-e.htm.

Indian and Northern Affairs Canada (INAC). 2000. "Fact Sheet—The Nisga'a Treaty." Northern Affairs Program. www.ainc-inac.gc.ca/pr/info/nit_e.html.

————. 2006. "Gains Made by Inuit in Formal Education and School Attendance, 1981–2001." *Inuit Social Trends Series*, No. 3. Strategic Research and Analysis Directorate. www.ainc-inac.gc.ca/pr/ra/gmif/gmif1_e.html.

————. 2008. "Frequently Asked Questions about the Tlicho Agreement." 28 Oct. www.ainc-inac.gc.ca/ai/mr/nr/j-a2005/02586bk-eng.asp.

Irlbacher-Fox, Stephanie, and Stephen J. Mills. 2008. *Devolution and Resource Revenue Sharing in the Canadian North: Achieving Fairness across Generations*. http://www.alternativesnorth.ca/Portals/0/2010%2006%2019%20Devolution%20and%20Resource%20Revenue%20Sharing%20in%20the%20Canadian%20North.pdf.

Inuit Tapiriit Kanatami (ITK). 2004. "Backgrounder on Inuit and Housing: For Discussion at Housing Sectorial Meeting, November 24 and 25th in Ottawa." www.aboriginalroundtable.ca/sect/hsng/bckpr/ITK_BgPaper_e.pdf.
Inuvialuit Regional Corporation (IRC). 2006. Annual Report 2005. www.irc.inuvialuit.com/publications/pdf/IRC%20Annual%20Report%202005.pdf.
————. 2007. "2007 Distribution Payments." www.irc.inuvialuit.com/beneficiaries/2006payments.html.
————. 2010. Annual Report 2009. www.irc.inuvialuit.com/publications/pdf/2009%20IRC%20Annual%20Report%20-%20Combined.pdf.
————. 2012. Annual Report 2010. http://www.irc.inuvialuit.com/publications/pdf/2010%20IRC%20Annual%20Report.pdf.
————. 2014. 2013 Annual Report. http://www.irc.inuvialuit.com/publications/pdf/2013%20IRC%20Annual%20Report%20+%20Financial%20Statement.pdf.
Kennett, S.A. 1999. A Guide to Impact Benefits Agreements. Calgary: Canadian Institute of Resources Law, University of Calgary.
Kitsaki. 2006. "Northern Resource Trucking Limited Partnership." www.kitsaki.com/nrt.html.
————. 2010. Newsletter (Summer). www.kitsaki.com/Kitsaki%20Updates%20&%20Brochure/Kitsaki%20Update%20Summer%202010.pdf.
Klinkig, Eileen. 2005. Presentation by Makivik Construction Division and Kativik Municipal Housing Bureau to Indian and Northern Affairs Canada. 23 Aug. Ottawa: Indian and Northern Affairs Canada.
Krolewski, A. 1973. Northwest Territories Statistical Abstract 1973. Ottawa: Department of Indian Affairs and Northern Development.
Kulchyski, Peter. 1994. Unjust Relations: Aboriginal Rights in Canadian Courts. Toronto: Oxford University Press.
Laidler, G.J., J. Ford, W.A. Gough, T. Ikummaq, A. Gagnon, A. Kowal, K. Qrunnut, and C. Irngaut. 2009. "Travelling and Hunting in a Changing Arctic: Assessing Inuit Vulnerability to Sea Ice Change in Igloolik, Nunavut." Climatic Change 94, 3 and 4:363–97.
Laidler, Gita. 2010. Inuit Siku (Sea Ice) Atlas. app.fluidsurveys.com/surveys/gjlaidler/siku-atlas-consultation-1/?code=X45wR&l=en.
Lac La Ronge Indian Band. 2014. Consolidated Financial Statements, March 31, 2014. http://llrib.com/wp-content/uploads/2014/09/Lac-La-Ronge-FS-for-AANDC.pdf.
Laliberté, Arlene, and Michel Tousignant. 2009. "Alcohol and Other Contextual Factors of Suicide in Four Aboriginal Communities of Quebec, Canada." Journal of Crisis Intervention and Suicide Prevention 30, 4: 215–21.
Land Claims Agreements Coalition. 2014. Modern Treaties. http://www.landclaimscoalition.ca/modern-treaties/.
Légaré, André. 2000. "La Nunavut Tunngavik Inc.: Un examen de ses activités et de sa structure adminis-trative." Études/Inuit/Studies 24, 1: 97–124.
————. 2001. "The Spatial and Symbolic Construction of Nunavut: Towards the Emergence of a Regional Collective Identity." Études/Inuit/Studies 25, 1 and 2: 143–68.
————. 2008a. "Canada's Experiment with Aboriginal Self-Determination in Nunavut: From Vision to Illusion." International Journal on Minority and Group Rights 15: 335–67.
————. 2008b. "Inuit Identity and Regionalization in the Canadian Central and Eastern Arctic: A Survey of Writings about Nunavut." Polar Geography 31, 3 and 4: 99–118.
———— and Markku Suksi. 2008. "Rethinking the Forms of Autonomy at the Dawn of the 21st Century." International Journal on Minority and Group Rights 15: 143–55.
MacDonald, Michael. 2001. "Innu of Davis Inlet Prepare for Year of Big Changes." Canadian Press, 21 Dec. http://hrsbstaff.ednet.ns.ca/waymac/Sociology/A%20Term%201/5.%20Research%20Methods/Davis%20Inlet.htm.
McNeil, Kent. 1998. "Defining Aboriginal Title in the 90s: Has the Supreme Court Finally Got It Right?" The John P. Roberts Professor of Canadian Studies Twelfth Annual Roberts Lecture, York University, 25 Mar. http://robarts.info.yorku.ca/files/lectures-pdf/rl_mcneil.pdf.
McKay, Raymond, A. 2004. "Kitsaki Management Limited Partnership: An Aboriginal Economic Development Model." Journal of Aboriginal Economic Development 4, 1: 1–5.

Makivik Corporation. 2011a. "The Information Tour Preceding the Referendum on the Final Agreement Starts." 14 Feb. www.makivik.org/the-information-tour-preceding-the-referendum-on-the-final-agreement-starts/.

————. 2011b. "Housing Development." www.makivik.org/building-nunavik/housing-development/.

Newhouse, David. 2000. "From the Tribal to the Modern: The Development of Modern Aboriginal Societies." In Ron F. Laliberte, Priscilla Settee, James B. Waldram, Rob Innes, Brenda Macdougall, Lesley McBain, and F. Laurie Barron, eds, *Expressions in Canadian Native Studies*, 395–409. Saskatoon: University of Saskatchewan Extension Press.

————. 2001. "Modern Aboriginal Economies: Capitalism with a Red Face." *Journal of Aboriginal Economic Development* 1, 2: 55–61.

Noble, Bram F. 2010. *Introduction to Environmental Impact Assessment*, 2nd edn. Toronto: Oxford University Press.

Northern Transportation Company. 2014. "Changes at NorTerra but Business as Usual for NTCL." 3 Apr. http://www.ntcl.com/changes-norterra-business-usual-ntcl/.

Nunavik Regional Government. 2010. *Timeline with Target Dates of Negotiation Activity Leading to the Final Agreement (FA), 2010–11*. 2 Nov. www.nunavikgovernment.ca/en/documents/NRG_Detailed_Timeline_2010-11.pdf.

————. 2014. *Towards a Representative Public Service*. 31 Dec. http://www.gov.nu.ca/sites/default/files/files/Finance/IEPstats/TRPS_December_2013_English.pdf.

Nunavimmiut Kavamanga. 2007. Map of Nunavik. www.nunavikgovernment.ca/en/archives/photos/map_of_nunavik.html and nunavikgovernment.ca/en/photos/map_carte1.html.

Nunavut Bureau of Statistics. 2012. *Nunavut Community Population Counts*. 8 Feb. http://www.stats.gov.nu.ca/Publications/census/2011/StatsUpdate,%20Nunavut%20Community%20Population%20Counts_2011%20Census.pdf.

Nunavut, Department of Finance. 2015. *Towards a Representative Public Service: Statistics as of December 31st, 2014*. At: http://www.gov.nu.ca/sites/default/files/final_dec_2014_trps_-_english_0.pdf.

O'Reilly, Kevin, and Erin Eacott. 1998. "Aboriginal Peoples and Impact and Benefit Agreement: Summary of the Report of a National Workshop." Canadian Arctic Resources Committee. www.carc.org/pubs/v25no4/2.htm.

Prno, Jason, and Ben Bradshaw. 2007. "Assessing the Effectiveness of Impact and Benefit Agreements in the Canadian North." Presentation at the annual meeting of the Canadian Association of Geographers, 31 May. www.usask.ca/geography/cag2007/final_program.pdf.

Quebec. 2002. "Complementary Agreement to the James Bay and Northern Québec Agreement—The Quebec Government Allocates an Additional $4.4 Million to the Income Security Program for Cree Hunters." News release, 22 May. communiques.gouv.qc.ca/gouvqc/communiques/GPQE/Mai2002/23/c8749.html.

————. 2010. *Annual Report of the Cree Hunting and Trappers Income Security Board, 2009–2010*. Departement de l'Emploiet de la Solidarité, 17 Dec. www.osrcpc.ca/images/osrcpc/rapportannuel/2009-2010.pdf.

Rodon, Thierry. 2014. "From Nouveau-Québec to Nunavik and Eeyou Istchee: The Political Economy of Northern Quebec." *Northern Review* 38: 93–112.

Roslin, Alex. 2001. "Cree Deal a Model or Betrayal?" *National Post*, 10 Nov., FP7.

Rynard, Paul. 2001. "Ally or Colonizer? The Federal State, the Cree Nation and the James Bay Agreement." https://www.questia.com/library/journal/1P3-99725068/ally-or-colonizer-the-federal-state-the-cree-nation.

Salisbury, Richard F. 1986. *A Homeland for the Cree: Regional Development in James Bay, 1971–1981*. Montreal and Kingston: McGill-Queen's University Press.

Schellenberger, Stan, and John A. MacDougall. 1986. *The Fur Issue: Cultural Continuity, Economic Opportunity*. Report of the House of Commons Standing Committee on Aboriginal Affairs and Northern Development. Ottawa: Queen's Printer.

Schultz, Stephanie. 2001. "A Simple Smile, a Scream Inside." In Jack Caufield, Mark Victor Hansen, and Kimberly Kirberger, eds, *Chicken Soup for the Teenage Soul: Letters of Life, Love & Leaving*, 56–8. Deerfield Beach, Fla: Health Communications.

Simeone, Tonina. 2014. "Resource Revenue-Sharing Arrangements with Aboriginal People." Library of Parliament Research Publications, Hill Note Number 2014-10-E, 26 Feb. http://www.parl.gc.ca/Content/LOP/ResearchPublications/2014-10-e.htm.

Slocombe, D. Scott. 2000. "Resources, People and Places: Resource and Environmental Geography in Canada." *Canadian Geographer* 44, 1: 56–66.

Stackhouse, John. 2001. "Increasing Raglan's Inuit Workforce." *Globe and Mail*, 14 Dec. www.globeandmail.com/series/apartheid/stories/20011214-5.htm.

Statistics Canada. 1998. "1996 Census: Education, Mobility and Migration." *The Daily*, 14 Apr. www.statcan.ca/Daily/English/980414/d980414.htm.

————. 2003. "Overview: Canadians Better Educated Than Ever." *Education in Canada: Raising the Standard.* www12.statcan.ca/english/census01/Products/Analytic/companion/educ/canada.cfm.

————. 2010. "Canada's Population Estimates." *The Daily*, 22 Dec. www.statcan.gc.ca/daily-quotidien/101222/t101222a2-eng.htm.

————. 2012a. "Visual Census. 2011 Census." Ottawa, 24 Oct. http://www12.statcan.gc.ca/census-recensement/2011/dp-pd/vc-rv/index.cfm?Lang=ENG&TOPIC_ID=1&GEOCODE=61.

————. 2012b. *Focus on Geography Series, 2011 Census.* Statistics Canada Catalogue no. 98-310-XWE2011004. Ottawa. Analytical products, 2011 Census, 24 Oct.

————. 2013a. Iqaluit, CY, Nunavut (Code 6204003) (table). National Household Survey (NHS) Aboriginal Population Profile. 2011 National Household Survey. Statistics Canada Catalogue no. 99-011-X2011007. 13 Nov. http://www12.statcan.gc.ca/nhs-enm/2011/dp-pd/aprof/index.cfm?Lang=E.

————. 2013b. Nunavik, Inuit region, Quebec (Code 640002) (table). 2011 National Household Survey (NHS), Aboriginal Population Profile. Statistics Canada Catalogue no. 99-011-X2011007. Ottawa, 13 Nov. http://www12.statcan.gc.ca/nhs-enm/2011/dp-pd/aprof/index.cfm?Lang=E.

————. 2014a. "Aboriginal Identity (8), Age Groups (20), Area of Residence: Inuit Nunangat (7) and Sex (3) for the Population in Private Households of Canada, Provinces and Territories, 2011 National Household Survey." 2011 National Household Survey: Data Tables. 3 Apr. http://www12.statcan.gc.ca/nhs-enm/2011/dp-pd/dt-td/Rp-eng.cfm?TABID=1&LANG=E&APATH=3&DETAIL=0&DIM=0&FL=A&FREE=0&GC=0&GK=0&GRP=1&PID=105397&PRID=0&PTYPE=105277&S=0&SHOWALL=0&SUB=0&Temporal=2013&THEME=94&VID=0&VNAMEE=&VNAMEF=.

————. 2014b. *Focus on Geography Series, 2011 Census: Nunavut.* 17 May. http://www12.statcan.gc.ca/census-recensement/2011/as-sa/fogs-spg/Facts-pr-eng.cfm?Lang=eng&GC=62.

Swidrovich, Cheryl. 2001. "Stanley Mission: Becoming Anglican but Remaining Cree." *Native Studies Review* 14, 2: 71–108.

Tait, Heather. 2008. *Inuit Health and Social Conditions.* Statistics Canada, Catalogue no. 89-637-XWE2008001. www.statcan.gc.ca/bsolc/olc-cel/olc-cel?catno=89-637-XWE2008001&lang=eng.

Trovato, Frank, and Anatole Romaniuk, eds. 2014. *Aboriginal Populations: Social, Demographic, and Epidemiological Perspectives.* Edmonton: University of Alberta Press.

Tshaukuesh (Elizabeth Penashue). 2010. "Miam Euauiat Tshekuan (It's Like a Circle)." In Ursula A. Kelly and Elizabeth Yeoman, eds, *Despite This Loss: Essays on Culture, Memory, and Identity in Newfoundland and Labrador*, 246–53. St John's: ISER Books.

Usher, Peter J. 1976. "Inuit Land Use in the Western Canadian Arctic." In M.M.R. Freeman, ed., *Inuit Land Use and Occupancy Project*, vol. 1, 21–31; vol. 3, 1–20. Ottawa: Department of Indian Affairs and Northern Development.

————. 1982. "Unfinished Business on the Frontier." *Canadian Geographer* 26, 3: 187–90.

————. 2002. "Inuvialuit Use of the Beaufort Sea and Its Resources, 1960–2000." *Arctic* 55 (supplement 1): 18–28.

————. 2003. "Environment, Race, and Nation Reconsidered: Reflections on Aboriginal Land Claims in Canada." *Canadian Geographer* 47, 4: 365–82.

————, F.J. Tough, and R.M. Galois. 1992. "Reclaiming the Land: Aboriginal Title, Treaty Rights, and Land Claims in Canada." *Applied Geography* 12, 2: 109–32.

———— and George Wenzel. 1987. "Aboriginal Harvest Surveys and Statistics: A Critique of Their Construction and Use." *Arctic* 40, 2: 145–60.

————— and —————. 1989. "Socio-Economic Aspects of Harvesting." In Randy Ames, Don Axford, Peter Usher, Ed Weick, and George Wenzel, eds, *Keeping On the Land: A Study of the Feasibility of a Comprehensive Wildlife Harvest Support Program in the Northwest Territories*, ch. 1. Ottawa: Canadian Arctic Resources Committee.

Warick, Jason. 2007. "Suicide Epidemic: Northern Community of Sandy Bay Rocked by Five Deaths, 12 Attempts in Recent Months." *Saskatoon Star-Phoenix*, 22 Feb.

Wenzel, George W. 1986. "Canadian Inuit in a Mixed Economy: Thoughts on Seals, Snowmobiles, and Animal Rights." *Aboriginal Studies Review* 2, 1: 69–82.

—————. 1995. "*Ningiqtuq*: Resource Sharing and Generalized Reciprocity in Clyde River, Nunavut." *Arctic Anthropology* 24, 2: 56–81.

—————. 2006. "Cultures in Collision: Traditional Knowledge and Euro-Canadian Governance Processes in Northern Land-Claim Boards." *Arctic* 59, 4: 401–14.

White, Marianne. 2011. "Crees to Ink Regional Government Deal with Quebec." *National Post*, 26 May, A3.

Whitelaw, Graham, and Daniel McCarthy. 2010. "Learning from the Victor Diamond Mine Comprehensive Environmental Assessment, Ontario." In Bruce Mitchell, ed., *Resource and Environmental Management in Canada: Addressing Conflict and Uncertainty*, 473–4. Toronto: Oxford University Press.

Williamson, Robert. 1974. *Eskimo Underground: Socio-cultural Change in the Canadian Central Arctic*. Uppsala: Institutionen for Allman.

- 8 -

Geopolitics, Climate Warming, and the Arctic Ocean

Setting the Stage

The Arctic Ocean has always held great symbolic importance for Canada. The search for the Northwest Passage, so deeply etched into Canadian psyche, remains one feature of Canada's northern image. Now that the Arctic Ocean is no longer an impenetrable frozen body of water, its geopolitical value has increased significantly. With an ever-warming Arctic, the prospect of an ice-free Northwest Passage for as long as three months in the summer is no longer a pipe dream. In September of 2013, the bulk carrier *Nordic Orion* loaded 73,000 tons of coal at Vancouver and took the southern route of the Northwest Passage (with a Canadian Coast Guard icebreaker escort) instead of the longer Panama Canal route to reach its destination in Pori, Finland. Rather than an isolated event, this successful voyage appears to have been a forerunner of things to come: a year later, in September 2014, a reinforced Polar 4 class cargo ship, the *Nunavik*, without an icebreaker escort, carried 23,000 tons of nickel concentrate from Deception Bay in Nunavik along a northern route of the Passage and on to China (Fednav, 2014).

The changing Arctic Ocean poses a dilemma for Canada. Until now, Canada has kept its investment in Arctic transportation to a minimum because of the high cost, but with the growing presence of domestic and foreign ships in Canada's Arctic waters, how will Ottawa manage the traffic? Properly managing an increase in Arctic Ocean transport can be a sign of Canada exerting its sovereignty over what could become internationally contested waters, but Canada to date does not have a strong marine presence in the Arctic and is thus ill-equipped to police such traffic. In fact, Canada lags far behind the other circumpolar nations. In 1985, Prime Minister Brian Mulroney recognized the need for a stronger Arctic presence. His government focused on a Polar 8 class icebreaker, but that project was cancelled in 1990 due to "budgetary difficulties." In 2008, the federal government revealed plans for a similar icebreaker plus patrol ships that would allow Canada to secure and manage its Arctic waters. Named for former Prime Minister John Diefenbaker, construction of this state-of-the-art icebreaker has yet to begin. The news is slightly better for the patrol vessels. Work on the first of six Arctic patrol ships, the HMCS *Harry DeWolf*, is scheduled to begin in September 2015 in Halifax.

Geopolitics has another dimension—the scramble for the last remaining unclaimed territory in the world, Arctic Ocean seabed. The **Arctic Five**—Canada, Denmark (Greenland), Norway, Russia, and the United States (Alaska)—have the inside track in this race. At a 29 March 2010 meeting, the Arctic Five countries confirmed their commitment to follow a co-operative approach by following the United Nations Convention on the Law of the Sea. Accordingly, each nation must base its claim on scientific evidence that its continental shelf is geologically linked to a portion of this seabed (Fisheries and Oceans Canada, 2010; see Figure 8.1).

A warmer Arctic has implications for Canada's Inuit. They are already facing the reality of an Arctic Ocean with less **pack ice**, if not entirely free of ice for part of the year. Their traditional hunting of seals and other mammals on land-fast ice is already compromised, and a shorter period of land-fast ice has implications for food security and hunter safety. But how do Inuit and other Arctic peoples play a role in this changing environment? The answers, such as they are, lie in territorial and federal governments

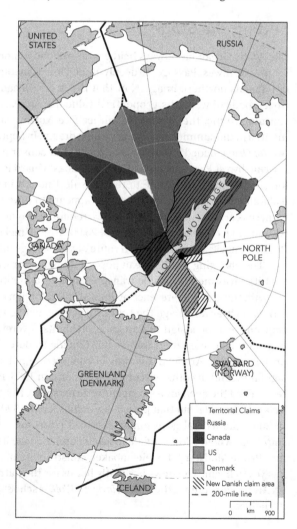

Figure 8.1 Disputed Waters and Seabed in the Arctic Ocean

Canada's claim to the seabed in the disputed area is likely to amount to around 40 per cent or 1.5 million km². Canada's submission is strongly dependent on the geomorphic connection of the Lomonosov Ridge to Canada's continental shelf (Fisheries and Oceans Canada, 2009). Denmark, in a surprise move, has claimed parts of the Arctic Ocean long associated with areas thought to belong to Canada and Russia. A diplomatic solution will likely play a role in the final outcome.

Source: Adapted from a map of "Maritime Jurisdiction and Boundaries in the Arctic Region" prepared by Durham University (2015).

as well as the Arctic Council, where Indigenous organizations, such as the **Inuit Circumpolar Council**, have permanent participation status.

Arctic Sovereignty

The question of sovereignty over the **Arctic Archipelago** was settled in 1880 when the British government transferred those islands to Canada,[1] but questions remain about Arctic waters and the seabed. Inuit use of ice to hunt remains a wild card in Canada's claim for sovereignty over the disputed waters of the Northwest Passage. In sum, except for the tiny Hans Island (see Figure 8.2) between Nunavut and Greenland (see Gray, 1997: 69; Byers, 2009: 23), territorial matters have been resolved, leaving questions about the marine boundaries, the Northwest Passage, and the international Arctic seabed as unresolved questions.

The Northwest Passage[2] is a top concern. Canada considers the Northwest Passage internal Canadian waters while foreign countries, especially the United States, argue

Figure 8.2   Hans Island—Part of Canada or Greenland?

In 1973, Canada and Denmark formally set the boundary as the midpoint in the waters separating Nunavut and Greenland. However, during those negotiations, no agreement was reached on Hans Island, which straddles the marine boundary. Located in Kennedy Channel of Nares Strait at a latitude of nearly 81°N, this tiny island is claimed by Canada and Denmark. For more on the legal case for each country and possible outcomes, see Byers (2009: ch. 2).

Source: Gary Clement

that the Northwest Passage lies in international waters. These issues were dormant until open water became more of a regular feature of the Northwest Passage.

Besides the Northwest Passage, two boundaries remain contested, one by the United States and the other by Denmark. Negotiations are taking place to resolve these two outstanding maritime **boundary** issues and Canada's claim for part of the seabed of the Arctic Ocean now lying in international waters is underway. The areas of contention are:

- With Denmark: a 200 km² section of the Lincoln Sea located north of Ellesmere Island and Greenland at latitude 84°N.
- With the United States: the position of the maritime boundary in the Beaufort Sea. Here the contested space is much larger, at 21,436 km².
- In 2013, Canada submitted its claim to part of these waters under the United Nations Convention on the Law of the Sea (UNCLOS). In 2014, Canada indicated that it wishes to make an additional claim to the area around the North Pole. At issue is the Lomonosov Ridge, which runs from Ellesmere Island and across the North Pole.

The Arctic Ocean and Its Changing Ice Cover

The Arctic Ocean is by far the smallest of the world's five oceans, with an area of just over 10 million km².[3] It is distinguished by several unique features, including the Arctic environment, an ice cap, and an encirclement by the land masses of North America, Eurasia, and Greenland. Within the Arctic Ocean, the principal seas are the Barents Sea, Beaufort Sea, Chukchi Sea, East Siberian Sea, Greenland Sea, Kara Sea, and the Laptev Sea, plus the many straits and channels separating the islands forming the Canadian Arctic Archipelago, including the Northwest Passage.

The change in ice conditions has been remarkably swift. In the late twentieth century, the Nanisivik and Polaris lead/zinc mines, the former on the northern tip of Baffin Island and the latter on Little Cornwallis Island near Resolute, were truly extreme outposts of the world economy, surrounded by thick ice for all but a short span during which the year's production was exported by the ice-strengthened MV *Arctic* to European markets. Fast forward to the twenty-first century and the Mary River iron mine on the northern tip of Baffin Island, where the company hopes to ship iron ore up to 10 months per year (*Nunatsiaq Online*, 2014).

The Arctic ice cap extends over most of the Arctic Ocean. It consists of both pack ice and **new ice**. Its geographic extent varies from winter to summer. In the winter, the Arctic ice reaches an area of nearly 15 million km²—in other words, it covers practically the entire ocean, including Hudson Bay; in late summer it is reduced to less than 5 million km². Formed from the freezing of sea water into pack ice and subjected to the centrifugal forces associated with drifting around the Arctic Ocean, its topography is formed by a combination of broken and refrozen ice creating a rough and irregular surface and large, flat, and relatively undisturbed ice sheets, which have served as ice islands for scientific research on the Arctic Ocean (Vignette 8.1). The major topographic features on the surface of the ice cap are pressure ridges caused by the collision of huge sheets of ice. While pack ice reaches three or more metres in thickness,

---

Vignette 8.1 The Nature of Pack Ice

Pack ice is the basic building material of the Arctic ice cap. Pack ice is formed in the following manner. As a crust of sea ice is frozen on the surface of the Arctic Ocean, this sea ice attaches itself to the existing pack ice, and this newly formed pack ice adds to the mass of the Arctic ice cap. Without a summer melt, pack ice gains in hardness and thickness, thus providing a challenge to icebreakers. Pack ice, in the form of small and large ice sheets, circulates around the Arctic Ocean.

---

pressure ridges reach heights two to four metres above the pack ice. In the summer, the Arctic ice cap is reduced in size and lies above 75°N; in the winter, the ice cap reaches its maximum extent with seasonal or young ice that extends from land to the permanent **multi-year ice**. Seasonal ice, with an average thickness of one metre, is much thinner than multi-year ice.

The Arctic ice cap, driven by polar currents and winds, slowly moves around the Arctic Ocean in a clockwise direction known as the Beaufort Gyre, while the **Transpolar Drift** carries ice from the Russian Arctic across the North Pole into the Greenland Current. Given these dynamic motions, the Arctic ice cap is not one solid sheet of ice but consists of huge sheets of pack ice separated by narrow patches of open water. The slow drifting of the ice cap causes it to bend, crack, and split. The cracking of the pack ice causes **leads** or open water (polynyas). In fact, as much as 10 per cent of the ice-covered Arctic Ocean consists of leads and open water. In the twenty-first century, the Arctic ice cap began to retreat each summer, resulting in the Arctic Ocean having record amounts of open water. Ice thickness varies through the year due to changes in wind and temperature.

The central question—still unresolved—is whether the Arctic Ocean—like Hudson Bay—will become ice-free each summer? One indicator is the decrease in young ice (less than one year old) from 60 per cent of the ice cover in 1983 to 40 per cent in 2014 (National Snow and Ice Data Center, 2014). Another indicator is the geographic extent of the ice cover. In 1978, the spatial extent of ice cover in September was 8.5 million km², but in 2014 it was 6.2 million km². In September 2007, the ice cap was reduced to its smallest size ever recorded—4.2 million km². This diminished geographic size translates into more open water in late summer.

Looking at the ice retreat from a longer perspective, leading scientists in the field of global warming have predicted that the Arctic Ocean will be virtually ice-free by 2050 (Stocker and Plattner, 2009: 4). Professor David Barber, the Canada Research Chair in Arctic System Science at the University of Manitoba, expressed an even more radical view, calling for an ice-free Arctic Ocean sometime between 2013 and 2030 (Campbell, 2010). For such predictions, the dynamics of annual ice loss are complicated, but the general thesis is that with more open water in the summer a greater amount of solar energy is absorbed by the Arctic waters, thus accelerating the rate of melting of the ice cap (Howell, Duguay, and Markus, 2009).

Other evidence supports this hypothesis. Glaciers, for instance, provide a valuable indicator of long-term temperature change. Thus, the retreat of glaciers in the northern hemisphere that began in earnest in the late twentieth century, followed by the shrinking of the Arctic ice cap, provides empirical evidence supporting the hypothesis of climate warming. In the case of the ice cap, the albedo effect is also at play (more open water equals more heat absorption). As discussed in Chapter 2, albedo refers to the amount of sunlight reflected by an object. As the Arctic ice cap melts, two events take place. First, the amount of sunlight reflected is reduced, so more warming of the ice cap surface takes place. Second, with more "dark" water exposed, the Arctic Ocean absorbs more solar energy, thereby warming its waters, which in turn increases ice melt.

With glaciers and ice caps melting, what will be the effect on ocean levels? Since the 1990s, sea levels have risen by two millimetres per year (National Ocean Service, 2014). While this slow but relentless annual increase seems destined to continue, the media often refer to the threat of massive flooding resulting from the melting of ice around the world. This threat could become a reality far in the future if both the Greenland and Antarctic ice sheets melted, thus causing sea levels to rise by more than 70 metres (National Snow and Ice Data Center, 2007). However, the disappearance of the Arctic ice cap alone would add little additional water to the oceans of the world because it is only about three metres thick on average and is already displacing ocean water. The loss of the ice cap alone, then, would only result in a negligible increase in the global sea water level, affecting only low-lying islands and coastlines where elevations are near sea level. The recent increases in sea level are explained in part by the melting of glaciers, with the rest a result of ocean thermal expansion (National Snow and Ice Data Center, 2007).

In spite of the solid science and evidence of global warming, a note of caution about predictions is appropriate. First, the complex interaction of the various natural forces that affect the Arctic ice cap, the Arctic Ocean, and its climate are not fully understood and therefore the rate of warming over the past 20 years could slow or even decrease. Second, an unexpected natural event could interrupt this recent warming trend, such as a modern version of the Mount Tambora volcanic eruption of 1815, which chilled the northern hemisphere for several years.[4] Nonetheless, the scientific evidence strongly suggests that global warming will continue and that the Arctic ice cap will disappear sometime in the twenty-first century.

Sovereignty in the Twenty-First Century

In the 1958 federal election, John Diefenbaker's "Northern Vision" captured the attention of the Canadian public and rallied voters to support his Progressive Conservative Party. But Arctic sovereignty goes beyond a vision that espouses economic development under programs like Diefenbaker's "Roads to Resources." Federal governments over the past 30 years have waffled on spending serious dollars in the North. Any serious threat to Canada's Arctic sovereignty, it seemed, soon disappeared from public sight—only to reappear years later in another form. As Coates, Lackenbauer, Morrison, and Poelzer (2009: 1) put it:

> Arctic sovereignty seems to be the zombie—the dead issue that refuses to stay dead—of Canadian public affairs. You think it's settled, killed and buried, and then every decade or so it rises from the grave and totters into view again.

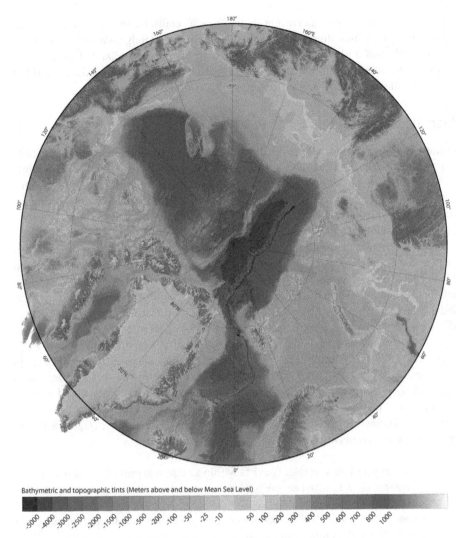

Bathymetric and topographic tints (Meters above and below Mean Sea Level)

-5000 -4000 -3000 -2500 -2000 -1500 -1000 -500 -200 -100 -50 -25 -10 50 100 200 300 400 500 600 700 800 1000

Figure 8.3 Bathymetric Map of the Arctic Ocean

Source: Jakobsson et al (2012).

Today, climate warming has changed the game, with the prospect of ocean shipping passing through the Northwest Passage and Canada's partial submission in 2013 for a piece of the unclaimed seabed of the Arctic Ocean. An understanding of the various claims for this last piece of world territory begins with the bathymetric (topographic) map of the floor of the Arctic Ocean. Its main features consist of three main landforms: shallow continental shelves, submarine mountains, and deep basins. While the waters of the Arctic Archipelago feature extensive continental shelves, a submarine mountain range extends some 3,000 metres above the ocean floor and stretches from the northern tip of Ellesmere Island and Greenland to the North Pole and then to the New Siberian Islands, a distance of 1,770 km. Known as the Lomonosov Ridge, this topographic feature plays a central role in Canada's claim for the seabed around the North Pole. Russia has already argued that the Lomonosov Ridge is an extension of its continental shelf and Denmark joined Russia in making a similar but overlapping claim. While late in the game, Canada began gathering oceanographic data in 2014 to make a similar argument for seabed around the North Pole.

Use It or Lose It

Ottawa is on the hot seat to provide both a vision and funding for its Arctic. In 2007, Prime Minister Harper declared that "Canada has a choice when it comes to defending our sovereignty over the Arctic. We either use it or lose it. And make no mistake, this Government intends to use it" (*Victoria Times Colonist*, 2007). One of Canada's leading Arctic experts, Michael Byers, puts it this way in his definitive book, *Who Owns the Arctic?* (2009: 6): "Sovereignty, like property, can usefully be thought of as a bundle of rights." Unfortunately, each right calls for substantial federal investments. With the exception of the Diefenbaker government of the 1950s, federal governments, both Conservative and Liberal, have made huge commitments but have failed to provide the funding because of conflicting demands from other regions of the country and/or the need to balance the federal budget. The Arctic is a costly region, and the Harper government faces a dilemma—pay the Arctic piper or, like so many previous governments, duck and run.

Maritime sovereignty, unlike territorial sovereignty, often does not contain a full bundle of rights. For example, under the UNCLOS regime, full sovereignty only extends to 12 nautical miles (22 km) from shore. Described as territorial seas, coastal states have full sovereign rights over these waters, including airspace, seabed, and subsoil. Beyond the 12 nautical miles, the bundle of rights diminishes. The diagram below indicates the complexity of determining the various maritime boundaries (Figure 8.4). The most important points are:

- *Territorial seas.* Full sovereignty exists within 12 nautical miles from shore.
- *Continental shelf.* Limited sovereignty occurs between 12 nautical miles and 200 nautical miles from shore.
- *Extended continental shelf.* Seabed sovereignty goes beyond 200 nautical miles to a limit of 350 nautical miles from shore.
- *High seas.* The so-called "high seas" lie beyond national sovereignty and are considered international waters under the control of the **International Seabed Authority**.
- *The area.* Seabed sovereignty belongs to the international community under the control of the International Seabed Authority.

Ottawa is confronted with a new Arctic reality. On 10 August 2010, the federal government announced its policy in a *Statement on Canada's Arctic Foreign Policy*. Canada's policy is based on three points:

1. resolve boundary issues;
2. secure international recognition for the full extent of our extended continental shelf;
3. address Arctic governance and related emerging issues, such as public safety.

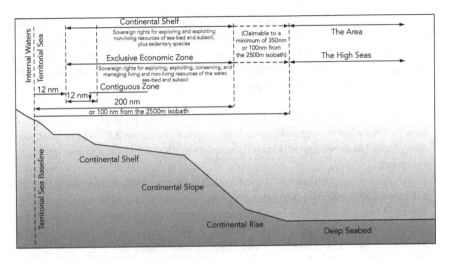

Figure 8.4 **Zones of Maritime Sovereignty**

Note: "The area" is the sea floor beneath the "high seas." nm = nautical miles

Source: Fisheries and Oceans Canada (2010). This is copy of the version available at http://www.dfo-mpo.gc.ca/oceans/
canadasoceans-oceansducanada/marinezones-zonesmarines-eng.htm. Reproduced with permission. This does not constitute an
endorsement by Fisheries and Oceans Canada of this product.

Such statements have a strong appeal to the Canadian public, but when it comes
to investing large sums outside of the Canadian voting ecumene, governments are
always cautious. While the government made a strong commitment in 2007 to beef up
its Arctic presence, the cost is high and construction of new ships readily capable of
plying Arctic waters remains in the future.

Vignette 8.2 Evolving Marine Sovereignty

With water covering about two-thirds of the Earth's surface, a basic question con-
fronts the nations of the world: How far does the sovereignty of coastal nations
extend into the ocean? While the answer has evolved over time, especially since
1945, the basic trend has seen coastal states gain more control over these waters.
The importance for Canada, with one of the longest coastlines in the world, is critical.
Canada's Arctic Archipelago and its Arctic continental shelf are playing a key role in
determining the geographic range of its ownership of the waters of Arctic Ocean
and its seabed. Based on the 1982 **United Nations Convention on the Law of the
Sea (UNCLOS)**, which describes the nature of sovereignty and the associated maritime boundaries that coastal states can set, Canada passed its Oceans Act (1991),
which delineates its maritime zones (Department of Justice, 2010). Equally important, Article 234 of UNCLOS—the **Arctic Clause**—allows Canada to enforce its pollution regulations in its **exclusive economic zone**. Another regulation that asserts
Canada's sovereignty over its waters is **NORDREG**, which requires that domestic and
foreign vessels formally register before entering Canadian ice-covered areas.
Unfortunately, Canada lacks the Arctic naval power and Arctic marine ports to
enforce such international regulations.

Given the cost and time horizon, how has Canada matched its stated goals with concrete action? The answer is "very slowly," because the federal government curtailed its Arctic expenditures, estimated at $20 billion, in order to reach a surplus in time for the 2015 election. As a result, its 2007 promises for Arctic projects have fallen off the rails. Here are the promises and, to date, the realities:

- To build three huge, armed icebreakers. This was later reduced to one, the CCGS *John G. Diefenbaker*, originally scheduled for completion in 2017 at the Vancouver Shipyards. The most recently stated completion date is 2021, but construction has yet to begin.
- To create a deepwater port at Nanisivik. Construction of Nanisivik Naval Port was to begin in 2011, but the physical site was not suitable[5] and the Defence Department replaced these plans with a much more modest refuelling station. In 2014, officials claimed the facility would be completed by 2017; in 2015, the project manager asserted it would be operational in 2018 (CBC News, 2015).
- To establish a small military training centre at Resolute Bay. This was later downgraded to sharing space with the Polar Continental Shelf Project.
- To modernize and expand the Canadian Rangers. The Rangers were scheduled to receive new rifles in 2015.
- To establish a High Arctic Research Centre at Cambridge Bay (Figure 8.5). Construction started in 2014.

In addition, Ottawa is heavily committed to Radarsat satellite surveillance of its Arctic waters (Vignette 8.3).

Drawing Lines in Water

As noted earlier, Canada has several outstanding international jurisdictional issues related to Arctic waters and the need to draw or redraw lines in the water—no easy task!—to indicate which nations own what. The principal issues here involve the Beaufort Sea, the Northwest Passage, and the Arctic seabed (see Figure 8.1).

The Beaufort Sea Dispute

The Beaufort Sea dispute between Canada and the United States revolves around a relatively small area of 21,436 km². Attempts at resolving the Beaufort Sea boundary are underway, but success may be elusive. At stake within this contested triangular-shaped area, besides management of the water and fish stocks, is ownership of potentially vast seabed petroleum deposits. The disputed area is tangled in history, with Canada arguing that the 1815 treaty between Great Britain and Russia set the boundary along the 141st meridian west not only for the **border** between Alaska and Yukon but also for the maritime boundary in the Beaufort Sea. The US, however, argues for a line equidistant from the coast of each country that would shift the boundary east of the 141st longitude and thus provide the United States with sovereignty rights over the contested area of the Beaufort Sea (Figure 8.6).

Figure 8.5 Proposed Canadian High Arctic Research Centre at Cambridge Bay, Nunavut

The location of the Canadian High Arctic Research Centre (lower right) is superimposed on this aerial view of Cambridge Bay. This world-class scientific facility is expected to open in the summer of 2017. It will contain space for research, technology development, and traditional knowledge as well as facilities for teaching, training, and community affairs. Its role in the Central Arctic will be to conduct environmental research and to protect the Arctic environment. In these ways, the centre will contribute to the regional importance of Cambridge Bay and employment opportunities for its residents.

Source: Canada (2014).

Vignette 8.3 Space Technology, Sovereignty, and Hard Political Decisions

One of the most controversial and expensive investments in Arctic sovereignty involves Radarsat satellite surveillance. This space technology provides very detailed daily surveillance of Canada's Arctic waters with the ability to monitor not just shipping but the state of the ice pack and other environmental variables. In 2008, the Canadian government was confronted with a dilemma: to approve or reject the sale of MacDonald Dettweiler Associates (MDA) to the US firm of Alliant Techsystems. Ottawa rejected the sale on the ground of national security, which translated into Arctic sovereignty. The value of this satellite surveillance cannot be underestimated (Vachon, 2010). To maximize the data flow from the three orbiting Radarsat satellites, Canada's first Arctic Satellite Station Facility (ISSF) was opened at Inuvik in 2010. While Inuvik benefits from economic spinoffs from the operation of the satellite station, by far the bulk of the economic benefits will flow to southern Canada, especially the Vancouver area where the MDA plant is located. Southern taxpayers should recognize that large public investments in the Arctic benefit industries in southern Canada. The explanation for the spatial pattern of economic benefits associated with northern projects lies in the concept of economic leakage.

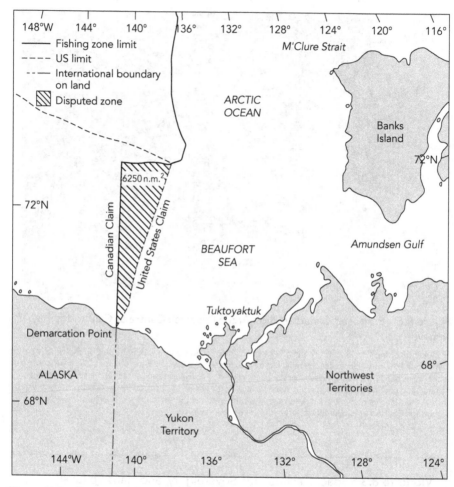

Figure 8.6 Contested Area of the Beaufort Sea

Source: Gray (1997). IBRU, Durham University, UK.

Canada, as with its other boundary disputes, is seeking a negotiated solution. One step in this direction has been the joint Canada/US seabed mapping program for the Beaufort Sea in the contested area. This political decision for a co-operative approach to hydrographic investigation may lead to a solution. The joint seabed mapping program involves the Canadian Coast Guard icebreaker *Louis S. St-Laurent* and the US Coast Guard cutter *Healy* working together. This effort to map the seabed has three objectives:

- to help resolve the maritime boundary between the two countries;
- to identify potential petroleum and other seabed resources;
- to assist the two countries in preparing their respective claims for a section of the international seabed of the Arctic Ocean.

The Northwest Passage: Internal Waters or International Strait?

For years, Canada's claim to the Northwest Passage as internal waters was solidified by the massive ice pack that occupied much of this waterway. With the Northwest Passage virtually impassable for commercial vessels, ice was Canada's friend. Added to the "ice as a barrier to commercial navigation" argument was the "ice as a platform for Inuit hunters to kill seals" argument. From both perspectives, ice reinforced Canada's position.

But with climate warming and a retreating ice pack, the Northwest Passage now is ice-free, albeit for a very short time in the summer. In 2009, Canada declared that ships could traverse the Northwest Passage without an escorting icebreaker. This decision opened the Northwest Passage to tourist cruisers and, in 2014, to the unescorted passage of the cargo ship *Nunavik*, carrying nickel concentrate to China from Deception Bay in Nunavik by way of a northern route through Viscount Melville Sound and Prince of Wales Strait.

Canada continues to argue that the waters between the islands of the Archipelago are internal seas. Countering that position, Washington and other foreign governments claim that the Northwest Passage is an international waterway because it connects two segments of the high seas—the Atlantic and Pacific oceans. Support for this position comes from UNCLOS: the right of passage exists "between one part of the high seas or an exclusive economic zone and another part of the high seas or an exclusive economic zone." While both perspectives accept that these are Canadian waters, the key difference is that if the Northwest Passage is designated as an international waterway, then foreign ships could pass freely. On the other hand, if the Northwest Passage was declared internal waters of Canada, then "foreign vessels must have Canada's permission [to pass through the Northwest Passage] and are subject to the full force of Canadian domestic law" (Byers, 2009: 43). In December 2010, Canada gained more credence for its "internal waters" position by announcing that Lancaster Sound will become a national marine conservation area and, in the future, may become a World Heritage Site. Lancaster Sound serves as the eastern entrance to the Northwest Passage and therefore is a critical piece of the ownership puzzle of the Northwest Passage. Along the southern route, Canada's claim to the waters of the Northwest Passage are strengthened by the declaration that the HMS *Erebus* and the yet-to-be-found HMS *Terror* from the ill-fated 1840s Franklin expedition are national historic sites.[6]

An ice-free Northwest Passage could alter world shipping patterns because it offers a much shorter sea route between Asia and Europe. In fact, an Arctic shipping route through the Northwest Passage would shave roughly 4,000 km off current maritime routes between Europe and Asia. As we saw in Chapter 3, this dream of a shortcut between Europe and Asia was stymied for centuries by the much colder climate of the Little Ice Age and an even more formidable and extensive Arctic ice cap.

The southern route through the Northwest Passage begins in Lancaster Sound but quickly veers south along the Peel Channel. The attraction of this route is the presence of open water and less chance of encountering hard, thick ice. This route is named after Roald Amundsen, the Norwegian polar explorer who, from 1903 to 1906, meandered his way across the waters separating the islands of Canada's Arctic Archipelago. His small but agile ship, the *Gjoa*, was able to follow leads in the ice until the ship became lodged in pack ice. After three summers, Amundsen became the first person to

Figure 8.7 The SS *Manhattan* Challenges the Northwest Passage

The reinforced Arctic class US supertanker SS *Manhattan* entered the Northwest Passage at Lancaster Sound in the summer of 1969. The question facing the oil producers at Prudhoe Bay, Alaska, was how to transport the oil to market. One possibility was to use the Northwest Passage. The *Manhattan* took the northern route but ran into impassable ice at M'Clure Strait. With the help of the Canadian icebreaker *John A. Macdonald*, the *Manhattan* freed itself and then changed course to the Prince of Wales Strait, Amundsen Gulf, and on to Prudhoe Bay.

Source: Dan Guravich/CORBIS.

complete the journey across the Northwest Passage. The *Gjoa* entered Lancaster Sound and then Barrow Strait. Facing a wall of ice, Amundsen veered his ship almost due south along Peel Sound to reach what was the settlement of Gjoa Haven on King William Island (also known as Uqsuqtuuq). At this site, Amundsen spent two years conducting research on the magnetic North Pole. In the summer of 1906, Amundsen sailed westward to Queen Maud Gulf and then to the Beaufort Sea and beyond.

The second crossing of the southern route of the Northwest Passage began in 1940, when the *St Roch*, under the command of Inspector Henry Larsen of the RCMP, set sail from Vancouver for Halifax across the Arctic Ocean. His orders were to assist the Canadian armed forces that were to occupy Greenland.[7] Taking two summers, the *St Roch* was the first ship to cross the Northwest Passage from west to east. Larsen followed Amundsen's route in reverse, with one exception. Instead of proceeding north along Peel Sound to Barrow Strait, Larsen took his small craft through the very narrow Bellot Strait that separates Southhampton Island from Boothia Peninsula and then sailed northward from Prince Regent Inlet to Lancaster Sound.

Over the last decade, a remarkable number of vessels of all shapes and sizes have crossed the Northwest Passage (Table 8.1). In sharp contrast, no ship crossed the passage in the nineteenth century and only Amundsen and Larsen were successful in the

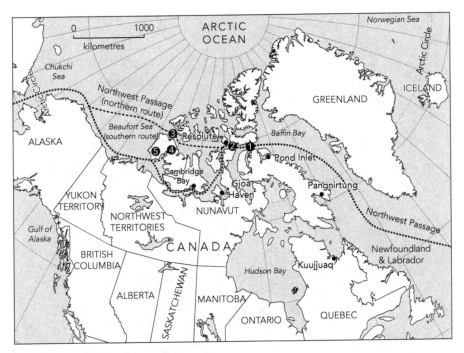

Figure 8.8   The Northwest Passage

The Northwest Passage is a rarely used sea route across the Arctic Ocean connecting Europe with the Pacific coast of North America and with Asia. In fact, there are two basic routes through the Arctic Archipelago. The northern route stretches from Lancaster Sound (1 on map) to M'Clure Strait (3) or Prince of Wales Strait (4), which separates Banks Island and Victoria Island, and west to the Beaufort Sea. The southern route veers south near Resolute after passing through Barrow Strait (2) and passes between Somerset Island (to the east) and Prince of Wales Island, which are north of the Boothia Peninsula. Until recently, ships found it impossible to navigate M'Clure Strait. This strait is named after Commander Robert M'Clure, whose ship, the HMS *Investigator*, approached M'Clure Strait from the west but was trapped in pack ice in the strait for three winters (1851–3) and finally was abandoned just off Banks Island (5). He and his crew were rescued by another expedition that had sailed to M'Clure Strait from Lancaster Sound. Both expeditions were part of the search for Sir John Franklin but only M'Clure and his crew were recognized as the first to cross the Northwest Passage. In 2010, Parks Canada discovered the HMS *Investigator* in shallow water near the northwest corner of Banks Island while Franklin's HMS *Erebus* was found in shallow water at the eastern end of Queen Maud Gulf, south of King William Island (where Gjoa Haven is located). Franklin and his crew followed the southern route; M'Clure attempted the northern route.

Source: Adapted from Dawson et al. (2008).

first half of the twentieth century. By the end of the first decade of the twenty-first century, more open water each summer allowed at least one crossing of the Northwest Passage each year. Other ships, including research vessels, have entered Arctic waters but have not travelled the entire length of the Northwest Passage.

As Arctic waters provide one of the world's last wilderness areas, cruise ships have found a niche market by scheduling Arctic voyages. Billed as "Arctic Adventures," these expensive voyages have grown in number. In 2009, Robert McCalla (2010b: Table 2) reported that 15 Arctic voyages by 10 cruise ships took place in Canada's Arctic Ocean with two ships crossing the entire southern Northwest Passage route. Since these waters are neither fully charted nor necessarily totally free of ice, navigating the Northwest

Figure 8.9 Approximate Location of Sir John Franklin's HMS *Erebus*

Parks Canada has kept the exact location of the wreck of HMS *Erebus* secret to prevent others from exploiting the site, though it has been reported the ship is near O'Reilly Island in the eastern waters of Queen
Maud Gulf not far from the Adelaide Peninsula. Sir John Franklin's two ships entered Lancaster Sound and
made a land camp at Beechey Island in 1845–6. Franklin and his crew are believed to have become trapped
in thick ice in Victoria Strait somewhere along the northwest coast of King William Island. The discovery of
the *Erebus* in 2014 confirmed the Inuit oral testimony, first recorded by John Rae in 1854, that a ship sank in
the area south of King William Island (Woodman, 1991). The captain of the HMS *Terror*, Francis Crozier, led
the desperate trek of men—hauling a boat loaded with provisions across the snow and ice—in their attempt
to reach the mouth of the Great Fish River. Based on a note left by Crozier in a cairn on King William Island
and on oral accounts from Inuit, the *Erebus* likely drifted south as a ghost ship along the western side of
King William Island for some distance until it jammed in the ice off the Adelaide Peninsula, was breached by
local Inuit scavenging for iron and wood, and, after the ice melt the following summer, sank near O'Reilly
Island, where it remained undisturbed in the frigid Arctic waters until its discovery in September 2014.

Source: Adapted from Royal Canadian Geographical Society, "The 2014 Victoria Strait Expedition: The Search for the Lost Ships
of the Franklin Expedition," at: http://www.canadiangeographic.ca/franklin-expedition/assets/map2/.

Table 8.1 Key Events: Crossing the Northwest Passage, 1969 to 2014

Date	Event
1969–70	The Exxon supertanker, the SS *Manhattan*, travels through the Northwest Passage twice to prove it is a viable commercial route for shipping oil from Prudhoe Bay, Alaska, to refineries along the US east coast. The hull of the supertanker is covered with a thick protective steel belt and its reinforced bow is designed to ride up on the pack ice, allowing the weight of the massive ship to break the ice apart. While Exxon had not asked for Ottawa's permission, Canada grants "unsolicited" permission and provides an icebreaker escort. While the two crossings (one from the east to Prudhoe Bay and the other from the west to the US east coast) are successful, damage to the *Manhattan* is considerable and plans are abandoned for using the Northwest Passage as a shipping route for Alaskan oil.
1984	The first cruise ship to navigate the Northwest Passage is the MS *Explorer*.
1985	The US Coast Guard icebreaker, *Polar Sea*, traverses the Northwest Passage without asking permission. Ottawa is deeply troubled and approaches Washington for clarification.
2000	The RCMP patrol vessel, *St Roch II*, escorted by the Canadian Coast Guard icebreaker, *Simon Fraser*, departs Vancouver and completes the voyage in one summer.
2006	Open ice becomes more extensive in the Northwest Passage. The cruise liner, MS *Bremen*, crosses the passage from east to west.
2009	Canada declares that cruise ships do not require an escort by a Canadian icebreaker. Nine small vessels and two cruise ships (MS *Bremen* and MS *Hanseatic*) make the crossing.
2010–14	Each year three to eight luxury cruise ships, as well as numerous small craft, travel the historic southern Northwest Passage route. More cruise ships commence their journey from Cambridge Bay, making that Inuit centre a summer tourist town.
2013	The coal-carrying *Nordic Orion* sails from Vancouver through the southern route of the Northwest Passage on its way to Finland, becoming the second cargo ship (after the *Manhattan*) to make this journey across the Arctic Ocean.
2014	The *Nunavik* becomes the first unescorted bulk carrier to traverse the Northwest Passage, carrying nickel concentrate from Deception Bay in Nunavik to China by way of the northern route and Prince of Wales Strait.

Passage is risky—as the captain of the cruise ship, *Clipper Adventurer*, discovered when his ship ran aground in 2010 on its way to Kugluktuk (formerly Coppermine), Nunavut. Fortunately, calm weather prevailed and the passengers were transferred to the Canadian icebreaker *Amundsen*, taken to Kugluktuk, and then flown to Edmonton and on to other destinations (Cohen, 2010). Given the short navigation season, a salvage operation began a few days later. The ship was refloated and towed to Cambridge Bay and then to Nuuk, Greenland, where it was repaired (Marinelog, 2010). This misadventure was a relatively minor one, but it underscores the perilous nature of such voyages as well as the need for Canada to expand its presence in the Arctic Ocean.

Canada's Arctic Seabed Submission

Canada and other circumpolar countries—the US, Denmark, Norway, and Russia—seek a share of the international part of the Arctic Ocean and, with the exception of the US,[8] these countries have ratified the United Nations Convention on the Law of the Sea (UNCLOS).[9] Each country's claim for a part of the Arctic seabed relies on the geology and geomorphology of the Arctic Ocean.

Under UNCLOS, the goal is to "delimit the area in which each state would have sovereign rights to seabed minerals" (Steinberg, Tasch, and Gerhardt, 2015: 12). Under the aegis of UNCLOS, the sharing of the international seabed, known as "the area," is heading for a peaceful and scientific-driven solution, though overlapping claims may add some spark to the negotiations. Each country has 10 years from the time that it signs the UNCLOS agreement to present its claim to a portion of the extended continental shelf (Figure 8.1). Canada ratified the Convention in 2003, and made its initial submission in 2013 with a proviso for an additional claim to the area near the North Pole at a later date (Foreign Affairs, Trade and Development Canada, 2013).

What can Canada expect to gain? Beyond Canada's 200 nautical mile boundaries, the areas of extended continental shelf shown in Figure 8.1 consist of approximately 1.5 million km^2, which is equivalent to the size of Quebec (Fisheries and Oceans Canada, 2010). In comparison to the possible additional territory to be gained from marine boundary disputes with the US and Denmark, this claim is far greater in geographic extent and has the potential to greatly add to Canada's mineral and petroleum reserves. Indeed, the motivating factor behind the race for the Arctic Ocean's seabed is its potential wealth of resources, especially petroleum. An estimate of oil and gas resources within Canada's 200 nautical miles is found in Table 8.2, but much more likely lies in the international seabed—hence the scramble to submit claims to the United Nations.

In a 2009 report, Fisheries and Oceans Canada described how the data for its submission to the UN was being gathered:

The Arctic is subdivided into two areas: the Eastern and Western Arctic. The Eastern Arctic required seismic surveys to show the natural prolongation of the Lomonosov and Alpha-Mendeleev Ridges, then bathymetric surveys to map the foot of slope and 2,500-metre depth contour. The Western Arctic required bathymetric surveys to map the foot of slope and seismic surveys to determine sediment thickness. There are few existing data sets in these Arctic areas relevant to the continental shelf submission so the Program must conduct all the work itself.

Conclusion

Geopolitics has again turned the world's attention to the Canadian Arctic. Without a doubt, the economic and social implications of a warmer (or less frozen) northern world are enormous—and to some degree frightening. Driven by climate warming, the gradual demise of the ice cap has wide-ranging implications for Canada, its Arctic sovereignty, and the peoples of the Arctic. Those implications will require a considerable adjustment to the realities of a more open Arctic Ocean with increased

Table 8.2 Oil and Gas Resources within Canada's Continental Shelf

OIL RESOURCES

Region	Discovered Resources		Undiscovered Resources		Ultimate Potential	
	10^6 m³	MMbbls	10^6 m³	MMbbls	10^6 m³	MMbbls
Northwest Territories and Arctic offshore	187.9	1,182.5	799.7	5,032.6	987.6	6,215.0
Nunavut and Arctic offshore	51.3	322.9	371.8	2,339.4	423.1	2,662.3
Arctic Offshore Yukon	62.5	393.8	412.7	2,596.8	475.2	2,990.6
Total	301.7	1,899.1	1,584.1	9,968.8	1,885.9	11,867.9

GAS RESOURCES

Region	Discovered Resources		Undiscovered Resources		Ultimate Potential	
	10^6 m³	trillion cubic feet	10^6 m³	trillion cubic feet	10^6 m³	trillion cubic feet
Northwest Territories and Arctic offshore	457.6	16.2	1,542.2	54.8	1,999.8	71.0
Nunavut and Arctic offshore	449.7	16.0	1,191.9	42.3	1,641.6	58.3
Arctic offshore Yukon	4.5	0.2	486.6	17.3	491.1	17.4
Total	911.8	32.7	3,220.7	114.3	4,132.6	146.7

Notes: MMbbls = million barrels (of oil).

The Arctic offshore includes marine areas offshore Yukon and the Northwest Territories in the Beaufort Sea, and offshore Nunavut in the High and Eastern Arctic. Resources within Yukon are not included.

Source: Courtesy of Aboriginal Affairs and Northern Development Canada

resource/tourist development, ocean shipping, and possibly even a larger Arctic population, along with the threat of pollution of the Arctic Ocean. Moving forward into this warmer northern world, Arctic sovereignty demands a more active Canadian Coast Guard and military presence; a more substantial marine infrastructure, including a deepwater port for Canadian naval vessels; and more Arctic ships to enforce Canada's laws and to respond to marine disasters.

With much greater access to world markets, resource and tourist development should flourish and, if crafted carefully through impact and benefits agreements, benefits will accrue locally. For example, besides the commercial traffic passing through the Northwest Passage, mining operations will greatly benefit from longer periods of ocean shipping. The change in ice conditions has been remarkably swift, and means that resource projects such as the Mary River iron mine may be able to ship their product for most of the summer, whereas the Nanisivik and Polaris lead/zinc mines were limited to a short window for shipping. The effect of global warming has extended to the tourist industry with luxury cruise ships now sailing in ice-free waters along the southern route of the Northwest Passage. The discovery of HMS Erebus along this southern route may well spark an influx of tourists to the area of Queen Maud Gulf and to Gjoa Haven.

With such developments close at hand, Canada is under pressure to protect its interest as custodian not only of the Northwest Passage, but also of its other Arctic waters and seabed. Environmental issues loom large in a more active shipping season. Fears of oil leaks and other forms of pollution are not unfounded—the issue of accidents and spills is not a question of "if" but of "when." The *Exxon Valdez* oil spill in Prince William Sound on the southern coast of Alaska in 1989 is a reminder of the danger of environmental disasters in the Far North: remote locations are difficult to access for quick and sufficient response; and northern ecosystems, because of the cold, cannot recover nearly as quickly as those in more temperate climates. Canadian resources must be put in place to manage such pollution and to charge the ship's captain and shipping company each time an incident occurs. That means having feet on the ground and men and women on the water in the Far North. Imagine an Arctic where armed Canadian Coast Guard vessels patrol these waters and, when necessary, force foreign ships to port to ensure that every ship entering the Northwest Passage has obtained a Canadian permit. The solution is close at hand. Canada's submission to UNCLOS, which doubtless will meet some challenges, especially from Russia, should settle ownership of waters and seabed. At the same time, Ottawa must invest considerable resources to ensure a secure Arctic and provide enough military muscle to enforce its regulations and to patrol this vast and changing region.

Challenge Questions

1. Why have the Arctic Ocean and the Northwest Passage moved from the outer margins of global geopolitics to become central concerns?
2. What is the physical difference between young ice and multi-year ice and what is the economic significance of this difference?

3. Why is M'Clure Strait, the western end of the northern route of the Northwest Passage, rarely free of ice while Lancaster Sound, at the eastern entrance of the Passage, is commonly free of ice?
4. Assuming that ocean shipping begins to use the Northwest Passage, why would the northern route be more attractive to super-size vessels than the southern route?
5. Why do predictions vary widely as to when the Arctic ice cap will disappear?
6. From an economic development perspective, why do the large investments designed to defend Canada's Arctic sovereignty largely benefit industries in southern Canada?
7. Climate warming has its downside but could there be upsides? For instance, while the Inuit have already found hunting on ice for seals more difficult, could new hunting opportunities emerge in a much warmer climate such as the Medieval Optimum (700 to 1,200 years ago), when a larger and more sedentary population existed in the Canadian Arctic?
8. Ottawa is faced with many demands for money. If you were Prime Minister, would you stick to your Arctic commitments, such as the very costly construction of the CCGS *John G. Diefenbaker* and the building of deepwater ports, or, like previous prime ministers, would you put such projects on hold to balance the budget?
9. Can you make a case that Arctic Canada and its Inuit residents could gain from the impact of climate warming? Explain.
10. Geopolitics is a rough game. The Northwest Passage debate pits two views against each other—are these Canadian internal waters or international straits? What evidence does Canada have to support its claim?

Notes

1. Canada's claim to the Arctic Archipelago was enhanced by an agreement in 1930 with Norway regarding the Sverdrup Islands located just west of Ellesmere Island. This agreement released any claim on these islands by Norway. Such a claim would have been based on the discovery and mapping of these islands by the famous Norwegian explorer, Otto Sverdrup. His expedition into the Arctic Archipelago took place from 1898 to 1902.
2. Ownership of the ice-covered waters of the Arctic Ocean between the islands of the Arctic Archipelago is not entirely resolved, especially for the waters of the Northwest Passage, where the distance between islands sometimes exceeds 12 nautical miles.
3. The definition of the Arctic Ocean may include Hudson Bay. By this definition, the geographic extent of the Arctic Ocean reaches 14 million km^2.
4. The eruption of Mount Tambora in Indonesia changed the world's climate for several years. In 1815, this volcanic activity filled the atmosphere with volcanic ash and dust, reducing the amount of sunlight reaching the Earth. Weather stations did not exist so temperature readings were not recorded, but the following summer of 1816 was so cool that it was dubbed "the year without summer." By 1991, weather stations existed around the world. Based on their recordings, the impact of the 1999 eruptions of Mount Pinatubo in the Philippines and Mount Hudson in Chile resulted in a mean world temperature decrease of about 0.5°C for the next two years (Rosenberg, 2010).
5. The site of the proposed Nanisivik port is settling because of a deep layer of clay under the wharf. This layer of clay is slowly compressing, thus making the site unsuitable for a major investment in an Arctic port facility.

6. In 1992, the federal government declared the HMS *Erebus* and HMS *Terror* "undiscovered" national historic sites. This political decision had sovereignty implications and it set in motion a well-funded annual search. In 2008, Ottawa assigned Parks Canada the task of leading the search as well as organizing a partnership of public and private organizations. In the summer of 2014, Parks Canada began its Victoria Strait Expedition. The objective was to concentrate the search in the waters of Victoria Strait, which lies between Victoria Island and King William Island. On 1 September, the Nunavut archaeology team discovered an iron fitting from a Royal Navy ship on Hat Island in Victoria Strait. Six days later, one of the vessels, the HMS *Erebus* on which Sir John Franklin sailed, was found in the waters just off O'Reilly Island near King William Island. In 2010, Parks Canada found M'Clure's ship, the HMS *Investigator*, near Banks Island.

7. The *St Roch* undertook the difficult task of crossing the Arctic Ocean because of British and Canadian concerns for Greenland. Nazi Germany had already conquered Denmark and had its eyes on Denmark's Arctic colony, which would have given it access to the east coast of North America. At the same time, the US was concerned about British/Canadian influence and even occupation of Greenland. According to Shelagh Grant (2010: 249–53), Ottawa sent the *St Roch* to Greenland from Vancouver through the Arctic Ocean to disguise its true mission to support Canadian occupation forces in Greenland.

8. The United States has not yet ratified the Law of the Sea Convention because it fears US national interests might be curtailed and its power in the international arena weakened. (This same approach has been reflected in the American refusal to adhere to other international conventions and agreements, such as the Ottawa Convention to Ban Landmines, the Kyoto Protocol, and the International Criminal Court.) UNCLOS would require that the US submit to the International Seabed Authority (ISA).

9. The United Nations Convention on the Law of the Sea (UNCLOS) trumps the previous doctrines. The first doctrine was known as "freedom of the seas," which limited rights and jurisdiction over the oceans by coastal nations to a narrow strip of three nautical miles. The freedom of the seas has its roots in the seventeenth century when European countries, especially Great Britain, "ruled the waves." By the twentieth century, this doctrine was challenged by countries seeking to control wider strips of the continental shelves. In 1945, the United States took the first step to extend its jurisdiction over all natural resources on its continental shelf. Other nations soon followed suit, but with varying distances from their shores. In 1967, the United Nations sought to bring order to this matter. In 1982, the United Nations adopted the Law of the Sea Convention, which regulates oceans by dividing the sea floor into zones of national and international jurisdiction. Nations have managerial control to a distance of 200 nautical miles offshore and, under special circumstances, beyond 350 nautical miles. On the other hand, international jurisdiction covers those waters and ocean floor beyond national control. These waters are called "high seas" and the term for the ocean floor is "the area." The United Nations created the International Seabed Authority to manage the use, such as mining, of the area.

References and Selected Reading

Abele, Frances, Thomas J. Courchene, F. Leslie Seidle, and France St-Hilaire, eds. 2009. *Northern Exposure: Peoples, Powers and Prospects in Canada's North, vol. 4, The Art of the State*. Montreal: Institute for Research on Public Policy.

————— and Thierry Rodon. 2007. "Inuit Diplomacy in the Global Era: The Strengths of Multilateral Internationalism." *Canadian Foreign Policy* 13: 3.

Arnold, Samantha. 2008. "Nelvana of the North: Traditional Knowledge and the Mythical Function of Canadian Foreign Policy." *Canadian Foreign Policy* 14 (Spring): 95–108.

Beauchamp, Benoît, and Rob Huebert. 2008. "Canadian Sovereignty Linked to Energy Development in the Arctic." *Arctic* 61, 3: 341–3.

Berton, Pierre. 1988. *The Arctic Grail: The Quest for the Northwest Passage and the North Pole, 1818–1909.* Toronto: McClelland & Stewart.

Birchall, S. Jeff. 2006. "Canadian Sovereignty: Climate Change and Politics in the Arctic." *Arctic* 59, 2: iii–iv.

Borgerson, Scott. 2008. "Arctic Meltdown: The Economic and Security Implications of Climate Change." *Foreign Affairs* 87, 2: 63–76.

Byers, Michael. 2009. *Who Owns the Arctic? Understanding Sovereignty Disputes in the North.* Vancouver: Douglas & McIntyre.

Campbell, Dean. 2010. "Walking to the North Pole? Hurry Up." *Globe and Mail*, 4 June. www. theglobeandmail.com/life/travel/walking-to-the-north-pole-hurry-up/article1592323/singlepage/.

Canada. 2014. "Canadian High Arctic Research Station." 10 Sept. http://www.science.gc.ca/default. asp?lang=En&n=A7775FC8-1.

Canadian Coast Guard. 2010. "The CCGS *John G. Diefenbaker* National Icebreaker Project." 28 Apr. www. ccg-gcc.gc.ca/e0010762.

Canadian Press. 2011. "Nanisivik, Nunavut Naval Facility Project Delayed for at Least Two Years." *Daily Commercial News and Construction Record*, 7 Aug. dcnonl.com/article/id44797.

CBC News. 2015. "Arctic Naval Facility at Nanisivik Completion Delayed to 2018." 4 Mar. http://www.cbc. ca/news/canada/north/arctic-naval-facility-at-nanisivik-completion-delayed-to-2018-1.2980312.

Charron, Andrea. 2005. "The Northwest Passage: Is Canada's Sovereignty Floating Away?" *International Journal* 60 (Summer): 831–48.

Coates, Kenneth S., P. Whitney Lackenbauer, Bill Morrison, and Greg Poelzer. 2009. *Arctic Front: Defending Canada in the Far North.* Toronto: Thomas Allen.

_____ and William R. Morrison. 2008. "The New North in Canadian History and Historiography." *History Compass* 6, 2: 639–58.

Cohen, Tobi. 2010. "Cruise Ship Runs Aground in Nunavut." *National Post*, 29 Aug. www.nationalpost. com/Cruise+ship+runs+aground+Canadian+Arctic/3457290/story.html.

Dawson, J., P.T. Maher, and D.S. Slocombe. 2007. "Climate Change, Marine Tourism and Sustainability in the Canadian Arctic: Contributions from Systems and Complexity Approaches." *Tourism in Marine Environments* 4, 2 and 3: 69–83.

Department of Justice. 2010. Oceans Act. 5 Nov. laws.justice.gc.ca/eng/O-2.4/page-2.html#anchorbo-ga:l_I.

Dodds, Klaus. 2010. "A Polar Mediterranean? Accessibility, Resources and Sovereignty in the Arctic Ocean." *Global Policy* 1, 3: 303–11.

Durham University. 2015. "Maritime Jurisdiction and Boundaries in the Arctic Region." https://www.dur. ac.uk/resources/ibru/resources/ibru_arctic_map_27-02-15.pdf.

Elliot-Meisel, Elizabeth. 1999. "Still Unresolved after Fifty Years: The Northwest Passage in Canadian–American Relations, 1946–1998." *American Review of Canadian Studies* 29 (Fall): 407–22.

Fednav. 2014. "Nunavik's Log Book." 19 Sept.–15 Oct. http://www.fednav.com/en/voyage-nunavik.

Fisheries and Oceans Canada. 2009. "Canada's Submission to the Commission on the Limits of the Continental Shelf under United Nations Convention on the Law of the Sea (The 'Continental Shelf Project')." 23 Dec. www.dfo-mpo.gc.ca/ae-ve/evaluations/09-10/6b060-eng.htm.

_____. 2010. *Canada's Ocean Estate: A Description of Canada's Maritime Zones.* 9 July. www.dfo-mpo. gc.ca/oceans/canadasoceans-oceansducanada/marinezones-zonesmarines-eng.htm.

Foreign Affairs and International Trade. 2007. "Defining Canada's Extended Continental Shelf." 31 July. www.international.gc.ca/continental/limits-continental-limites.aspx?lang=eng.

_____. 2010a. "Arctic Ocean Foreign Ministers' Meeting." 29 Mar. www.international.gc.ca/polar-polaire/arctic-meeting_reunion-arctique-2010_index.aspx.

_____. 2010b. *Statement on Canada's Arctic Foreign Policy.* 10 Aug. www.international.gc.ca/POLAR-POLAIRE/ASSETS/PDFS/CAFP_BROCHURE_PECA-ENG.PDF.

Foreign Affairs, Trade and Development Canada. 2013. "Partial Submission of Canada to the Commission on the Limits of the Continental Shelf Regarding Its Continental Shelf in the Arctic Ocean—Executive Summary." 9 Dec. http://www.international.gc.ca/arctic-arctique/continental/summary-resume.aspx?lang=eng.

Gerhardt, Hannes, Philip E. Steinberg, Jeremy Tasch, Sandra J. Fabiano, and Rob Shields. 2010. "Contested Sovereignty in a Changing Arctic." *Annals, Association of American Geographers* 100, 4: 992–1002.

Globe and Mail. 2010. "Theories Differ, but Climatologists Concur on Need for Action," 14 Dec., CC1 (Climate Change, a special information feature).

Gray, David H. 1997. "Canada's Unresolved Maritime Boundaries." *IBRU Boundary and Security Bulletin* 5, 3 (Autumn): 61–70: www.dur.ac.uk/resources/ibru/publications/full/bsb5-3_gray.pdf.

Grant, Shelagh D. 2010. *Polar Imperative.* Vancouver: Douglas & McIntyre.

Griffiths, Franklyn, ed. 1987. *Politics of the Northwest Passage.* Montreal and Kingston: McGill-Queen's University Press.

————. 2004. "Pathetic Fallacy: That Canada's Arctic Sovereignty Is on Thinning Ice." *Canadian Foreign Policy* 11 (Spring): 1–16.

Hall, C.M., and J. Saarinen, eds. 2010. *Tourism and Change in Polar Regions: Climate, Environment and Experiences.* Milton Park, Abingdon, Oxon, UK: Routledge.

Harper, Stephen. 2007. "Prime Minister Stephen Harper Announces New Offshore Arctic Patrol Ships." News release: Prime Minister of Canada, 9 July. http://pm.gc.ca/eng/news/2007/07/09/prime-minister-stephen-harper-announces-new-arctic-offshore-patrol-ships.

————. 2010. "PM Announces High Arctic Research Station Coming to Cambridge Bay." News release: Prime Minister of Canada, 24 Aug. http://pm.gc.ca/eng/news/2010/08/24/pm-announces-high-arctic-research-station-coming-cambridge-bay.

Heininen, Lassi, and Chris Southcott. 2010. *Globalization and the Circumpolar North.* Fairbanks: University of Alaska Press.

Howell, S.E.L., C.R. Duguay, and T. Markus. 2009. "Sea Ice Conditions and Melt Season Duration Variability within the Canadian Arctic Archipelago: 1979–2008." *Geophysical Research Letters* 36, L10502. doi: 10.1029/2009GL037681.

Huebert, Rob. 2006. "Canada–United States Environmental Arctic Policies: Sharing a Northern Continent." In Philippe Le Prestre and Peter Stoett, eds, *Bilateral Ecopolitics: Continuity and Change in Canadian–American Environmental Relations,* 115–32. Aldershot: Ashgate.

————. 2007. "Renaissance in Canadian Arctic Security?" *Canadian Military Journal.* www.journal.forces.gc.ca/vo6/no4/north-nord-eng.asp.

————. 2008. "Walking and Talking Independence in the Canadian North." In Brian Bow and Patrick Lennox, eds, *An Independent Foreign Policy for Canada? Challenges and Choices for the Future,* 118–36. Toronto: University of Toronto Press.

————, ed. 2009. *Thawing Ice, Cold War: Canada's Security, Sovereignty, and Environmental Concerns in the Arctic.* Winnipeg: University of Manitoba Press.

————. 2010. *The Newly Emerging Arctic Security Environment.* Prepared for the Canadian Defense and Foreign Affairs Institute, Mar. www.cdfai.org/PDF/The%20Newly%20Emerging%20Arctic%20Security%20Environment.pdf.

Honderich, John. 1987. *Arctic Imperative: Is Canada Losing the North?* Toronto: University of Toronto Press.

Indian and Northern Affairs Canada. 2010. *Northern Oil and Gas Annual Report 2009.* 6 May. www.ainc-inac.gc.ca/nth/og/pubs/ann/ann2009/ann2009-eng.pdf.

Jakobsson, M., L.A. Mayer, B. Coakley, J.A. Dowdeswell, S. Forbes, B. Fridman, H. Hodnesdal, R. Noormets, R. Pedersen, M. Rebesco, H.-W. Schenke, Y. Zarayskaya A, D. Accettella, A. Armstrong, R.M. Anderson, P. Bienhoff, A. Camerlenghi, I. Church, M. Edwards, J.V. Gardner, J.K. Hall, B. Hell, O.B. Hestvik, Y. Kristoffersen, C. Marcussen, R. Mohammad, D. Mosher, S.V. Nghiem, M.T. Pedrosa, P.G. Travaglini, and P. Weatherall. 2012. "The International Bathymetric Chart of the Arctic Ocean (IBCAO) Version 3.0." *Geophysical Research Letters,* doi: 10.1029/2012GL052219.

Kirton, John, and Don Munton. 1992. "The *Manhattan* Voyages, 1969–70." In Don Munton and John Kirton, eds, *Canadian Foreign Policy: Selected Cases.* Toronto: Prentice-Hall.

Lackenbauer, P. Whitney. 2008. "The Canadian Rangers: A 'Postmodern' Militia That Works." *Canadian Military Journal* 6, 4. www.journal.forces.gc.ca/vo6/no4/index-eng.asp.

Lajeunesse, Adam. 2008. "The Northwest Passage in Canadian Policy: An Approach for the 21st Century." *International Journal* 63 (Autumn): 1037–52.

Lasserre, Frédéric. 2007. "La souveraineté canadienne dans le passage du nord-ouest." *Policy Options* 28 (May): 34–41.

McCalla, Robert J. 2010a. "Arctic Shipping: Possibilities and Challenges in the Canadian Sector." Presentation at the Prairie Summit annual meeting of the Canadian Association of Geographers, Regina, 2 June.

_____. 2010b. "Adventure Cruise Shipping in Polar Regions: An Overview." In Aldo Chircop, Scott Coffen-Smout, and Moira McConnell, eds, *Ocean Yearbook*, 359–75. Boston: Brill.

Marinelog. 2010. "Resolve Marine Group Starts *Clipper Adventurer* Salvage." 3 Sept. www.marinelog.com/DOCS/NEWSMMIX/2010sep00033.html.

National Ocean Service.2014. "Is the Sea Level Rising?" *Ocean Facts*, 10 Apr. http://oceanservice.noaa.gov/facts/sealevel.html.

National Snow and Ice Data Center. 2007. "The Contribution of the Cryosphere to Changes in Sea Level." *State of the Cryosphere*, 12 Mar. https://nsidc.org/cryosphere/sotc/sea_level.html.

_____. 2010. "Sea Ice." *State of the Cryosphere*, 27 Oct. http://nsidc.org/cryosphere/sotc.

_____. 2014. "2014 Melt Season in Review." *Arctic Sea Ice News & Analysis*, 7 Oct. http://nsidc.org/arcticseaicenews/2014/10/2014-melt-season-in-review/.

Nicol, Heather N. 2014. "Discourses of Power: Sovereignty and the Canadian Arctic." In K. Dodds and M. Nuttall, eds, *A Northern Nation: Canada's Arctic Policies and Strategies*. Edmonton: University of Alberta/CCI Press.

_____ and Lassi Heininen. 2014. "Human Security, the Arctic Council and Climate Change." *Polar Record* 50, 1: 80–5. doi: 10.1017/S0032247412000666.

Nunatsiaq Online. 2014. "Nunavut Iron Producer Proposes Big Changes for Mary River." 5 Nov. http://nunatsiaqonline.ca/stories/article/65674nunavut_iron_producer_proposes_big_changes_for_mary_river/.

Richards, S. 1999. *Glaciers: Clues to Future Climate?* USGS General Interest Publications. pubs.usgs.gov/gip/glaciers/glaciers.pdf.

Rosenberg, Matt. 2010. "The Volcanic Mount Pinatubo Eruption of 1991 That Cooled the Planet." 17 Nov. geography.about.com/od/globalproblemsandissues/a/pinatubo.htm.

Steinberg, Philip E., Jeremy Tasch, and Hannes Gerhardt. 2015. *Contesting the Arctic: Politics and Imaginaries in the Circumpolar North*. London: Tauris.

Stewart, E.J., A. Tivy, S.E.L. Howell, J. Dawson, and D. Draper. 2010. "Cruise Tourism and Sea Ice in Canada's Hudson Bay Region." *Arctic* 63, 1: 57–66.

Stocker, Thomas, and Gian-Kasper Plattner. 2009. "The Physical Science Basis of Climate Change: Latest Findings To Be Assessed by WG I in AR5." United Nations Climate Change Conference, COP 15 IPCC Side Event, "IPCC Findings and Activities and Their Relevance for the UNFCCC Process." Copenhagen, 8 Dec.

United Nations Convention of the Law of the Seas (UNCLOS). 1982. "Straits Used for International Navigation: Transit Passage." Part III, Section 2. www.un.org/Depts/los/convention_agreements/texts/unclos/part3.htm.

_____. 2007. "The United Nations Convention on the Law of the Sea (A Historical Perspective)." *Oceans and Law of the Sea*. www.un.org/Depts/los/convention_agreements/convention_historical_perspective.htm.

Vachon, Paris W. 2010. "New Radarsat Capabilities Improve Maritime Surveillance." 18 Oct. geos2.nurc.nato.int/mrea10conf/reports/pdf/Vachon,%20New%20RADARSAT%20Capabilities%20Improve%20Maritime%20Surveillance.pdf.

Van Aken, Hendrik M. 2007. *The Oceanic Thermohaline Circulation: An Introduction*. New York: Springer Science & Business Media.

Victoria Times Colonist. 2007. "Harper on Arctic: Use It or Lose It." 10 July. www.canada.com/topics/news/story.html?id=7ca93d97-3b26-4dd1-8d92-8568f9b7cc2a&k=73323.

Weber, Bob. 2011. "Treaty to Boost Arctic Search and Rescue." *Globe and Mail*, 5 Jan., A5.

Woodman, David. 1991. *Unravelling the Franklin Mystery: Inuit Testimony*. Montreal and Kingston: McGill-Queen's University Press.

- 9 -

Looking to the Future

Rapid change remains the common denominator in the Canadian North. Arctic sovereignty and the Northwest Passage have emerged as critical issues. These topics—as well as Aboriginal and environmental issues—are now front and centre, and together they continue to reverberate in the halls of Parliament and in the boardrooms of corporate Canada. Looking to the end of the twenty-first century, it is safe to say that climate change will transform the North's physical and economic world—it already is. This new North should see an ice-free Arctic Ocean with unprecedented resource development taking place in Canada's Arctic Archipelago due to low-cost ocean shipping to world markets; a substantial federal presence to manage and regulate international shipping; and larger and more prosperous communities along the Northwest Passage due to resource-sharing, a global tourist industry, and the emergence of a coastal-based fishing industry.

In this new world, a frontier version of "development" is no longer acceptable. Such an image of the North, typified by the Klondike gold rush and the poems of Robert Service, represents an earlier time. The old dream of the Northwest Passage has revived, but less as a challenge and more as a commercial venture (Table 9.1). In addition, the federal government will have had to abandon its long-standing laissez-faire approach to the North and take a more active role in northern affairs. Not since Diefenbaker's "Northern Vision" of more than a half-century ago will the North once again gain the attention it has needed. A sign of this today is the search by Parks Canada for the lost Franklin ships, which signalled a renewed federal interest in the Northwest Passage and its prominent place in our claim to the Arctic Archipelago. The stirring chorus of the 1981 Stan Rogers song, "The Northwest Passage," provides a measure of our earlier, romantic image of the North but also indicates how deeply Canadians feel about the North and these Arctic waters. As a powerful symbol of Canada's northern character, sailing through the Northwest Passage could become a "passage of pride" for Canadians.

> Ah, for just one time I would take the Northwest Passage
> To find the hand of Franklin reaching for the Beaufort Sea
> Tracing one warm line through a land so wild and savage
> And make a northwest passage to the sea

Table 9.1   Recent Events Associated with Arctic Sovereignty

Date	Event
1985	Canada declares a 200-mile exclusive economic zone, extends its three-mile territorial waters to 12 nautical miles, and passes the Arctic Waters Pollution Prevention Act 1985.
1988	The Arctic Co-operation Agreement between Canada and the United States is signed. One term of the agreement states that the US government will seek Canadian approval to traverse the Northwest Passage.
1997	Ottawa passes the Oceans Act. In doing so, Canada extends its national jurisdiction over most of the Canadian continental margin, marine resources, and seabed.
2003	Canada ratifies the United Nations Convention on the Law of the Sea, which provides Ottawa 10 years to file its claim to the Arctic seabed.
2007	Radarsat-2, Canada's second Earth observation satellite, has the capacity to track and identify ships in the Canadian Arctic.
2013	Canada files ownership claim for Arctic seabed beyond its 200-mile economic zone, but signals that it wishes to extend its claim towards the North Pole—already claimed by Russia and Denmark. Canada's claim to the Arctic seabed—potentially an area similar in size to the three Prairie provinces—depends on "proof" that the ocean floor adjacent to Canada's Arctic continental shelf is geologically linked; its claim to the North Pole depends on its physical connection to the Lomonosov Ridge.
2014	Parks Canada expedition discovers Franklin's ship, the *Erebus*, in Queen Maud Gulf near King William Island. With the lost ships of the Franklin expedition designated as national historic sites in 1992, Canada's claim to the waters of the Northwest Passage can only be strengthened by this discovery and that of the HMS *Investigator* in 2010.

While the central issues facing this region and its people will be tested, recalibrated, and tested again, what distinguishes the North from other regions of Canada is its geography, its Aboriginal peoples, and most importantly, its ongoing and necessary search for a more equitable resolution of land/resource issues between Aboriginal peoples and large corporations. As the twenty-first century unfolds, the future Aboriginal face of the North will be shaped by:

- *Supreme Court of Canada decisions.* Aboriginal title to traditional lands, as a result of Supreme Court rulings, now requires resource companies to consult and accommodate First Nations before proceeding with their projects, thus pushing the boundaries of resource-sharing and adding another element towards a fairer division of the North's wealth.
- *Aboriginal readiness.* Aboriginal business organizations are eager to partner with private and public companies and their workforce is better equipped than ever before to participate in such employment.
- *Population dominance.* In the Territorial North and in parts of the Provincial North, Aboriginal peoples form a majority—and their numbers and share of the northern population continue to increase.

In sum, the North's future—and that of Canada—is bound to have a more prominent place for Aboriginal peoples. Already, the nature of resource development has shifted towards resource-sharing and the concept of partnerships. If Enbridge's Northern Gateway pipeline still has a chance, Enbridge will have to fully engage First Nations along the proposed pipeline route. Rapid cultural, economic, and political change is the order of the day. Within this context, the Northwest Passage, Arctic sovereignty, and resource-sharing have seized the spotlight and they are driving political events. A few years ago, Frances Abele (2009: 19) succinctly summarized the North, and her view is still valid:

> The Canadian North is the site of impressive and sometimes daring social and political innovations. It is also, increasingly, the focus of gathering international economic and political pressure, as well as new economic opportunity. Northern development will be an increasingly important aspect of Canadian political, social and economic vitality, and it should be a priority for public discussion and debate.

Northern Vision at End of the Twenty-First Century

By the end of this century, what will the physical geography of the North look like? With a high degree of certainty, the North will be a warmer place: the Arctic or Tundra biome will occupy a smaller geographic area; an ice-free Arctic Ocean will occur each summer; and much of the North is likely to be permafrost-free. In another sense, the geographic size of Canada's North will increase with the addition of a portion of the seabed of the Arctic Ocean.

A lower degree of certainty exists for economic and social changes. Nevertheless, the prospects for more resource projects in the Arctic are likely while commercial shipping through the Northwest Passage is almost assured. The vast oil and gas reserves in the Beaufort Sea and the Sverdrup Basin might come into play. As for demographic change, the likelihood that Aboriginal peoples in the North will maintain an increasing majority is promising, and the North's population should exceed 1.5 million; by how much remains unclear. Well-being in some Aboriginal communities should improve because of resource-sharing. Most remote communities, however, will likely remain dependent on government, and their nagging social problems could increase due to population pressures.

At that time, will Canadians still agree that the purpose of northern development is to harness the region's natural resources, strengthen and broaden its economy, and improve the living conditions of its residents? From a conceptual perspective, will the popular resource-driven model of regional development (see Chapter 1) first conceived by Harold Innis (with the added wrinkle by Mel Watkins of a staple trap outcome) be discarded for a fresh one? Also, will the hidden costs of resource projects to the environment and to the socio-economic well-being of Aboriginal peoples disappear or worsen by the end of the twenty-first century? Although existing environmental legislation and monitoring of resource projects, the rulings of the Supreme Court of Canada, and resource-sharing have gone some distance to address environmental and Aboriginal issues, the job is not done—especially as governments, led by the Harper government in Ottawa, have sought to limit the number of projects requiring assessment and to streamline assessments and approvals through new legislation.

At this point, are we are on the right road to closing these gaps in the next round of resource megaprojects? Canada's northern energy/pipeline policy on the Northern Gateway project poses the first test. The next major test may involve managing and regulating traffic through the Northwest Passage. Under this scenario, Canada's readiness to regulate shipping in the Northwest Passage and its capacity to enforce Canada's marine pollution laws on domestic and foreign vessels will be sorely tested. Much further into the twenty-first century, the vast oil and gas reserves in the Beaufort Sea and **Sverdrup Basin** may come into play. Well beyond the twenty-first century, Canada's share of the international seabed might be ripe for exploitation. Such resource development and oil tanker traffic come with risks, making Canada's preparedness to protect its Arctic environment of the utmost importance.

Question 1: Can the Resource Economy Support the Northern Labour Force?

The resource economy can support only a small portion of the workforce in the North. In fact, even with an air-commuting option, only a few communities are fortunate enough to benefit from resource development. The geographic reality is that the vast majority are beyond the economic and hiring orbit of resource projects. In addition, a serious mismatch exists between the skilled labour required by resource companies (as well as by governments) and the qualifications of Aboriginal workers (Vignette 9.1). One consequence is that skilled workers from the south work at these resource projects and in government offices. Impact and benefits agreements do help individual communities. For instance, Baker Lake's IBA focuses on preferential treatment for service contracts and employment opportunities (George, 2010). One IBA benefit was Inuit employment in building the 110-km road connecting Baker Lake with the mine site. Another benefit came in the form of Inuit companies obtaining the service contracts for the road construction and maintenance. Yet the future does not look bright because the demographic circumstances indicate the number of potential members of the Aboriginal labour force increasing very rapidly while the demand for workers in the resource industry fluctuates with the pace of construction work and the demand from the global economy.

Vignette 9.1   Mismatch: Jobs and Inuit Workers

The government of Nunavut has a preferential hiring practice for Inuit. The goal of increasing the percentage of Inuit in Nunavut's government to 50 per cent is not an easy task. This goal is complicated because of the limited number of Inuit who qualify for jobs in Nunavut's public sector. Even at the lowest entrance level, the Nunavut public sector normally requires a high school degree.

Question 2: How Can Government Ensure That the Resource Industry Limits Its Impact on the Environment?

Environmental impacts are kept to a minimum by government regulations and monitoring of resource projects. The principal means of identifying and regulating environmental damage derive from an environmental impact study, which is prepared by the project proponent and then assessed by a review panel, while restoration of the site, a requirement today, falls to government regulations and monitoring of the resource industry (see Chapter 6 for a fuller discussion). Mine closures go through four stages: shut-down; decommission; remediation/reclamation; post-closure monitoring (Miningfacts.org, 2012). Understandably, companies are concerned about their costs of complying with these regulations, but the cost to the public of unregulated resource development is enormous. For example, the Giant gold mine restoration alone is costing Ottawa over a billion dollars.

Even with these safeguards, resource development does make a footprint on the environment. The most serious case is found in Alberta's oil sands, a major source of Canada's greenhouse gas emissions. Environment Canada reports that, while most economic sectors are expected to reduce their emissions significantly by 2020, a major increase in total emissions is predicted to come from the oil sands. This industry, unless it can drastically curb its release of greenhouse gases, is projected to nearly double its already massive emissions between 2010 and 2020, from 49 to 92 **Mt CO_2e** (Environment Canada, 2011a: Table 5). Yet, companies operating in the oil sands have been slow to accept their social responsibilities in caring for the environment and, equally troubling, Alberta-led environmental monitoring has not been up to the task, thus creating a false sense of acceptable impact. For over a decade, the oil sands monitoring has remained badly flawed, suggesting that the Alberta and federal governments turned a blind eye because of the economic importance to Alberta and Canada of the oil sands projects.

The damage to the environment is not in dispute—and all industrial projects have negative environmental impacts. But what was not appreciated was the degree of damage, the geographic range of the impact, and the time horizon to restore the mined areas. In July 2011, the federal government made a sharp U-turn and insisted on a world-class monitoring system for the oil sands (Environment Canada, 2011b). It remains to be seen what difference this will make, but the fact is that so long as huge projects continue to exploit the oil sands, the local, regional, and global environments will continue to be compromised.

Question 3: Is There a Place for Aboriginal Corporations in the Resource Economy?

To be sure, Aboriginal development corporations have a role to play in resource projects. In fact, this model of resource development offers the potential for widespread and lasting benefits to specific Aboriginal corporations and their larger communities through direct employment and business opportunities. The growing presence of

Aboriginal corporations in the market economy represents a remarkable shift of economic power. Aboriginal leaders have chosen to participate in order to create jobs and business opportunities for their members. Modern land claim agreements have provided cash payments and a business structure that allow Aboriginal organizations the option to participate in major projects and to develop their own companies to take advantage of various opportunities related to northern development. The first two agreements, the James Bay and Northern Quebec Agreement in 1975 and the Inuvialuit Final Agreement in 1984, created the Cree Board of Compensation[1] and the Makivik Corporation in northern Quebec and the Inuvialuit Regional Corporation in the Northwest Territories. These organizations were responsible for receiving and managing the funds flowing from the two agreements. In each instance, funds were invested in the marketplace.

Partnership agreements that include resource-sharing mark the wave of the future for resource development in the North, thereby assuring benefits to Aboriginal peoples (Coates, 2015).

Future Directions

The key question remains: What is the purpose of northern development? While the North's economy will remain focused on energy and mineral extraction for world markets, aspects of that economy require greater involvement of Aboriginal peoples and greater respect for the environment. But oil companies in particular show little sign of respect and, after pressuring the federal government to open the Beaufort Sea for oil wells, were disappointed with the report that documented the difficulty of an oil spill cleanup in Arctic waters (Vanderklippe, 2011; Weber, 2011). For the Inuvialuit, the Arctic environment remains a precious feature of their culture and economy and a major oil spill would have devastating impacts on the sea and coastal environments.

Economic solutions do not come easily or without cost. Mega resource development alone cannot solve the perplexing problem of poverty and chronic housing shortages in Aboriginal communities. Ironically, both are products of the federal relocation program of the 1950s to sites with little economic promise, and prospects for stronger local economies are not bright—unless the federal government intervenes. Such intervention could involve support for small business, such as wind power generators, a hunter/trapper subsidy, an expansion in the number of Canadian Rangers, even a militia of local Inuit part-time soldiers who serve as an integral part of Canada's surveillance system and assist in search-and-rescue missions (Lackenbauer, 2008).

Breakthrough solutions require visionary leadership from Ottawa and a long-term commitment by all levels of governments and industry. Recognizing that the North has limited opportunities for economic diversification compared to other regions of Canada, how can its small population and the great distance to markets work to its advantage? One way is to make more use of resources within the North; another is to harness unused resources. Could wind power development achieve several seemingly elusive goals, such as making sustainability of northern communities closer to reality?

Currently, each community and remote mine must import diesel fuel at great cost to produce electricity. In time, these costs can only become more of a burden because of the increasing global price of oil, not to mention that energy from fossil fuels is being phased out throughout the world. One step in that direction was taken when the G7

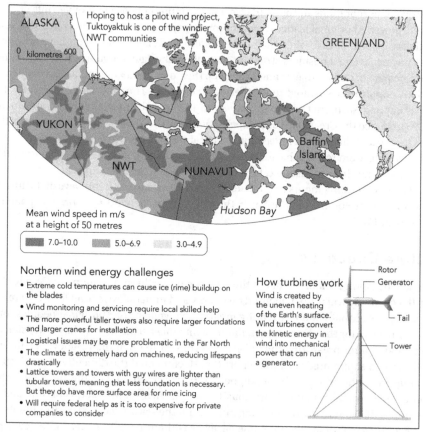

Figure 9.1 Wind Energy: Pipe Dream or Clean Energy Solution?

Source: O'Neill (2008). Licensed from *The Globe and Mail* for republication.

industrial nations—including Canada—agreed to cease using fossil fuels by the end of the century. Wind power, as a supplementary source of energy to reduce dependency on fossil fuels, could utilize the extensive wind energy found in the North, but it requires organization—perhaps an innovative combination of private industry, governments, and Aboriginal organizations. The wind energy exists but it needs to be harnessed. Canada's Arctic coast, for instance, registers some of the highest average annual wind speeds in North America. In an Arctic environment, design and maintenance are issues, but they are resolvable. Producing electricity from wind energy has several advantages:

- It is a renewable resource.
- Training of local utility workers would be required, and they would increase the size of the local workforce.
- Wind power technology could be the basis of a new scientific industry in Canada where a wind turbine technology centre could undertake research design and product development for small northern communities and isolated industrial

sites; offer advanced training in the operation and maintenance of this technology to northerners; and, with the help of northern leaders, government officials, and company officials, sell this technology to similar communities and industrial sites around the circumpolar world.

- Substituting wind energy for fossil fuel energy would help Canada reduce its greenhouse gas emissions.
- As a joint venture led by resource companies but strongly supported by governments and Aboriginal organizations, such a project could serve as a model for other joint efforts to search for breakthrough solutions to northern problems. Resource companies operating in a hinterland will find it to their advantage to go beyond their economic goal of profit-making and harness their vast power to tackle social and environmental issues of concern to northerners. Small-scale wind power generation could be one such challenge for the stakeholders in the Canadian North: resource companies, governments, and Aboriginal organizations.

The Northwest Passage

Global warming has already created more open water along routes of the Northwest Passage than ever before. Could an ice-free Arctic Ocean become a reality by the end of this century? In anticipation of ocean shipping using the Northwest Passage, the federal government has announced major public investment in marine infrastructure beginning with the Arctic patrol ships and later an icebreaker. A deepwater port is desirable along the Northwest Passage, but the first site selected, Nanisivik, proved unsuitable. Another site along the Northwest Passage should be selected soon and work begun. Whether this will happen is another matter.

Few Arctic coastal communities have a wharf (Figure 9.2). The construction of a wharf is far less costly than ferrying goods from a ship, and a wharf at each Arctic coastal community would benefit these places in three ways: first, the loading and unloading costs of goods transported by sea would be substantially reduced; second, local tourism would benefit from crews and passengers of small vessels and sailing boats docking and spending time in the community; and third, a local fishing industry could take root. Warmer water has seen an expansion in fish stocks, with Arctic char, flounder, clams, crabs, and scallops especially flourishing. With the recent resurgence of northern cod, Labrador and Nunavut fishers have an opportunity to harvest cod. Such harvesting in northern waters would involve community-based inshore fisheries. Processing fish would be a great boon to Inuit communities that face extremely high unemployment rates.

Summing Up

The North has broken away from the narrow perspective of a resource frontier. Arctic sovereignty, the Northwest Passage, and resource-sharing/partnerships have emerged as critical themes. Looking to the end of the twenty-first century, it is highly likely that climate change will transform the North's physical and economic world. This new North in all likelihood will include an ice-free Arctic Ocean and greater resource development with ocean shipping to world markets, which has already begun. At the same time, the

Figure 9.2   Cambridge Bay Wharf

During the September 2014 search for the two ships from the Franklin expedition, Cambridge Bay's harbour provided a secure place for the tugboat, M/V *Martin Bergmann* (foreground, at the Cambridge Bay wharf), and the Canadian Coast Guard icebreaker, *Sir Wilfrid Laurier* (at anchor in the background).

Source: Parks Canada/Jonathan Moore, 2014.

federal government will be pressed to bolster its northern presence, especially in the Arctic to manage and regulate shipping and to assist northern communities that, because of the geographic luck of the draw, are not in a position to prosper from resource-sharing, increased adventure tourism related to the Northwest Passage, and opportunities in a coastal fishing industry. Indeed, government-led relocation of northern peoples may not be only a thing of the past; in addition, significant voluntary migrations of people to locales of opportunity might occur in the decades to come.

By accepting the challenges of the twenty-first century, Canada could, with federal leadership, redirect the purpose of northern development, greatly increase its control of the Arctic Ocean as well as the extent of its control over the seabed of the Arctic Ocean, and, at the same time, create a more prosperous and equitable North.

Challenge Questions

1. Climate change has affected two of the three critical issues currently facing Canada's North. Of these three issues (resource-sharing, Arctic sovereignty, and the Northwest Passage), which one is not directly affected by climate change? Explain.
2. By the end of this century, the physical geography of the North is expected to change because of global warming. List as many of these changes as you can.

3. Canada's claim to its Arctic waters is not recognized by many countries, including the United States. Why would the Parks Canada discovery of the wreck of the *Erebus* strengthen Canada's claim to the waters along the Northwest Passage?
4. The North has always been an important element of the national psyche. Do you think that an ice-free Northwest Passage, even if just for a short time each summer, has increased or decreased its symbolic role?
5. The author suggests that the future Aboriginal face of the North will be shaped by three forces. What are these three forces?
6. Is it likely that resource-sharing will take hold in the next decade and bring a higher level of prosperity to some Aboriginal communities?
7. Protecting the environment from industrial pollution is now widely accepted and companies today accept their environmental responsibilities. If this statement is true, why did it take so long to create a "world-class" monitoring system for Alberta oil sands projects?
8. Canada has an opportunity to obtain part of the international seabed under the Arctic Ocean where huge petroleum deposits exist. While their commercial development lies far in the future, what evidence must Canada present to the United Nations to secure its claim to the international seabed?
9. With very high average annual wind speeds along the Arctic coast, what physical factors are holding back the development of wind-generated power (as a supplement to diesel power) in Arctic communities?
10. As the CEO of a major oil company plotting its future investment strategy, which offshore oil base would you plan to develop first—the Beaufort Sea, the Sverdrup Basin, or the oil in the seabed near the North Pole? Explain.

Note

1. "The Board of Compensation (managing funds from the 1975 James Bay Agreement) finances economic ventures directly and manages Air Creebec, Cree Construction, Valpiro, and Cree Energy through its holding company, Creeco. Eeyou Corporation manages the funding received under the La Grande 1986 Agreement and invests in community development and economic development ventures. The Cree Development Corporation set up under the New Agreement [Paix des Braves] with Quebec (February 7, 2002) is a vehicle for investment in economic ventures in the Territory using the funding from the New Agreement." From www.gcc.ca/cra/economicdevelopment.php.

References and Selected Reading

Abele, Frances. 2009. "Northern Development: Past, Present and Future." In Frances Abele, Thomas J. Courchene, F. Leslie Seidle, and France St-Hilaire, eds, *Northern Exposure: Peoples, Powers and Prospects in Canada's North*, 19–65. Montreal: Institute for Research on Public Policy.

————, Katherine A. Graham, and Allan M. Maslove. 1999. "Negotiating Canada: Changes in Aboriginal Policy over the Last Thirty Years." In Leslie A. Pal, ed., *How Ottawa Spends 1999–2000*, 251–92. Toronto: Oxford University Press.

Alberta. 2011. "Aboriginal People." *Alberta's Oil Sands*. www.oilsands.alberta.ca/aboriginalpeople.html.

Bell, Jim. 2007. "Tapardjuk: GN Won't Change Language Laws." *Nunatsiaq News*, 14 Dec. nunatsiaqnews.com/archives/2007/712/71214/news/nunavut/71214_780.html.

Bell, Patricia. 2007. "Harper Announces Northern Deep-Sea Port, Training Site." CBC News. http://www.cbc.ca/news/canada/harper-announces-northern-deep-sea-port-training-site-1.644982.

Berger, Thomas R. 1977. *Northern Frontier, Northern Homeland: The Report of the Mackenzie Valley Pipeline Inquiry.* Ottawa: Department of Supply and Services.

Byers, Michael. 2009. *Who Owns the Arctic?* Vancouver: Douglas & McIntyre.

Birchall, S.J. 2006. "Canadian Sovereignty: Climate Change and Politics in the Arctic." *Arctic* 59, 2: iii–iv.

Cairns, Alan C. 2000. *Citizens Plus: Aboriginal Peoples and the Canadian State.* Vancouver: University of British Columbia Press.

Campbell, Colin. 2006. "Saving Inuktitut: Nunavut Has Passed Legislation Intended to Keep the Inuit Language from Fading." *Maclean's.* 20 Sept. www.macleans.ca/canada/national/article.jsp?content= 20060925_133597_133597.

Canada. 2007. "Prime Minister Harper Bolsters Arctic Sovereignty with Science and Infrastructure Announcements." International Polar Year 2007–2008. www.ipy-api.gc.ca/me/nr/05-10-07_e.html.

—————. 2010. "Investments in Forest Industry Transformation Program." *Action Plan.* http://www.actionplan.gc.ca/en/initiative/investments-forest-industry-transformation-program.

—————. 2011. Budget 2011, 6 June. www.budget.gc.ca/2011/home-accueil-eng.html.

Canadian American Strategic Review (CASR). 2007. "Carving Up the Arctic Seabed—Two Options: The 'Meridian' Method or the 'Sector' Solution." Oct. www.sfu.ca/casr/id-arctic-empires-4.htm.

Carnaghan, Matthew, and Allison Goody. 2006. *Canadian Arctic Sovereignty.* Ottawa: Library of Parliament. http://www.parl.gc.ca/Content/LOP/researchpublications/prb0561-e.htm.

CBC News. 2011. "Oilsands Mine Approval Condemned." 28 Jan. www.cbc.ca/news/business/story/2011/01/28/total-oilsands-mine-approval.html.

Chase, Steven, and Bertrand Marotte. 2011. "Halifax, Vancouver Win $33-Billion in Shipbuilding." *Globe and Mail,* 20 Oct., A1.

Coates, Ken S. 2015. *Sharing the Wealth: How Resource Revenue Agreements Can Honour Treaties, Improve Communities, and Facilitate Canadian Development.* Macdonald-Laurier Institute. At: http://www.macdonaldlaurier.ca/files/pdf/MLIresourcerevenuesharingweb.pdf.

Coates, Kenneth, P. Whitney Lackenbauer, William R. Morrison, and Greg Poelzer. 2008. *Arctic Front.* Toronto: Thomas Allen.

Dawson, Jackie, Emma J. Stewart, Patrick T. Maher, and D. Scott Slocombe. 2008. "Climate Change, Complexity and Cruising in Canada's Arctic: A Nunavut Case Study." In Robert B. Anderson and Robert M. Bone, eds, *Natural Resources and Aboriginal People in Canada: Readings, Cases and Commentary.* Toronto: Captus Press.

Department of National Defence, 2011. "Canadian Rangers." 27 June. www.army.dnd.ca/land-terre/cr-rc/index-eng.asp.

Environment Canada. 2011a. *Canada's Emissions Trends.* July. www.ec.gc.ca/Publications/E197D5E7-1AE3-4A06-B4FC-CB74EAAAA60F%5CCanadasEmissionsTrends.pdf.

—————. 2011b. "An Integrated Monitoring Plan for the Oil Sands." 21 July. www.ec.gc.ca/default. asp?lang=En&n=56D4043B-1&news=7AC1E7E2-81E0-43A7-BE2B-4D3833FD97CE.

Fortier, M., and L. Fortier. 2006. "Canada's Arctic, Vast, Unexplored and in Demand: Canadian-led International Research in the Changing Coastal Canadian Arctic." *Journal of Ocean Technology* 1, 1: 1–8.

George, Jane. 2010. "A Nunavut Milestone: Meadowbank's First Gold Bar." *Nunatsiaq Online,* 1 Mar. www.nunatsiaqonline.ca/stories/article/98767_a_nunavut_milestone_meadowbanks_first_gold_bar/.

Grant, Shelagh D. 2010. *Polar Imperative.* Vancouver: Douglas & McIntyre.

Griffiths, F. 2004. "Pathetic Fallacy: That Canada's Arctic Sovereignty Is on Thinning Ice." *Canadian Foreign Policy* 11, 3: 1–16.

Hassol, Susan Joy. 2004. "Impacts of a Warming Arctic: Arctic Climate Impact Assessment." amap.no/acia.

Harrison, C. 2006. "Industry Perspectives on Barriers, Hurdles, and Irritants Preventing Development of Frontier Energy in Canada's Arctic Islands." *Arctic* 59, 2: 238–42.

Hicks, Jack, and Graham White. 2000. "Nunavut: Inuit Self-Determination through a Land Claim and Public Government?" In Jens Dahl, Jack Hicks, and Peter Jull, eds, Inuit Regain Control of their Lands and Lives, 30–117. Skive, Denmark: Centraltrykkeriet Skive A/S.

Huebert, R. 2003. "The Shipping News: How Canada's Arctic Sovereignty Is on Thinning Ice." *International Journal* 58, 3: 295–308.

Jull, Peter. 2000. "A Different Place." In Jens Dahl, Jack Hicks, and Peter Jull, eds, Inuit Regain Control of their Lands and Lives, 118–36. Skive, Denmark: Centraltrykkeriet Skive A/S.

Koperqualuk, Lisa, ed. 2001. "Premiere Issue: Our Land, Our Future." *Nunavik* 1: 1–17.

Lackenbauer, P. Whitney. 2008. "The Canadian Rangers: A Postmodern 'Militia' That Works." *Canadian Military Journal*, 14 Aug. www.journal.forces.gc.ca/vo6/no4/north-nord-03-eng.asp.

Laghi, Brian, Karen Howlett, and Rhéal Séguin. 2008. "Charest, McGuinty to Push PM on Economy: Harper Government to Unveil $1-Billion in Aid for One-Industry Towns." *Globe and Mail*, 10 Jan. http://www.theglobeandmail.com/news/national/charest-mcguinty-to-push-pm-on-economy/article18441841/.

Lalonde, Suzanne. 2007. "Le passage Nord-Quest: zone canadienne ou internationale?" *Le Multilatéral*: 1–13.

Miningfacts.org. 2012. "What Happens to Mine Sites after a Mine Is Closed?" Fraser Institute. http://www.miningfacts.org/environment/what-happens-to-mine-sites-after-a-mine-is-closed/

National Post. 2002. "Nunavut's Bill 101," 4 Feb., A15.

National Snow and Ice Data Center (NSIDC). 2007. "Arctic Sea Ice News Fall 2007." http://nsidc.org/arcticseaicenews/2007/10/589/.

Nunavik Commission. 2001. *Amiqqaaluta—Let Us Share: Mapping the Road toward a Government for Nunavik*. Quebec: Nunavik Commission.

Nunavut Tunngavik Inc. 2010. "Nunavut Harvest Support Program." www.tunngavik.com/documents/beneficiaryProgramForms/NHSP%20Community%20Harvest%20Program%20Description%20ENG.pdf.

O'Neill, Katherine. 2008. "Territories Hope Arctic Winds Pack Power." *Globe and Mail*, 3 Jan. www.theglobeandmail.com/servlet/story/RTGAM.20080103.wind03/BNStory/National/home.

Parks Canada. 2014. "The Franklin Expedition—Photo Gallery—Discovery Photos: The M/V Martin Bergmann at the Cambridge Bay wharf . . ." http://www.pc.gc.ca/eng/rech-srch/clic-click.aspx?/cgi-bin/MsmGo.exe?grab_id=0&page_id=88870&query=cambridge bay wharf&hiword=CAMBRIDGES WHARFS bay cambridge wharf.

Pharand, Donat. 2007. "The Arctic Waters and the Northwest Passage: A Final Revisit." *Ocean Development and International Law* 38, 1 and 2: 3–69.

Salisbury, Richard F. 1986. *A Homeland for the Cree: Regional Development in James Bay, 1971–1981*. Montreal and Kingston: McGill-Queen's University Press.

Vanderklippe, Nathan. 2011. "Oil Industry Outlines Cleanup Strategy for Arctic Spill." *Globe and Mail*, 9 June, B1.

Weber, Bob. 2011. "Arctic Oil Spill Cleanup Impossible One Day in Five, Energy Board Report Finds." *Globe and Mail*, 1 Aug., B1.

Appendix I

The Method of Calculating Nordicity

While the division of the North into the Arctic and Subarctic demonstrates the existence of two physical environments (the Tundra and Boreal biomes), nordicity permits us to measure differences between places and to create human/physical regions. But what exactly does nordicity measure? A Canadian geographer, Louis-Edmond Hamelin, created this term to quantify "northernness" into a single value measured as polar units. Nordicity is a quantitative measure based on 10 variables found at a particular northern place. It is based on both physical elements, such as annual cold, permafrost, and ice cover of water bodies, and human elements, such as population size, accessibility by land, sea, and air, and degree of economic activity. These 10 physical and human elements seek to represent all facets of the North. Hamelin set numeric values (polar units) for each variable. At a particular place, the sum of polar values for the 10 variables results in a single numeric value. This value represents the nordicity of that particular place. The North Pole, for example, has a nordicity of 1,000 polar units. Isachsen, the northernmost weather station in Canada, has 925 polar units while Vancouver has only 35. Hamelin determined that the boundary between northern and southern Canada followed a line representing 200 polar units. Hamelin is describing the Canadian North from a southern perspective. Northerners may not have the same mental map of Canada. For them, southern Canada is a distant and different place, while the terms Middle North, Far North, and Extreme North have no meaning for them.

The method of calculating the number of polar units for a place is summarized below. For example, the number of polar units assigned to varying degrees of latitude ranges from zero units for latitudes of 45°N or less to 100 units at the North Pole (90°N).

List of 10 Variables and Their Polar Units

Variable		Polar Units
1. Latitude	90°	100
	80°	77
	50°	33
	45°	0

Variable			Polar Units
2. Summer Heat		0 days above 5.6°C	100
		60 days above 5.6°C	70
		100 days above 5.6°C	30
		>150 days above 5.6°C	0
3. Annual Cold		6,650 degree days below 0°	100
		4,700 degree days below 0°	75
		1,950 degree days below 0°	30
		550 degree days below 0°	0
4. Types of Ice	Frozen ground	Continuous permafrost 457 m thick	100
		Continuous permafrost <457 m thick	80
		Discontinuous permafrost	60
		Ground frozen for less than 1 month	0
	Floating ice	Permanent pack ice	100
		Pack ice for 6 months	36
		Pack ice <1 month	0
	Glaciers	Ice sheet >1,523 m thick	100
		Ice cap 304 m thick	60
		Snow cover <2.5 cm	0
5. Annual Precipitation		100 mm	100
		300 mm	60
		500 mm	0
6. Natural Vegetation		Rocky desert	100
		50% tundra	90
		Open woodland	40
		Dense forest	0
7. Accessibility	Land or sea	No service	100
		For two months	60
		Up to four months by both land and sea	15
		Continuous service by either land or sea	0
8. Accessibility	Air	Charter only	100
		Weekly regular service	25
		Daily regular service	0
9. Population	Settlement size	None	100
		About 100	85
		About 1,000	60
		>5,000	0
	Population density	Uninhabited	100
		1 person per km²	50
		4 persons per km²	0
10. Economic Activity		No production	100
		Exploration	80
		20 hunters/trappers	75
		Interregional centre	0

Source: Hamelin (1979: ch. 1).

Appendix II

Population by Northern Census Divisions

Northern Census Divisions	2001	2011	% Change
Yukon	28,674	33,900	18.2
Northwest Territories*	37,360	41,460	11.0
Fort Smith, NWT	28,824	32,405	12.4
Inuvik, NWT	8,536	9,055	7.7
Nunavut	26,745	31,900	19.3
Baffin, Nunavut	14,372	16,935	17.8
Keewatin, Nunavut	7,557	8,955	18.5
Kitikmeot, Nunavut	4,816	6,010	24.8
Territorial North	92,779	107,275	15.6
British Columbia	198,289	197,747	–0.1
Bulkley-Nechako	40,856	39,208	–4.0
Fraser-Fort George	95,317	91,879	–3.6
Peace River	55,080	60,082	9.1
Stikine	1,316	1,000	–24.0
Northern Rockies	5,720	5,578	–2.5
Alberta	100,479	129,020	28.4
16	42,971	67,516	57.1
17	57,508	61,504	6.9
Saskatchewan	32,029	36,557	14.1
18	32,029	36,557	14.1
Manitoba	66,622	70,906	6.4
21	22,556	21,393	–5.2
22	35,077	40,923	16.7
23	8,989	8,590	–4.4

Northern Census Divisions	2001	2011	% Change
Ontario	416,475	400,656	-3.8
Cochrane	85,246	81,122	-2.3
Algoma	118,567	115,870	-2.2
Thunder Bay	150,860	149,063	-3.2
Kenora	61,802	57,607	-6.8
Quebec	515,126	504,100	-2.1
Rouyn-Noranda	39,621	41,012	3.5
Abitibi-Ouest	21,984	21,003	-4.4
Abitibi	24,613	24,354	-0.1
La Vallée-de-l'Or	42,375	42,896	1.2
La Tuque	15,862	15,130	-4.6
Le Domaine-du-Roy	32,839	31,870	-3.0
Maria-Chapdelaine	26,900	25,279	-6.0
Le Saguenay-et-son-Fjord	166,780	165,211	-0.9
La Haute-Côte-Nord	12,894	11,546	-10.5
Manicouagan	33,620	32.012	-4.8
Sept-Rivières-Caniapiscau	38,931	39,500	1.5
Minganie-Le Golfe-du-Sainte-Laurent	12,321	11,708	-5.0
Nord-du-Québec	38,575	42,579	10.4
Newfoundland & Labrador	47,735	43,515	-8.8
9	20,091	16,785	-16.5
10	25,230	24,110	-4.4
11	2,414	2,620	8.5
Provincial North	1,376,755	1,382,456	0.0
Territorial North	92,779	107,310	15.7
Canadian North	1,469,534	1,489,766	0.1

*The 2011 census divisions for the Northwest Territories have changed, with Inuvik equivalent to what formerly were Regions 1 [6,710] and 2 [2,345]; and Fort Smith equivalent to Regions 3 [2,810], 4 [3245], 5 [6,905], and 6 [19,445].

Source: Population and dwelling counts, for Canada, provinces and territories, and census divisions, 2001, 2011.

Appendix III

Demographic Tables

Death Rates in the Territorial North (per 1,000 persons)

Territory	2000	2001	2002	2003	2004	2005	2006	2007	2008	2009	2010	2011
Yukon	5.1	4.4	4.9	4.4	5.4	5.3	5.7	5.9	6.0	6.0	5.7	5.5
NWT	3.8	4.0	4.1	4.8	3.6	3.5	4.3	4.0	4.6	4.3	4.2	4.3
Nunavut	4.7	4.4	4.4	4.6	4.1	3.8	4.2	4.1	4.7	5.0	4.0	4.9
Canada	7.1	7.1	7.1	7.1	7.1	7.1	7.0	7.1	7.2	7.1	7.0	7.0

Source: Statistics Canada. 2014. "Deaths and Mortality Rates, by Age Group and Sex, Canada, Provinces and Territories." Table 102-0504. http://www5.statcan.gc.ca/cansim/a26?lang=eng&retrLang=eng&id=1020504&pattern=death+and+mortality+rates&tabMode=dataTable&srchLan=-1&p1=1&p2=49.

Birth Rates in the Territorial North (per 1,000 persons)

Territory	2000	2001	2002	2003	2004	2005	2006	2007	2008	2009	2010	2011
Yukon	12.1	11.4	11.3	11.0	11.8	10.3	11.7	10.8	11.3	11.4	11.0	12.2
NWT	16.5	14.9	15.3	16.6	16.3	16.7	16.2	16.7	16.5	16.3	16.0	15.6
Nunavut	26.5	25.3	25.3	26.0	25.2	23.3	24.6	25.4	25.5	27.2	25.2	24.9
Canada	10.7	10.7	10.5	10.6	10.5	10.6	10.9	11.2	11.3	11.3	11.1	11.0

Source: Statistics Canada. 2014. "Crude Birth Rate, Age-Specific and Total Fertility Rates (Live Births), Canada, Provinces and Territories." Table 102-4505. http://www5.statcan.gc.ca/cansim/pick-choisir?lang=eng&p2=33&id=1024505.

Rate of Natural Increase in the Territorial North (%)

Territory	2000	2001	2002	2003	2004	2005	2006	2007	2008	2009	2010	2011
Yukon	0.70	0.70	0.64	0.66	0.64	0.5	0.60	0.50	0.53	0.54	0.53	0.67
NWT	1.27	1.09	1.12	1.18	1.27	1.32	1.19	1.27	1.19	1.20	1.18	1.13
Nunavut	2.18	2.09	2.09	2.14	2.11	1.95	2.04	2.13	2.08	2.22	2.12	2.00
Canada	0.36	0.36	0.34	0.35	0.34	.035	0.39	0.41	0.41	0.41	0.41	0.40

Source: Derived from tables for birth and death rates.

Fertility Rate, Canada and Territories

Territory	2000	2001	2002	2003	2004	2005	2006	2007	2008	2009	2010	2011
Yukon	1.60	1.56	1.56	1.52	1.67	1.48	1.69	1.58	1.64	1.66	1.61	1.73
NWT	2.00	1.82	1.89	2.05	2.03	2.11	2.07	2.11	2.08	2.06	1.98	1.97
Nunavut	3.16	3.03	3.04	3.10	2.96	2.74	2.84	2.97	2.98	3.24	3.00	2.97
Canada	1.49	1.51	1.50	1.53	1.53	1.54	1.59	1.66	1.68	1.67	1.63	1.61

Source: Statistics Canada. 2014. "Crude Birth Rate, Age-Specific and Total Fertility Rates (Live Births), Canada, Provinces and Territories." Table 102-4505. http://www5.statcan.gc.ca/cansim/pick-choisir?lang=eng&p2=33&id=1024505.

Appendix IV

Aboriginal Population by Census Divisions, 1986 and 2011

Territory/ Province	Census Division	1986 Population	% 1986	2011 Population	% 2011
Yukon		4,995	21.4	7,710	23.1
Northwest Territories		14,680	49.2	21,165	51.9
	Fort Smith	9,165	37.7	14,130	44.4
	Inuvik	5,515	65.8	7,035	78.5
Nunavut		15,850	84.7	27,360	85.8
	Baffin	8,120	81.5	13,745	81.8
	Keewatin	4,415	88.8	8,150	91.5
	Kitikmeot	3,315	88.5	5,465	91.4
Territorial North		35,525	46.9	56,235	52.4
British Columbia		27,440	12.2	40,550	20.5
	Kitimat-Stikine	9,210	23.4	12,660	34.2
	Fraser-Fort George	6,525	7.3	10,915	12.1
	Peace River	5,890	10.3	9,365	14.5
	Buckley-Nechako	5,400	14.5	7,335	18.8
	Stikine	415	20.5	275	45.8
Alberta		22,735	23.3	32,000	24.8
	16	8,260	17.0	7,595	11.3
	17	14,475	29.7	24,405	40.0
Saskatoon	18	18,975	74.9	31,960	87.4

Territory/ Province	Census Division	1986 Population	% 1986	2011 Population	% 2011
Manitoba		38,385	51.9	64,725	74.6
	19	8,105	88.9	15,630	95.9
	21	7,160	30.0	10,765	50.9
	22	18,180	59.6	31,545	77.5
	23	4,940	48.2	6,785	79.2
Ontario		31,850	7.3	60,125	15.0
	Algoma	7,360	5.6	13,145	11.5
	Cochrane	4,195	4.5	9,860	12.5
	Kenora	10,865	20.7	19,990	36.4
	Thunder Bay	9,430	6.1	17,130	12.0
Quebec		28,560	6.0	58,305	11.6
	Abitibi (84)	3,420	3.7	9,060	6.4
	Lac-Saint-Jean-Ouest (90)	2,280	3.7	4,395	7.9
	Chicoutimi (94)	1,810	1.0	4,255	2.6
	Saguenay (97)	6,120	5.9	13,385	14.4
	Territoire-du-Nouveau Québec (98)	14,930	40.2	27,210	64.2
Newfoundland and Labrador		5,810	10.6	12,770	29.6
	9	350	1.4	1,190	7.2
	10	5,460	19.0	11,580	43.6
Provincial North		173,755	12.2	300,435	21.7
North		209,280	14.3	356,670	23.9
Canada		737,035	2.8	1,400,685	4.3

Notes:
1. BC: Northern Rockies census division did not exist in 1986 and it is included in the 2011 Peace River census division.
2. Quebec: Quebec altered the internal boundaries and renamed some of its northern census divisions: Abitibi (84) became, in 2011, Abitibi, Abitibi-Ouest, Rouyn-Noranda, La Vallée-de-l'Or, and La Tuque; Lac-Saint-Jean-Ouest (90) became Maria-Chapdelaine and Le Domaine-du-Roy; Chicoutimi was renamed Saguenay-et-son-Fjord; Saguenay (97) now consists of La Haute-Côte-Nord, Manicouagan, Sept-Rivières-Caniapiscau, and Mingainie-Basse-Côte-Nord.
3. Newfoundland and Labrador: In 1986, census division 10 represented Labrador; in 2011, Labrador was divided into two census divisions (10 and 11). In the table above, census division 11 is included in the population for census division 10.

Sources: Bone (1992: 190, 244); Statistics Canada. 2014. NHS Aboriginal Population Profile, 2011. 16 Jan. http://www12.statcan.gc.ca/nhs-enm/2011/dp-pd/aprof/index.cfm?Lang=E>.

Glossary

||||||||||||||||||||||||||||||||||||||||||||||||||||||||||||||||||||||||||||||||||||||||||||||||||||||||||||||||||||||||||||||||||||||||||||||||||||||||||||||||||

Aboriginal peoples: The original inhabitants of North America. Canada's Constitution Act, 1982 defines Aboriginal peoples as including "the Indian, Inuit and Métis peoples."

Aboriginal title: A legal right to land; the claim depends on documenting current and traditional occupancy and use of the land. *See also Calder; Delgamuukw; Tsilhqot'in.*

Acid rock drainage: The acidic water created when sulphide minerals, often taking the form of tailings from mining operations, are exposed to air and water and, through a natural chemical reaction, produce sulphuric acid. Once the acid enters the surface water, its toxicity poses a threat to fish stocks and to wildlife/humans that drink the water.

Active layer of permafrost: A thin layer of immature soil (cryosol) above the permanently frozen ground. The active layer allows plant growth, water drainage, and soil-forming to occur. With warmer Arctic summers, the active layer has thickened, which is another indication of the retreat of permafrost.

Age-dependency ratio: The ratio of persons in the "dependent" ages (under 15 and over 64 years) to those in the "economically productive" ages (15–64 years) in a population; used as an indicator of the economic burden that the productive portion of a population must carry.

Air mass: A large body of air that has similar horizontal temperature and moisture characteristics.

Alaska Current: A warm ocean current flowing northward along the coast of British Columbia and Alaska. It warms the adjacent land, causing northern British Columbia and southern Alaska to experience relatively warm weather for their latitudes.

Albedo: Proportion of solar radiation reflected by the Earth's surface back into the atmosphere. The more radiation reflected away from the Earth by light-coloured or white surfaces, the higher the albedo. Ice and snow may reflect as much as 90 per cent of incoming solar radiation back to outer space.

Arctic Archipelago: The group of islands in the Arctic Ocean representing a total land area of 1.3 million km². This Canadian archipelago consists of nearly 37,000 islands, many of which are tiny. Some of the islands are very large—Baffin Island is the fifth largest in the world, and other large islands include Victoria, Ellesmere, Banks, Devon, Axel Heiberg, Melville, and Prince of Wales.

Arctic Circle: An imaginary line, like latitudes, that extends in an east/west direction at 66° 33′N. At the time of the summer solstice (21 June), the sun at this location on the Earth's surface does not set below the horizon for the entire day; similarly, at the winter solstice (21 December), the sun does not rise above the horizon for 24 hours. Still, twilight does exist until much higher latitudes. At that point, the total darkness is called polar night.

Arctic Clause: Article 234 of UNCLOS, which allows Canada to enforce its Arctic Waters Pollution Prevention Act in its exclusive economic zone (200 nautical miles from shore).

Arctic Council: Group of eight nations (the Arctic Five plus Finland, Iceland, and Sweden) that play an advisory role on Arctic issues, especially related to the environment, but not those related to boundaries or resource development. Representatives of Aboriginal organizations have observer status at Council meetings.

Arctic Five: The five countries (Canada, Denmark, Norway, Russia, and the United States) with coasts bordering the Arctic Ocean and that therefore can make sovereign claims to the Arctic seabed under UNCLOS. Denmark's claim is through its possession of Greenland.

Arctic ice pack: Ice that covers much of the Arctic Ocean throughout the year, although in recent years this ice pack has reduced in size during the late summer. *See also* pack ice.

Arctic Messenger: A metaphor of warning for the world that the unprecedented speed with which the Arctic environment is under stress will have major global consequences, such as extreme weather and rising ocean levels.

Beaufort Gyre circulation system: The principal ocean currents in the Beaufort Sea, which have a clockwise direction (anticyclonic) in the winter months and counter-clockwise direction (cyclonic) in the summer, at which time cold and relatively fresh water and ice flow into Baffin Bay.

Berger Inquiry: A public inquiry (1974–7) into a proposed Mackenzie Valley gas pipeline in which, for the first time, public hearings were held in the small Aboriginal communities where the proposed pipeline project would be constructed. This novel approach gave a voice to local people who focused their comments on local issues, such as the impact on their culture and their ancestral lands.

Berger Report: The 1977 report of the Mackenzie Valley Pipeline Inquiry headed by Justice Thomas Berger. The report emphasized environmental issues and the potential clash between resource projects and the Aboriginal way of life. Berger recommended that no pipeline be built through the northern Yukon for environmental reasons and that a pipeline through the Mackenzie Valley should be delayed for 10 years, thus allowing time for the settling of Aboriginal land claims. Because of deteriorating economic conditions, this pipeline was never built.

Beverly-Qamanirjuaq Caribou Management Board: A co-management board concerned with resource management of caribou over several jurisdictions in the territories and provinces and among several Aboriginal user groups that must deal with the contentious issue of caribou harvesting by Dene and Inuit communities. The Board was formed in the early 1980s in response to the apparent decline of caribou. It consists of Aboriginal members from communities that harvest caribou and government biologists and resource managers. Their primary function is to exchange information and to ensure the survival of caribou by protecting the herd's habitat and, as needed, limiting the harvest by persuasion. While the Board has only an advisory role, the various governments responsible for caribou tend to accept its recommendations.

Biodiversity: The variety of life forms on Earth or that inhabit a particular ecosystem or geographic region; biological diversity.

Biomagnification: The increase in concentration of toxic substances in organisms as contaminants are passed up food webs from lower forms, such as plankton and fish, to higher forms, such as whales and seals.

Biomes: Large and naturally occurring communities of vegetation and wildlife adapted to the climatic and geological areas they inhabit, such as the Tundra biome. World biomes circle the globe and are controlled by climate. These large geographic zones contain natural communities of flora and fauna that, formed over a long period of time, have adapted to climatic conditions of the area as well as other natural conditions.

Birth rate: The number of births per 1,000 population in a given year. Also known as crude birth rate.

Bitumen: A thick, black, viscous oil consisting of naturally occurring hydrocarbons mixed with other substances such as sand and clay.

Bitumen-Royalty-in-Kind program: An Alberta program, initiated in 2009, to ensure a bitumen supply for Alberta's existing and proposed upgraders by having royalty payments in bitumen rather than cash.

Boom-and-bust cycle: A business cycle often associated with an economy based heavily on a single commodity. During the boom cycle, demand and prices for the commodity increase quickly and the region experiences a boom; during the bust cycle, demand and prices fall even faster, leaving the region's economy in tatters.

Border: Demarcation between two political units (countries, provinces, municipalities) as applied to land.

Boreal biome: A broad, circumpolar vegetation zone found in northern latitudes consisting of coniferous forest. It is one of the world's largest and most important biogeoclimatic areas. Black and white spruce dominate the boreal biome along with pine, larch, poplar, fir, and birch. This biome is noted for its many lakes, rivers, and wetlands. Permafrost exists, especially the discontinuous and sporadic forms.

Boundary: Demarcation of political jurisdiction between two countries as applied to a body of water.

Brent: The global benchmark for global spot oil prices. Brent is a light crude produced in the North Sea oil fields. Its price per barrel is normally higher than that for West Texas Intermediate oil.

Calder **decision:** *Calder v. Attorney General of British Columbia*, a 1973 Supreme Court of Canada decision that determined "Indian title" is a legal but undefined right, based on Aboriginal peoples' historic "occupation, possession and use" of traditional territories.

Canadian Environmental Assessment Act: Legal foundation for EIA under federal jurisdiction. The CEAA was passed in 1992, and in 1995 replaced the Environmental Assessment Review Process. The original CEAA was replaced in 2012 by a new Act, CEAA 2012, as part of a federal budget bill, despite protests from Aboriginal and environmental groups. The federal government argues that the new law streamlines the environmental process and reduces the time required to review major resource projects. Opponents claim the new legislation narrows the scope of the assessment process and puts the health of the environment at risk.

Canadian Environmental Assessment Agency: Agency that oversees federal EIA and implementation of the Canadian Environmental Assessment Act. Formed in 1994, this agency replaced the Federal Environmental Assessment Review Office.

Carrying capacity: The maximum population size in a given ecosystem that will allow sustainability of that system.

Churchill Falls hydro project: Hydroelectric project in Labrador begun in the late 1960s, when the government of Newfoundland badly wanted to develop the enormous water resources of the Churchill River. Located 200 kilometres from the "disputed" Quebec border, the project had only two potential markets—Quebec and New England. In 1967, construction began even though Hydro-Québec was the only utility to consider investing in what it viewed as a risky venture. In return for Hydro-Québec's investment, Newfoundland agreed to sell the power generated at Churchill Falls to Quebec at a fixed price for 65 years. A few years later, in the 1970s, the price of energy skyrocketed, and Quebec could resell Newfoundland's power at the much higher market rates while still purchasing it at the agreed low price, an issue that has riled the Newfoundland and Labrador government to the present.

Circumpolar World: Those lands and peoples found in the higher latitudes of the northern hemisphere; also the countries bordering on the Arctic Ocean whose environments consist of Arctic and Subarctic biomes. In a more narrow sense, it refers to the Arctic and its original inhabitants.

Clean energy policy: US policy announced by President Barack Obama on 15 June 2010. The policy is linked to US efforts to reduce greenhouse gas emissions. Later in 2010, the question of the US importing oil from Alberta's oil sands became a political issue that saw the term "dirty oil" become a rallying point against approval of the extension of the Keystone XL pipeline to heavy oil refineries on the Gulf of Mexico.

Climate change: Changes measured over a period of time by significant variations in global temperatures and precipitation due to natural factors or human activity. Climate is normally thought of as stable, but since the Earth was formed many climate changes have taken place. The most recent climate change involved the end of the Late Wisconsin glacial period and the beginning of the current climate known as the Holocene. The term is also used in connection with global warming.

Clyde River Protocol: An agreement that set the working relations between the government of Nunavut and Nunavut Tunngavik Inc.

Cohort: An age group, e.g., those born between 2010 and 2014 would fall into the 0 to 4 age cohort.

Colonialism: The imposition of one society/country on another society in another part of the world; associated with cultural, economic, and political domination. One common feature is the hierarchical relations of power between two different populations (one from away and one Indigenous to the land) where the local population is subordinated to satisfy the needs of the dominant newcomers.

Continental effect: Climate in the middle of a continent far from the moderating influence of oceans, so that there is a greater range of seasonal temperatures. Summers are hot and winters are cold.

Core/periphery model: A theoretical model that provides a geographic framework for interpreting the North, which is seen as a resource hinterland dominated and exploited by an industrial core (southern Canada, as well as wider North American and global economic forces). In reality, the flow of wealth from the North is somewhat offset by the injection of public funds through transfer payments.

Corporate memory: Acknowledgement by corporations operating in the North, based on past efforts to protect the environment and to ensure a place for Aboriginal workers, that they must be committed to these values in the present and future.

Country food: Food obtained by hunting, fishing, and gathering. While country food is no longer the chief source of food for many Native northerners, it remains important both nutritionally and culturally, particularly for those in more remote communities.

Cree Hunters and Trappers Income Security Program: A subsidized provincial program for Quebec Cree hunters and trappers that allows them to stay on the land for part of the year harvesting furs and country food for Cree communities.

Crude birth rate: *See* birth rate.

Crude death rate: *See* death rate.

Cryosolic soils: Thin soils formed in the continuous permafrost zone. These soils have active or thawed layers less than one metre thick.

Cyclic steam stimulation: Method of extracting bitumen whereby steam is injected into a well for a period of several weeks or months. Once the bitumen is heated, the hot bitumen is pumped out of the well.

Death rate: The number of deaths per 1,000 population in a given year. Also known as crude birth rate.

Delgamuukw decision: *Delgamuukw v. British Columbia*, a case that reached the Supreme Court of Canada, which in its 1997 judgment defined how Aboriginal title may be proved and outlined the justification test for infringements of Aboriginal title.

Demographic transition theory: A sequence of demographic changes in which populations progressively move over time from high birth and death rates to low birth and death rates.

Dene Declaration: A bold and defiant 1975 manifesto of the Dene Nation that declared: "We the Dene of the Northwest Territories insist on the right to be regarded by ourselves and the world as a nation." Their proposed state, called Denendeh, would be separate from but coexist with non-Dene living in the western half of the pre-Nunavut Northwest Territories.

Denendeh: The western half of the Northwest Territories (prior to the hiving off of Nunavut). These lands would form the territory of the proposed state for the Dene as stated in the Dene Declaration, while the eastern half (now Nunavut) would belong to the Inuit.

Dependency: A corollary of dominance; a situation where a region or people must rely on other regions or people for their economic well-being. Dependency can also mean that external capital and technology play a paramount role in the regional economy and that local politicians have little power relative to higher levels of government.

Dependency ratio: The ratio of the economically dependent part of a population to its productive part. Dependent members of a population are defined as ages less than 15 and over 64. Productive members fall into the age group of 15 to 64, the so-called working age group.

Dilbit: Diluted bitumen, a product resulting from adding a light petroleum liquid to bitumen. The purpose is to reduce the viscosity of bitumen so it flows more easily through a pipeline.

"Dirty oil": Derogatory expression for oil extracted from Alberta's oil sands that adds to global warming; a term used by opponents of oil sands development because this pollutes the environment and increases global temperatures by adding more greenhouse gases to the atmosphere each year. *See also* "ethical oil."

Drumlins: Elongated hills composed of glacial deposits formed by massive subglacial flooding; the long axis of a drumlin parallels the direction of glacier flow.

Duty-to-consult doctrine: Supreme Court of Canada determination in 2004 that the Crown has a duty to consult with affected Aboriginal communities about potential impacts of developments and, where appropriate, accommodate them. *See also Tsilhqot'in* decision.

Economic cycle: The business cycle, which consists of periods of economic expansion and contraction. This cycle is irregular and thus cannot be predicted.

Ecumene: Portion of the land that is permanently inhabited by humans and that contains the national transportation system.

Eeyou Istchee: Cree territory in Subarctic Quebec whose political body is the Grand Council of the Crees.

Environmental Assessment and Review Process: The first EIA review process affecting federal lands, created in 1973 by the federal government.

Environmental impact assessment (EIA): Assessment to identify shortcomings in a proponent's project, as detailed in the proponent's environment impact statement. EIA has been seen as one way to ensure that the environmental mistakes of the past are not repeated.

Epidemiological transition: A sequence of health changes in a human population that progressively moves over time from high infant mortality to low infant mortality, with death coming at higher ages. Control of infectious diseases and improved public health measures (e.g., reliable water and sewerage systems) account for the decline in infant mortality rates while control of degenerative diseases has resulted in longer lifespans.

Erratics: Boulders moved by glacial ice and deposited in another place.

Eskers: Ridges of sand and gravel, sometimes many kilometres long, deposited by streams beneath or within a glacier.

"Ethical oil": A slogan used by supporters of oil sands development and adopted by both the federal and Alberta governments to counter the term "dirty oil." The term "ethical oil" first appeared in the Canadian media in September 2010 with the release of Ezra Levant's book, *Ethical Oil: The Case for Canada's Oil Sands.*

Ethnocentrism: The belief in the superiority of one's own culture or people, which results in judging other cultures or groups using criteria specific to one's own group.

Ethnosphere: Term coined by Wade Davis to encompass the cultures and languages of past and present human populations; the global social web of life.

Exclusive economic zone: The territorial sea that extends 200 nautical miles from shore as measured from its baseline, i.e., at low tide. Within this area, coastal nations have sole rights over all natural resources.

Federal Environmental Assessment Review Office: Federal agency created in 1973 that later evolved into the Canadian Environmental Assessment Agency. Its responsibility was to implement the Environmental Assessment and Review Process.

Fertility: The actual reproductive performance of a population, such as the number of live births in a given year.

Fertility rate: The number of live births per 1,000 women between the ages 15 to 44 in a given year.

Gelifluction: The movement of the thawed active layer of permafrost downslope forming a series of distinct lobes; occurs in permafrost area. *See also* solifluction.

Geopolitics: A branch of political geography focusing on the interplay of geography, national power, and international relations; often takes the political form of national strategies to maximize favourable resolutions to territorial disputes.

Giant Mine Remediation Project: Billion-dollar remediation project involving two key steps to contain the arsenic trioxide waste. First, most waste will be frozen and stored in sealed underground chambers and vaults. Second, close to 100 buildings, tailings ponds, and contaminated soil will be removed and placed in a secure place. After remediation is complete, the site will be monitored to ensure no arsenic poses a threat to the environment and human life.

Glacial till: Unsorted and unstratified material deposited by ice sheets; also known as drift.

Glaciofluvial: Deposits or landforms produced by glacial meltwater streams.

Glaciolacustrine: Ancient lake bottoms, formed from the draining of glacial lakes.

Gleysolic soils: Water-saturated soils formed in marshy areas of the Subarctic.

Global circulation system: The general movement of air in the atmosphere and of water in the oceans from tropical areas to polar areas.

Global energy balance: The equilibrium resulting from incoming energy from the sun and outgoing heat from the Earth; its principal role is to regulate climate. Global warming is altering the natural equilibrium by increasing the level of greenhouse gases in the atmosphere, thus increasing global temperatures and affecting climate.

Global warming: The increasing temperature of the Earth due to the burning of fossil fuels, which adds greenhouse gases to the atmosphere. These gases act as a blanket keeping solar radiation from escaping to outer space. This process began with the Industrial Revolution but has accelerated temperature increases as the developing world industrialized, often using coal to drive their industries.

Governor-in-Council: Cabinet. If the Review Board determines that a proposed project will have significant adverse environmental impacts and the Minister of the Environment agrees, then the matter is referred to the Governor-in-Council (cabinet), who decide if the likely significant adverse environmental effects are justified in the circumstances.

Great Bear Rainforest: Area of old-growth forest between the Pacific Ocean and the Coast Mountains along the northern and central British Columbia coast. In 2006, the BC government and a wide range of conservation groups signed the Great Bear Rainforest Agreement. The aim of this agreement was to preserve this rainforest from various forms of resource development.

Greenhouse effect: The effect on the Earth's temperature of certain atmospheric gases known as greenhouse gases that trap energy

from the sun. These gases consist of water vapour, carbon dioxide, nitrous oxide, and methane.

Growth rate: The rate, expressed as a percentage, at which a population is increasing (or decreasing) in a given year. See natural rate of growth.

Hazmat suit: Hazardous materials gear providing varying levels of protection for workers involved in environmental cleanup. Total containment suits give protection from all forms of chemicals: solids, liquids, and gases/vapours. Other suits are not airtight and do not protect against vapours or gases. Also worn by medical personnel dealing with highly contagious and infectious disease and by laboratory workers dealing with hazardous materials and pathogens.

Heavy oil: Crude oil with a higher density than light crude oil so that it does not flow easily; product of upgraders.

Holocene Epoch: The most recent geological epoch that began some 10,000 years ago, marking the beginning of an interglacial phase and warmer climates than those experienced in the Late Wisconsin glacial period.

Homelands: Regions where the inhabitants have both a strong attachment and a commitment to their social and political institutions; a sense of place.

Horizontal drilling: A type of well drilling that deviates from a vertical drill hole and travels horizontally through a producing layer of natural gas or oil.

Hydraulic fracturing: Pumping a fluid into the well at very high pressure to create cracks in the reservoir rock and thus releasing natural gas or oil trapped in the reservoir rock; fracking.

Hydraulic mining: A pressurized stream of water directed at gold-bearing gravel terraces to wash away overburden and expose the gold-bearing gravel. In Bonanza Creek in Yukon following the Klondike gold rush, hydraulic mining also thawed the frozen terraces.

Ice road: Frozen ground, lakes, and rivers cleared of snow to create a road. The longest ice road in Canada serves diamond and gold mines. It extends from just north of Yellowknife for some 600 km to the northern end of Contwoyto Lake. Ice roads are also known as winter roads.

Ice shelf: A floating ice sheet attached to the coast that extends for some distance over the sea; from time to time, large blocks of ice fall into the ocean.

Indians: A legal term in Canada for those whose names are on the band list of any Indian community in Canada or on the central registry list in Ottawa. The main Indian linguistic groups in the Canadian North are Algonkian (e.g., Cree, Innu) and Athapaskan (e.g., Dene).

Infant mortality rate: The number of deaths to infants under one year of age per 1,000 live births in a given year. Often used as a proxy for the general health of a population.

In-migration: The relocation of a resident from one province/territory to take up residence in another province/territory.

Intergovernmental Panel on Climate Change: Group created in 1988 by the World Meteorological Organization and the United Nations Environment Programme with the objective of assessing published scientific literature and, every six years, publishing its findings. The most recent report, in 2013, is available at https://www.ipcc.ch/report/ar5/wg1/.

International Seabed Authority: Autonomous international organization established under the 1982 UN Law of the Sea Convention (UNCLOS) that organizes and controls activities of the ocean floor in international waters (known as "the area").

Inuit: Aboriginal people whose homeland is the Arctic and who, traditionally, were nomadic hunters, especially of sea mammals such as seal, whale, and walrus; they comprise about 85 per cent of the population of Nunavut and also live in Arctic Quebec, northern Labrador, and the western Arctic.

Inuit Circumpolar Council: Non-governmental agency founded in 1977 that represents approximately 150,000 Inuit of Alaska, Canada, Greenland, and Russia (Chukotka).

Inuit Qaujimajatuqangit: Knowledge, such as that related to the natural environment, that has been long known by Inuit and that is based on Inuit principles and values; traditional knowledge (TK).

Inuit Tapiriit Kanatami: The national advocacy organization of Inuit living in 53 communities across the Arctic. Inuit call this vast region Inuit Nunangat. Founded in 1971, ITK represents and promotes the interests of Inuit nationally on a wide variety of environmental, social, cultural, and political issues. ITK does not deliver or fund programs. Formerly known as Inuit Tapirisat of Canada.

Isostatic uplift: Rebound in the Earth's crust to a state of balance (isostasy) after massive ice sheets that depressed the crust have melted.

James Bay and Northern Quebec Agreement: Agreement made in 1975 by Quebec, the federal government, and the Aboriginal peoples (Cree and Inuit) of northern Quebec to allow the first phase of the James Bay Hydroelectric Project, La Grande Rivière, to proceed. The Canadian government recognized its obligations to northern Aboriginal peoples that went back to the transfer of Rupert's Land to Canada in 1880. The key point is that under this land transfer, Ottawa must obtain the surrender of Indian title to the lands that formerly composed Rupert's Land—and that included northern Quebec. The JBNQA preceded other Canadian modern land claim settlements, known as comprehensive land claim agreements, but the JBNQA was modelled after the Alaska Native Claims Settlement Act (1971).

Joint review panels: Under the Canadian Environmental Assessment Act, panels formed by the federal Minister of the Environment to conduct socio-economic assessments of industrial projects.

Keystone XL: Controversial pipeline proposal of TransCanada Corporation to transport diluted bitumen from Alberta's oil sands to the US Gulf coast of Texas for refining and shipment to world markets. The Keystone XL is caught between the need of oil sands producers to find refineries for their product and the concerns of environmentalists over global warming. A bill to approve the project was vetoed by US President Obama in February 2015 and the US Senate did not succeed in an attempt to override the President's veto, but the powerful lobbying efforts of the oil industry especially, and of Canadian governments, mean the issue has not disappeared from the policy agenda. Oil production in the United States is rising rapidly due to fracking technology, so the argument that the Keystone XL pipeline is a critical infrastructure project for American energy security and for strengthening the American economy no longer is valid.

Labrador Current: A cold current originating in Baffin Bay of the Arctic Ocean and flowing southward to meet the warm Gulf Stream Current in the waters offshore of the Maritime provinces. One effect of this current is to chill the summer temperatures along the Labrador coast and thereby extend the Tundra biome southward.

Labrador Trough: Geological structure, 1,600 km long and 160 km wide, that extends south from Ungava Bay through Quebec and Labrador and southwestward into central Quebec. The trough comprises early Proterozoic sedimentary and volcanic rocks with banded iron formations that have been mined since 1954. The deposits are unusually rich, with nearly 40 per cent of the ore consisting of iron.

Land-fast ice: Newly formed sea ice attached to shore but extending for some distance into the sea. In the winter, land-fast ice merges with the Arctic ice cap; in the late summer, this ice becomes detached from shore and floats into open water as ice floes. Unlike the Arctic ice pack, land-fast ice is attached to the land. Also known as fast ice.

Land mass: A large continuous area of land, such as a continent or a country.

Late Wisconsin glacial period: The last glacial phase in North America, also known as the ice age, when huge ice sheets covered practically all of present-day Canada and northern portions of what is now the United States. Its advance ended some 15,000 years ago. As the Earth began to warm again, the southern edges of the ice sheets began to melt, exposing barren land. The Late Wisconsin glaciation spanned a period from 25,000 BP to 10,000 BP.

Leads: Open water in the Arctic ice cap created by cracks resulting from its movement. The cracks cause smaller areas of solid ice with water between them; these leads can open and close quickly.

Life expectancy: The average number of additional years a person would live if current mortality trends were to continue; most commonly cited as life expectancy at birth.

Little Ice Age: A period of cool temperatures that lasted some 400 years from around 1450 to 1850.

Long Lake Project: Oil sands project purchased from Nexen by China's CNOOC in February 2013. The Long Lake commercial SAGD project facilities include a steam SAGD production field, central production facility, and two cogeneration plants integrated with an upgrading facility designed to significantly reduce reliance on natural gas. By June 2015, technical problems kept the project from reaching its maximum production.

Mackenzie Valley Gas Pipeline Project: Proposed project in the early 1970s, with various proponents seeking the go-ahead to build a pipeline to transport natural gas from vast gas reserves along the Arctic coast, especially at Prudhoe Bay and in the Mackenzie Delta, across Yukon to the mouth of the Mackenzie River and then up the Mackenzie River Valley to northern Alberta, where the pipeline would connect with the national/international pipeline grid. This project and its potential impacts were examined by the Berger Inquiry.

Malthusian population trap: Concept introduced by the English economist Thomas Malthus (1766–1834) that populations, if unchecked, tend to increase at a geometric rate while the means of subsistence (the food supply) increases at an arithmetic rate. The trap occurs when a country or region has a higher rate of population increase than its rate of economic growth.

Marine effect: The effect of oceans on adjacent land bodies causing mild temperatures throughout the year. Further inland, the marine effect weakens and summers become hotter and winters colder. This is known as the continental effect.

Maunder Minimum: The period roughly spanning 1645 to 1715 when sunspots became exceedingly rare, which may have been a factor in cooling global temperatures and the Little Ice Age.

Mean annual temperature: Calculated by adding the average of the 12 mean monthly temperatures and then dividing by 12.

Mean daily temperature: Calculated by adding the daily minimum and maximum temperatures and dividing by 2.

Mean monthly temperature: Calculated by adding the mean daily temperatures for that month and then dividing that sum by the number of days in the month.

Megaprojects: Large-scale industrial undertakings, which, because of their enormous size, dominate the local and regional economy during the construction phase. Construction costs usually exceed $1 billion and the start-up phase can extend for several years.

Métis: Originally understood to mean the offspring of fur traders and Indian (primarily but not exclusively Cree and Ojibwa) women who developed a separate culture and history focused on the buffalo hunt and centred around the Red River colony in present-day Manitoba. In 1982, the Métis gained official recognition as one of the three Aboriginal peoples of Canada, and the definition of Métis has been expanded by some groups and scholars to include all Canadians of mixed Aboriginal–European heritage who pursue a common local or regional culture and lifestyle.

Milankovitch theory: Theory positing that some 15,000 years ago the Earth's orbit shifted, causing the angle of incoming solar radiation to increase, which, in turn, strengthened the intensity of solar radiation and thus began the warming of the Earth that brought the Late Wisconsin glacial period to an end.

Modernization theory: Social scientific theory, popular in the twentieth century, that attempts to explain human progress over time from less advanced (i.e., non-Western) to more advanced (i.e., Western) societies, stressing those social elements that help or hinder social advancement.

Mt CO^{2}e: Megatonnes of CO^2 equivalent; unit used to report the amount of greenhouse gas emissions or reductions.

Multi-year ice: Ice that does not melt for one year or more, which is much thicker and harder than new ice.

Muskeg: A Cree term for bogland covered by sphagnum moss; found primarily in the Subarctic.

National Energy Board: An independent federal agency that regulates several parts of Canada's energy industry. Its purpose is to regulate pipelines and energy development and trade in the Canadian public interest.

Native peoples: Those Canadians of Indian, Inuit, and Métis ancestry; an older term describing Aboriginal peoples.

Natural increase: The surplus (or deficit) of births over deaths in a population over a given time period. See also rate of natural increase.

New ice: Sea water that freezes and then thaws within one year; also referred to as seasonal or young ice. *See also* multi-year ice; land-fast ice; pack ice.

Nexen: Canadian oil sands firm sold to the Chinese National Offshore Oil Company (CNOOC) after considerable discussion within the federal government, which finally approved the sale on 25 February 2013. At the same time, Ottawa barred further takeovers of Canadian oil sands firms, such as Suncor Energy and Cenovus Energy, by state-owned companies.

Non-governmental organizations: Organizations, focused on any of a host of issues, that represent what they view as the general public interest. Environmental NGOs often target a controversial resource theme, such as the Sierra Club's emphasis in recent years on the threat of global warming, the need for clean energy alternatives, and the impact of oil sands development and proposed pipelines on the environment.

Non-status Indians: Those Canadian Indians who by birth, marriage, or choice have no legal status, under the Indian Act, to benefit from reserve lands and special federal programs.

Nordicity: Concept created by Louis-Edmond Hamelin to measure the degree of "northernness" of a place. Nordicity provides a quantitative definition of the southern boundary of the North and is based on 10 variables, such as latitude, degree of isolation, and annual cold, that are supposed to represent all facets of the North.

NORDREG: Canada's Northern Canada Vessel Traffic Services Zone Regulations. Mandatory since 2010, NORDREG requires vessels to report and receive permission to enter Canadian ice-covered waters.

Northern benefits: Requirement of resource companies to ensure employment and business opportunities are specifically allocated to northerners. Without such encouragement by governments, employment and business opportunities would flow out of the region and the North would benefit very little (if at all) from resource development.

Northern Gateway: Enbridge proposal involving two pipelines. The eastbound pipeline would import natural gas condensate and the westbound pipeline would export diluted bitumen from the Athabasca oil sands to the marine terminal in Kitimat for transport to the Asian markets by oil tankers. This $6.5 billion pipeline project has run into strong opposition from environmentalists, First Nations, and much of BC society, who place a high value on their land and ocean environments. The National Energy Board has approved this project subject to over 200 concerns, but the 2014 *Tsilhqot'in* decision by the Supreme Court has meant that Enbridge will need to gain the explicit approval of various First Nations along the proposed pipeline route—a tall order.

North Rankin nickel mine: Nickel mine that, from 1957 to 1962, was the first Arctic resource developer to train and then employ large numbers of Inuit miners who at best had a rudimentary command of English and no experience of wage employment. These Inuit miners had just resettled in centres along the west coast of Hudson Bay, including this mine site. They formerly were hunters and trappers.

Northwest Territories Devolution Act: Federal statute passed in 2014 that simultaneously decentralizes ultimate control over environmental issues from Ottawa to the NWT and centralizes control in one resource co-management board (under the Mackenzie Valley Resource Management Board) by eliminating five regional co-management boards that dealt with water and wildlife issues. The Act thereby has removed local input and knowledge from the management of NWT resources.

Nunangat: The current homeland of Inuit in Canada, composed of the four Inuit regions found in the Canadian Arctic: Nunatsiavut (northern coastal Labrador); Nunavik (northern Quebec); the territory of Nunavut; and the Inuvialuit region of the Northwest Territories. These regions were legally determined by the comprehensive land claims process and encompass only part of the area traditionally occupied by Inuit in what is today Canada.

Nunataks: Unglaciated mountain peaks that stood above the ice sheets.

Nunavik: Inuit territory in Arctic Quebec that is slowly moving towards a regional government within Quebec. See also Eeyou Istchee.

Nunavummiut: A person from Nunavut; a person of Inuit ethnicity.

Nunavut: Territory in Canada's eastern Arctic, formally established in 1999, that was hived off from the present Northwest Territories; means "our land" in Inuktitut, and represents a political expression of the vision of an Indigenous homeland.

Nunavut Tunngavik Inc.: Corporation that receives and manages the payments from the Nunavut Land Claims Agreement, including royalties from mining operations on settlement lands. Nunavut Tunngavik interfaces with the government of Nunavut over the management of land, water, and wildlife on behalf of the Inuit. Formerly called Tungavik Federation of Nunavut.

Pack ice: Floating ice of varying age, size, and thickness that makes up the Arctic ice cap.

Patterned ground: Stones and pebbles arranged by frost action in a geometric pattern, e.g., circles or polygons. It is widespread in Arctic environments where frost action is the dominant geomorphic force.

Physiography: The study of landforms and the processes that shape them. Physical geography or geomorphology.

Pingos: Ice-cored hills found in permafrost areas. Pingos range in height from a few metres to 50 metres, and expand in size as water seeps into the core and freezes. Most pingos are found in the Mackenzie Delta.

Place: The cultural and physical elements that form the special character of an urban or rural area. In terms of a "pull" factor in migration theory, place draws migrants back to their original place of birth and long-term residence where strong associations with family and friends as well as the combination of cultural and physical aspects are very appealing. In brief, it is like "going home."

Pleistocene Epoch: A geological epoch associated with the last ice age; included at least four major ice advances, including the Wisconsin; began some 1.6 million years ago and ended some 10,000 years ago when the Holocene Epoch began.

Podzolic soils: Acidic soils found in the boreal forest.

Polar night: A period of absolute winter darkness when twilight does not occur. Polar night occurs first at 72° 33′N when there is no twilight. The duration of polar night increases until the North Pole, which has six months of total winter darkness from around 25 September to 18 March. In the zone between the Arctic Circle and polar night, twilight exists.

Polynya (north water): Large area of open water in the Arctic Ocean surrounded by sea ice.

Population density: A measure of the number of people in a given area. In Canada, for instance, density is measured by population per km² for all of Canada's land mass.

Population explosion: An enormous and rapid population growth.

Population increase: The total increase or decrease in a population resulting from births, deaths, and in-migration in a population within Canada. *See also* natural increase.

Potential labour force: The total adult population as defined as falling between the ages of 15 and 64. The actual labour force is a much smaller proportion of the adult population and refers to those currently employed or seeking employment.

Push–pull migration theory: A model explaining the movement of people from economically depressed, conflict-ridden, or environmentally damaged areas or countries to regions or countries with more economic opportunity, less conflict, or a more supportive environment. Factors in the home area push migrants from that location and factors in the receiving area pull or attract migrants to that location.

Rate of natural increase: The rate of a population increasing or decreasing in a given year due to the difference between births and deaths occurring in that year.

Remediation: The process involved in cleaning up or remedying a highly contaminated former industrial site.

Resource revenues: All monies flowing to governments from resource development. Some are fees and taxes but most revenue comes from royalties.

Royal Proclamation of 1763: A pronouncement of the British Crown that Aboriginal approval had to be gained before any lands could be alienated from them, that only the Crown could alienate Indian lands, and that lands from the Ohio Valley westward were set aside as Indian lands. After the Seven Years War with France, the victorious British faced two conflicting issues: the desire for territorial expansion on the part of American settlers along the eastern seaboard, and the need to accommodate the various Indian tribes to the west. Settlers desperately wanted to cross the Appalachian Mountains and occupy the fertile lands of the Ohio Valley. At that time, these lands were considered Indian Territory. The Indian tribes, led by Pontiac, wanted those lands left in their control. The British, in the Royal Proclamation, sided with Pontiac, but the defeat of the British by the American rebels in the American Revolution (1775–83) set a different course. Pontiac's dream came to nothing and the Indian tribes of the Ohio Valley soon were displaced as American settlers occupied these lands. The Royal Proclamation remains an important document in Canadian jurisprudence and legal scholarship.

Royalties: The price that the owner of a natural resource charges for the right to develop the resource. In a sense, royalty is not a tax. Taxes refer to costs of providing a service, such as filing a claim.

Sense of place: The intense feeling of belonging and loyalty to a region.

Settlement area: The lands selected by Aboriginal negotiators in their comprehensive land claim negotiations with the federal government. They are divided into two categories: Category A includes both surface and sub-surface ownership; Category B includes the surface area only.

Sierra Club: Organization founded in the United States in the late nineteenth century that has transformed itself from a conservation group to one of the most powerful and influential environmental organizations with chapters in many countries, including Canada.

Slave Geological Province: An extremely rich mineral area in the Northwest Territories portion of the Canadian Shield.

Solifluction: The slow downslope movement of waterlogged soil. *See also* gelifluction.

Status Indians: Canadian Indians who are registered under the Indian Act. They have a right to use reserve lands held by their band and access to federal funding for programs such as housing and education. Those status Indians whose ancestors signed a treaty also have treaty rights.

Steam-assisted gravity drainage: An oil recovery technology for extracting bitumen by a pair of wells drilled into the oil sands, one a few metres above the other. Low-pressure steam is continuously injected into the upper wellbore to heat the bitumen and reduce its viscosity, causing the heated bitumen to drain into the lower wellbore, where it is pumped out.

Subarctic: A natural region of North America distinguished by its natural vegetation, the boreal forest; also the traditional homeland of northern Indians.

Subsistence economy: An economic system of relatively simple technology in which people produce most or all of the goods to satisfy their own and their family's needs by hunting, gathering, and/or subsistence farming; little or no exchange occurs outside of the immediate or extended family.

Sustainable development: According to the 1987 World Commission on Environment and Development (Brundtland Commission), "development that meets the needs of the present without compromising the ability of future generations to meet their own needs."

Sverdrup Basin: Sedimentary basin in Canada's High Arctic, on Axel Heiberg Island and Ellesmere Island, with vast reserves of hydrocarbon.

Tailings: Waste generated from a mine processing plant, often containing toxic material used in extracting the mineral. Since tailings are a cost to the mine operator, efforts are made to store the waste in the most inexpensive manner but within the environmental regulations.

Tailings ponds: Huge ponds containing a slurry-like water/oil mixture laced with toxic chemicals, a by-product of bitumen extraction. The process of sedimentation separates the fine particles suspended in the water, causing the particles to settle at the bottom of the pond. In 2010 Suncor developed technology involving polymers to hasten the sedimentation process, which otherwise might not have been completed in this century.

Taliks: Permanently unfrozen ground above the permafrost but below the active layer in regions of permafrost. First observed in Russia, taliks often occur underneath thermokarst lakes, where the deep water does not freeze entirely, thus keeping the soil underneath from freezing.

Tar sands: Original name for deposits of bitumen, a viscous mixture of oil, sand, and clay, such as the deposits in northern Alberta; oil sands.

Thermokarst landscapes: Irregular landscapes that occur in permafrost areas when melting of ice contained in the frozen ground causes subsidence of the ground.

Trade-off: A difficult choice whereby you must accept something that you do not want in order to have something that you want.

Traditional ecological knowledge: Knowledge about a local environment or ecosystem based on personal observation and experience, especially as it has been passed on from one generation to another living in a specific area.

Trans Mountain Expansion: Proposal by Kinder Morgan calling for an expansion of its current 1,150-km oil pipeline between Strathcona County (near Edmonton), Alberta, and Burnaby, BC. The original pipeline was completed in 1953. The proposed expansion, if approved, would create a twinned pipeline that would increase the nominal capacity of the system from 300,000 barrels per day to 890,000 barrels per day. However, opposition from BC residents is particularly strong. Unlike the Northern Gateway Project, the National Energy Board has yet to hold its hearings on this proposal.

Transpolar Drift: The movement of water and ice from the shores of Siberia across the North Pole, where it joins the East Greenland current that flows into the North Atlantic Ocean.

Tribal groups: Groups of Indians united by language and customs and belonging to the same bands.

Truth and Reconciliation Commission: Commission formed in June 2008 as a result of the Indian Residential Schools Settlement Agreement (2007). Its immediate goal was to allow victims of abuse to acknowledge their residential school experiences and the consequences of those experiences. Through this process, Canadians were made more aware of this dark page in Canadian history, now understood as cultural genocide. The Commission tabled its report in June 2015.

***Tsilhqot'in* decision:** June 2014 decision of the Supreme Court of Canada in *Tsilhqot'in Nation v. British Columbia* in which the Court ruled in favour of the Tsilhqot'in claim to its ancestral lands and its control of development on these lands. For First Nations without treaty, this ruling confirms Aboriginal title over lands that were historically used for hunting, fishing, and other activities but adds that proposed development on these lands will require the consent of the First Nation.

Tundra biome: One of the world's major biogeoclimatic areas, with a cold climate that inhibits most plant growth. Its natural vegetation is limited to dwarf shrubs, sedges, grasses, mosses, and lichens. Formed at the end of the last ice age, this biome is found in the highest latitudes.

Tungavik Federation of Nunavut: Now called Nunavut Tunngavik Inc.

United Nations Convention on the Law of the Sea (UNCLOS): An international agreement, first established in 1982, that defines the rights and responsibilities of nations in their use of the world's oceans, including the ownership of coastal waters and seabeds. With the exception of the United States, most countries have ratified this agreement.

Upgrader Alley: The industrial heartland of Alberta, extending from Edmonton to Fort Saskatchewan, where most bitumen upgraders outside of Fort McMurray are located.

Upgrader plants: Operations that convert bitumen into synthetic crude oil, which then can be transported by pipelines to refineries where the synthetic crude is refined into various commercial products. Bitumen that has not been processed by an upgrader is thinned by adding lighter oil. This product, known as dilbit, can flow through pipelines to refineries.

Ward Hunt Ice Shelf: At 400 km² in size, the largest ice shelf in Canada's North. It was formed in the twentieth century when the Ellesmere Ice Shelf slowly separated into six shelves. Since 2000, the Ward Hunt Ice Shelf has diminished in size and large blocks have fallen into the sea. The British Arctic Expedition of 1875–6 recorded the existence of the Ellesmere Ice Shelf which was already known to the Inuit.

West Texas Intermediate: The price of oil set at Cushing, Oklahoma, a major trading hub for North American crude oil. A bottleneck in the transport system prevents oil from quickly accessing Gulf coast refineries, causing the WTI price to fall below the Brent price.

Young population: A population with a high proportion under the age of 15. Such a population is often associated with developingcountries and parts of the Canadian North. See also dependency ratio.

Index